Food and Agriculture

Food and Agriculture

Edited by **Laura Vivian**

R CALLISTO REFERENCE

New York

Published by Callisto Reference,
106 Park Avenue, Suite 200,
New York, NY 10016, USA
www.callistoreference.com

Food and Agriculture
Edited by Laura Vivian

International Standard Book Number: 978-1-63239-664-8 (Hardback)

Printed in the United States of America.

Contents

Preface

The purpose of the book is to provide a glimpse into the dynamics and to present opinions and studies of some of the scientists engaged in the development of new ideas in the field from very different standpoints. This book will prove useful to students and researchers owing to its high content quality.

This book contains some path-breaking studies in the fields of food and agriculture. While understanding the long-term perspectives of the topics, the book makes an effort in highlighting their impact as a modern tool for the growth of this discipline. Earlier only traditional methods of farming were used but nowdays modern techniques such as agrochemicals, plant breeding, etc. have been employed to increase yield and improve the quality of produce. The aim of this book is to present researches that have transformed this discipline and aided its advancement. Some of the diverse topics covered in this book address the varied branches that fall under this category. The extensive chapters of this book have been contributed by experts from across the globe. This book will serve as a reference to a broad spectrum of readers such as agriculturists, botanists, biologists, academicians and students related with the field of agriculture at various levels.

At the end, I would like to appreciate all the efforts made by the authors in completing their chapters professionally. I express my deepest gratitude to all of them for contributing to this book by sharing their valuable works. A special thanks to my family and friends for their constant support in this journey.

Editor

Agricultural Sector of Bosnia and Herzegovina and Climate Change—Challenges and Opportunities

Ognjen Zurovec *, Pål Olav Vedeld and Bishal Kumar Sitaula

Department of International Environment and Development Studies (Noragric), Norwegian University of Life Science (NMBU), Universitetstunet 3 1430 Ås, Norway; E-Mails: pal.vedeld@nmbu.no (P.O.V.); bishal.sitaula@nmbu.no (B.K.S.)

* Author to whom correspondence should be addressed; E-Mail: ognjen.zurovec@nmbu.no

Academic Editor: Terence Centner

Abstract: Half of Bosnia and Herzegovina's (BH) population lives in rural areas. Agricultural production is a backbone of the rural economy and generates significant economic value for the country. BH is highly vulnerable to climate change, which poses a significant development challenge given the climate-sensitivity of the agricultural sector, the share of agriculture in the total economy, the number of people employed in the sector, and the closely related socio-economic issues of food security. BH has experienced serious incidences of extreme weather events over the past two decades, causing severe economic losses. Based on available data and currently available climate projections, exposure to threats from climate change will continue to increase. The review paper presents the current state of the BH agricultural sector and the impact of potential climate change on agricultural systems. It proposes policy options to optimize opportunities and mitigate consequences of possible climate change in the agricultural sector. Development of policy and research capacity should include harmonisation and centralisation of domestic agricultural policies, carrying out a vulnerability assessment and strengthening the public and private extension systems. Further technological development should include improvements in weather and climate information systems, crop development, irrigation and water management.

Keywords: Bosnia and Herzegovina; agriculture; climate change; adaptation

1. Introduction

Bosnia and Herzegovina (BH) belongs to a group of countries considered highly vulnerable to climate change [1,2]. Agriculture is an important and vulnerable economic sector in BH, given the climate-sensitivity of the sector, the share of agriculture in the economy (7% of the GDP), the number of people employed in the sector, and the closely related socio-economic issues of food security [3].

Climate change leads to adaptation among farmers and their agricultural production in the affected areas. However, adaptation does not occur independently, but rather as a process influenced by socio-economic, political, cultural, geographical, ecological and institutional factors [4,5]. Adequate responses will depend on the ability of decision makers, from the farm level to the national policy level, to perceive climate change and to take relevant action. The current state of politics in Western Balkan countries (WBC), where the public sector is mostly silent and non-transparent, and where the scientific contributions on climate change analyses are scarce, has led to limited development of activities in this field. It therefore comes as no surprise that climate change policy issues are not visible at any level of the policy-making agenda in these countries. As a result, climate change is not treated seriously in current published key strategic documents. One of the main problems comes from inadequate social and human capital when it comes to the introduction and implementation of measures and policies. However, there is a gradual rise of awareness in WBC about climate change, its importance and impact on all spheres of life [3,6–8].

The general objective of this review paper is to present the current state of the agricultural sector in BH and the impact of climate change on agricultural systems in BH. The first section of this paper gives an overview of BH's agricultural sector and its significance, together with the challenges and possible opportunities. The second section assesses the impact of climate change to BH's agricultural sector based on both current conditions and future predictions. The first two sections are based on data derived from official statistical releases, national and international reports and other relevant literature. In the last section, we propose policy options based on the international literature to optimize opportunities and mitigate consequences of climate change in the agricultural sector, in order to increase productivity and adapt agriculture in BH to changing climate.

2. Overview of Agricultural Sector in Bosnia and Herzegovina

BH was considered as a raw material-energy providing region, of the former Yugoslavia throughout the major part of the last century [9]. In the constitutional order of Yugoslavia, BH was part of the federation of six autonomous republics, ruled by a strong central government under the control of the Communist Party. Therefore, the development path and policies cannot be attributed to individual republics only, but external decisions taken at higher levels were crucial in policy formulation and outcome processes [10]. The natural resources of BH are the country's great fortune and misfortune at the same time, which was historically recognised by a large number of occupier exploiting these resources throughout the course of history. Prior to the Second World War, BH was a particularly undeveloped agrarian country compared to its western neighbours. Agriculture was the main sector of the economy during post-war reconstruction. However, at the same time, the foundations of the industrial development, exclusively related to potential in raw materials, were established. Industry employed only 2% of the total population prior to this period [11].

The main characteristics of the former BH economy and national planning is economic development based on the example of post-revolutionary Soviet Russia [9,12], which preferred development of heavy industry as a prerequisite for the development of light industry, transport and agriculture [13]. The leading industries were in metallurgy and chemical industry. The industry employed 54.3% of the entire population in 1961, reached its peak in 1981 (58.4%), and was 44% in 1991, after which comes a new period of war. Industrialization was the key cause of de-agrarian processes, which left deep impacts on social structures in rural areas. The share of agricultural population has decreased 76% in a 40-year period (1948–1981) [14]. BH experienced industrialization that initiated an urban development and migration to urban areas, de-populating rural areas. Agricultural resources as a public good have not been seen in accordance with general social interest—large areas of arable land have been left abandoned and uncultivated [15]. This, among other factors, led to a situation where the country could meet barely 50% of its needs for food [16].

One of the main revolutionary convictions of the newly established socialist state was that the inherited capitalist model of ownership and property rights was seen as a cause of social injustice and inequality. The new government attempted to achieve their vision of social equality and justice through the introduction of common ownership [17]. This was accomplished through adoption of laws and regulations that abolished private properties as the predominant form. At this time, important economic and industrial facilities were converted to state property through confiscation, sequestration, agrarian reforms and nationalization. Land as a common ownership was acquired primarily by confiscation of assets from persons convicted as "enemies of the state" and then significantly increased through agrarian reforms in 1945 and 1953. Agrarian reforms abolished large private land holdings and limited them to a maximum of 10 ha per private entity [17,18]. The confiscated land was awarded to landless peasants and farmers with insufficient land. This led to the emergence of a large number of small and medium sized farms with a tendency of further fragmentation. The agrarian reforms set back agricultural production almost to a scenario of natural or subsistence economy [9]. In addition, the remaining agricultural production was plagued by weak capital equipment of family farms [19] and obsolete technologies [20]. A shift in agricultural policies was recorded in the 70s, where much attention was paid to the development of agriculture and rural areas. The plan was to increase the intensity of production through higher yields and general increased productivity. Investment in land amelioration was one of the focus areas of this master plan. In the 80s, Yugoslavia was plunged into a deep economic crisis, which has affected investments in agriculture and the effective implementation of the planned investment programs.

Like all other sectors, the agricultural sector has suffered enormous damage during the war period (1992–1995). The programs of reconstruction and restoration of international donors focused on basic rural infrastructure and housing, purchase of agricultural machinery and inputs, seeds and fertilizers for the reconstruction and rehabilitation of crop production. It was more of a social aid to local people than a serious investment in the revitalization of agriculture, with the main objective to return displaced population in rural areas [21].

Today, BH is still a predominantly rural country. It is estimated that about 61% of the population live in rural areas [22]. Although the share of agriculture in GDP is constantly decreasing (11% in 2003 to

7% in 2013), agricultural production is a backbone of the rural economy, employing 20% of workforce. The economy of BH demonstrated considerable vitality by achieving high growth rates, especially in 2009, but it was not enough to significantly approach the level of medium developed countries. Actual GDP per capita in 2011 was only 30% of the EU 27 average and reached only 80% of GDP of which it had in 1989 [23]. BH's decentralized political and administrative structure is very complex. This unique constitutional order involves two entities: Federation of BH (FBiH) and Republika Srpska (RS), as well as the Brcko District of BH (BD), as separate administrative units. In addition, FBiH is divided into 10 Cantons. This complex governance structure also has a great impact on management competence and capacity in the agricultural sector. The situation in the agricultural sector in BH is featured by different regulations at different levels, legislative overlaps, limited capacities and communication channels, as well as a lack of clear vision and failure to implement necessary reforms. The legacy of the past socio-political system, coupled with the current complex political structure, have significant consequences for agricultural development, facing many challenges.

2.1. Agricultural Productivity

The main problem of the agricultural sector is low productivity, both per unit of production, and per farm [8]. The main feature is small-scale, subsistence agriculture oriented production rather than a more commercial or market oriented agro-food system. It is a main cause of low competitiveness, particularly in the domestic market. Low agricultural productivity is often a consequence of the absence of clear specialization, primarily in crop production, low technology levels of farms and extreme dependence on weather conditions. Shifts in terms of improving productivity are apparent, however, these processes are very slow [24]. The main reasons for a slow process of improving productivity are difficult and risky market access, and insufficient capacity for storing and processing, especially vegetables. In addition, production of seed material, nurseries and seedlings is underdeveloped and the production depends on imported seeds, often with questionable quality and without adequate control [8]. The level of technological and marketing knowledge among producers is low, which certainly has a negative effect on the productivity of the sector. The main cause for low production of basic agricultural products is that in the previous years, existing agricultural capacities have not been used intensively. Agricultural land covers 2.1 million hectares, of which 46.5% is arable (Table 1) and as much as half of arable land remains unused (Figure 1).

Table 1. Structure of agricultural land in Bosnia and Herzegovina. Source: [25].

Category	Area (000 ha)	%
Arable land	1004.9	19.6
Orchards	99.4	1.9
Vineyards	5.6	0.1
Meadows	460.2	9.0
Pastures	588.2	11.5
Total Agricultural Land	**2158.3**	**42.2**
Forests	2,795.1	54.6
Other	166.3	3.2
Total	**5119.7**	**100.0**

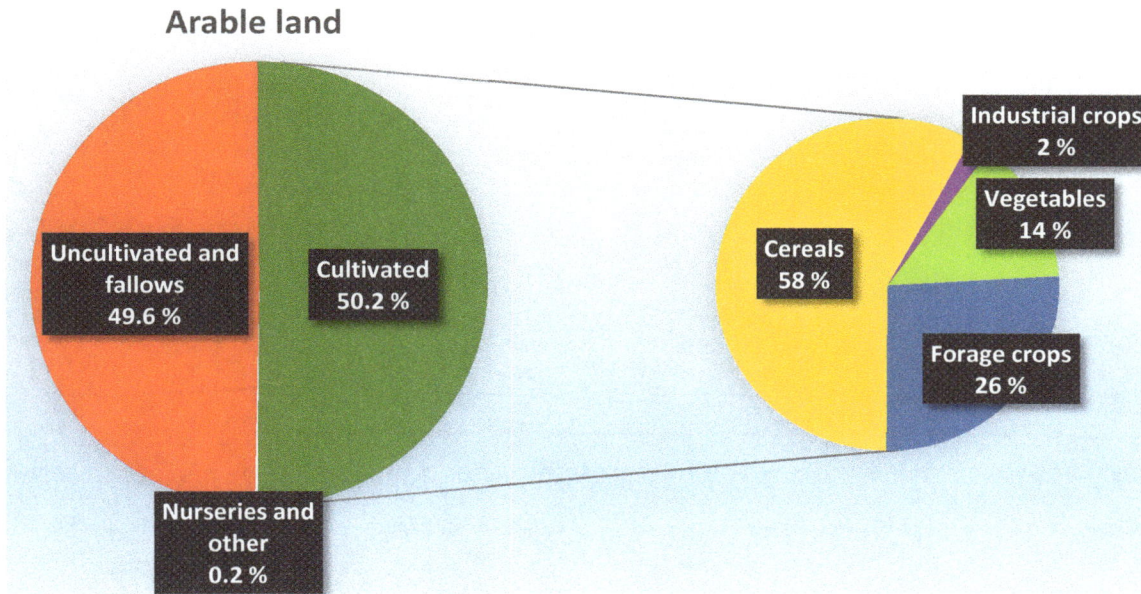

Figure 1. Structure of arable and cultivated arable land. Source: [26].

2.2. Livestock Production

Livestock production has a great significance for BH agriculture considering the available natural resources and the number of people engaged in this production. The key products from livestock involve milk and dairy products, meat and meat products. Since the number of livestock stagnated or declined in recent years, except in the case of poultry (Figure 2), growth in production of meat and milk is explained by improvements in yields and breed composition, but productivity is still low compared with countries in the region [22]. Low productivity is certainly partly due to still inadequate breed structure, inefficient breeding and selection work, but mostly because of the duality in production [8]. Extensive production on small farms is prevailing in livestock production, while on the other hand a small part of the production is organized on the modern, technologically well-equipped farms. While there has been some progress in exports and productivity in recent years, the overall competitiveness of livestock production in the international markets is still weak. BH currently achieves only a small share of imports in its major export destinations, mainly the Western Balkan countries [27]. One of the largest problems in livestock production is the banned on export of Products of Animal Origin to EU. For many years, EU has been requesting the establishment and reorganisation of control system for food and animal stock feed, based on the principle "From Farm to Fork". However, this issue has not been resolved, due to disagreements and lack of coordination and cooperation between the state and entity level institutions in the food safety system and it remains unclear when it will be addressed [28].

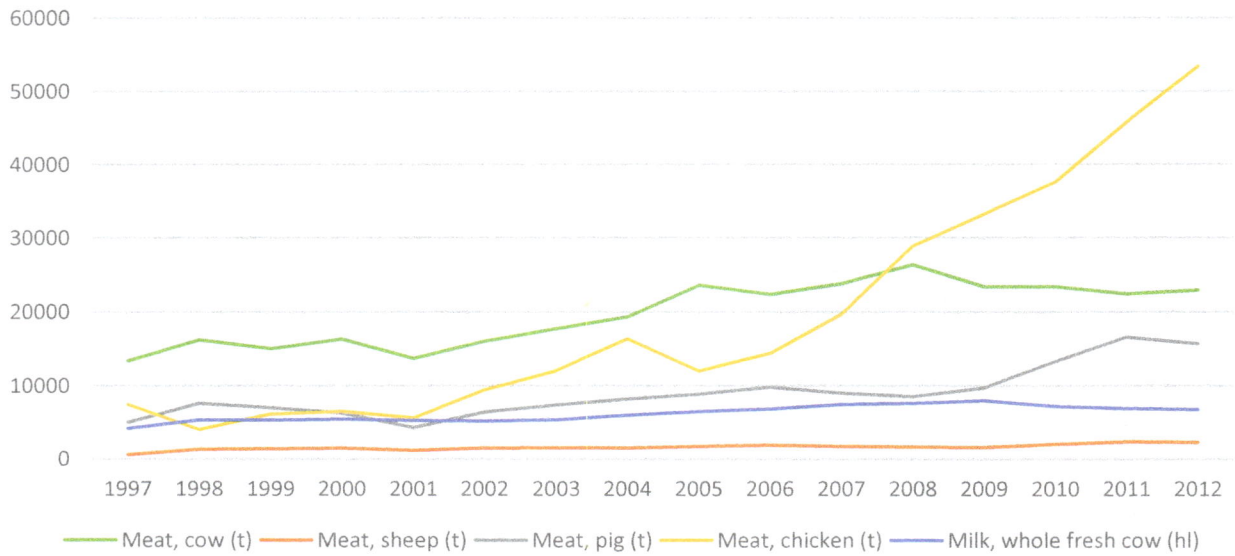

Figure 2. Meat and milk production in Bosnia and Herzegovina. Source: [29].

2.3. Crop Production

Despite relatively favourable natural conditions, crop production is facing many challenges. Frequent adverse weather conditions in key stages of crop growth (high or low temperatures, late spring or early autumn frosts, deficit or surplus rainfall) are further aggravated by lack of farm investment, high prices and poor quality of inputs (such as seeds, fertilizers, and pesticides), subsistence agriculture and traditional extensive farming practices. The result is low productivity and significantly lower yields compared with the rest of the region (Figure 3).

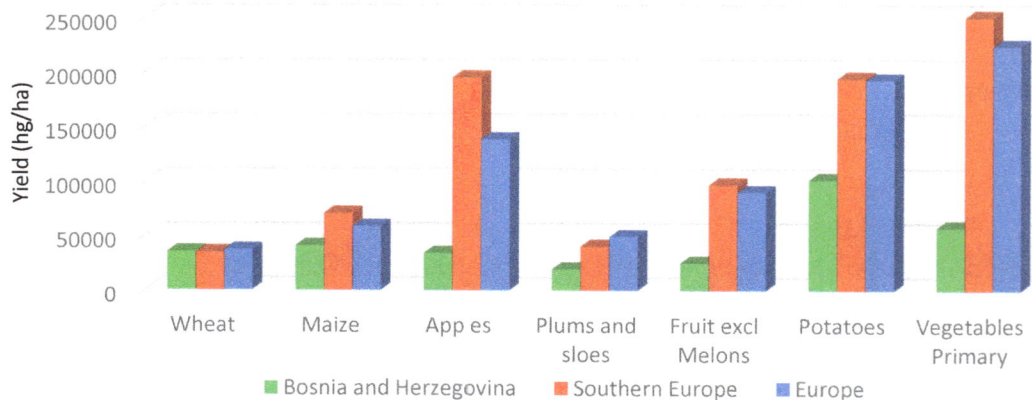

Figure 3. Yield comparison for most important crops in Bosnia and Herzegovina with the regional production. Source: [29].

Cereals dominate agricultural crop production areas (Figure 1), with maize as the main crop (more than 60% of total cereal production). Domestic production of cereals in BH is not sufficient to meet the total consumption demands [22]. The yields of forage crops are low and well below their genetic potential. Resulting feed has a low protein and high cellulose percentage, despite the usage of high quality legumes and their mixtures [30]. Hay yields on grasslands are even lower since most of them are not managed at all. Vegetable production is mainly focused on local markets and takes place in mixed farming systems, often as a supplementary source of income. Orchards and vineyards in BH suffered enormous damage during the war and are currently going through a phase of consolidation and regeneration. Plums, apples and pears dominate the production [26].

2.4. Farm Structure

The exact number of agricultural holdings, particularly family farms, is not known, since a census of agriculture that would provide a comprehensive picture of the structure of agricultural holdings has not been conducted since 1960. This includes determining and typologically categorising farms in relation to the utilized agricultural area (UAA), and determining the size of farms according to UAA. The general characteristic of the Western Balkans' agriculture is that the majority of the producers are smallholders. The main reason for this situation is the former Yugoslavian agricultural policy (farm size limit 10 hectares) [17,18]. Beside its low size (4.7 ha/farm), in most cases farms are formed from small parcels, often dislocated from each other, which makes the production more costly and less efficient., Some 250.000 farms (50%) are less than 2 hectares, and 400.000 or 80% are less than 5 hectares, while only 4% have areas beyond 10 hectares [31].

2.5. Agrarian Policy

The budget allocation to agriculture is quite low compared to developed and even developing countries amounting to about 2% of total public spending in 2007 [32]. By looking at the structure of agricultural support by groups of measures, it is noticeable that the largest amount of support during the analysed period was within the framework of direct payments to producers, which refers to payments based on output and the payment per unit area/cattle. The current BH policy resembles the EU policy in the mid-80s, which will not directly promote productivity increases in BH's agriculture. Budgetary transfers related to direct support to producers and payments based on output form more than 40% of the total agricultural budget [8]. This means that the composition of subsidies in BH is heavily oriented toward direct production rather than investments. Taking into account that most of the smallholder farmers orient their production towards subsidized crops or livestock in order to have access to an additional source of income, this form of support has the function of a social transfer rather than providing productive support to the agriculture sector [32]. The biggest difference between the agrarian policy of BH and the EU's Common Agricultural Policy (CAP) is that CAP has moved away from subsidies tied to production-unbound payments per unit area/cattle/revenue/income and is gradually moving toward investment grants (through Pillar II).

2.6. Agricultural Markets

Inadequate market access reduces the motivation of farmers to specialize and improve their performance and to increase production. Because of the fragmented agricultural production, one of the most important channels for agricultural outputs, the food industry, is not motivated to link up with agricultural producers on long-term contracts [22]. In addition, the current weak food industry is not able to take the role of being the main driver or vehicle of vertical connectivity for an efficient value chain. Consequently, the industry is not interested in improving the present value chain; it uses its negotiating powers and often does not respect agreements made. On the other hand, horizontally disjointed and disorganized producers, who since the time of socialism have expectations that the food industry has an additional social task—the purchase of produced foods—do not offer/respect contract quantities, and often do not deliver adequate quality [8]. The producers do not use the opportunity to develop its regional identity through the production of products with geographical origin, originality or traditional products. In other words, the product range is very narrow, and all companies operate within a narrow segment of the market, while the rest of the niche markets are left to import companies. That is why the food industry companies are faced with stiff competition within the domestic market, and a large number of companies are not able to independently adopt the standards required for foreign markets.

2.7. Agricultural Extension

The networks of public and private institutions, organizations and research institutions, which rely on a system of life-long education, efficient transfer of knowledge, technology and information, form a potential base for innovation and modernization, but they are not established or are severely underdeveloped. The informal sector and NGOs activities, which are implementing different, usually internationally funded, development projects, had a significant role in the transfer of technology in the previous period. These projects have launched initiatives for the establishment of private advisory services, as well as building a portal for the exchange and dissemination of various types of information, which upon completion of the project are either forgotten or unsustainable, because the system cannot become part of the technology transfer system [8]. The core of knowledge and technology transfer services is meant to go through the public extension service. The public extension services, which are located at the regional or cantonal ministries, depending on the entity, mainly perform administrative work and devote very little time to field related work. Coordination between regional extension services do not exist, nor does a systemic approach to their strengthening (especially education) and equipping. This results in a very slow technological progress in the sector compared to the agricultural sectors in neighbouring countries, which ultimately leads to a deepening of the technological and knowledge gap.

2.8. Trade Imbalance

Generally, relatively negative macroeconomic trends are the result of high foreign trade imbalance and high trade deficits, which are the main causes of the negative current account balance. Although the relative share of the negative trade balance in GDP is decreasing—from 31.9% (2006) to 28.4% (2011)—it is still unsustainably high [8]. Such a high negative trade balance is significantly affected by import of agri-food products (18.3% in 2012). Despite that BH has signed a large number of bilateral and multilateral agreements with neighbouring countries, existing inefficient trade policy mechanisms have not lead to a significant increase in exports of agriculture and food products [22]. At the same time, the agricultural sector faces many other challenges, especially in the part of the fulfilment of obligations towards the European Union, since joining inevitably requires adjustment and reform of the agricultural sector in line with EU requirements.

3. Bosnia and Herzegovina and Climate Change

BH is a mountainous country with lowlands along the banks of major rivers. Moving from north to south, the flat landscape gradually becomes wide foothills, arising from 200−600 m above sea level, and gradually turns into a mountainous region. The rest of the area is dominated by the Dinaric Alps, which extend across the whole country, from the western border with Croatia towards the southeast. The central part consist of hills among which are relatively broad river valleys and basins. Karst (barren rocky) terrains cover most of the south-western territories of BH.

As part of the general circulation of the atmosphere over the Balkans, and thus over BH, there are frequent shifts of tropical air masses during summer and the inflow of cold arctic air during winter. All these processes are largely modified by relief that occurs as a major climate modifier. For this reason, the territory of BH is split between three main types of climate: (1) Continental and moderate-continental (2) Mountain and mountain-depression; and (3) Mediterranean and modified Mediterranean climate [3,33]. Continental climate occurs in the north, the Mediterranean to the south, and the line that separates these two regions is dominated by high mountains, plateaus and cliffs which are, depending on the altitude, affected by the mountain climate.

3.1. Future Climate Change Scenarios

BH has experienced serious incidences of extreme weather events in the past two decades, causing severe economic losses. Based on available data and currently available climate projections, exposure to threats from climate change will continue to increase [33]. Observed climate changes are reflected through an increase in average temperatures in BH. For the last hundred years, the average temperature has increased by 0.8 °C (which is in line with global trends), with a tendency to accelerate—the decade of 2000–2010 is warmest in the last 120 years. According to IPCC SRES's scenarios based on SINTEX-G and ECHAM5 climate models (Figure 4), the mean seasonal temperature changes for the period 2001–2030 are expected to range from +0.8 °C to +1.0 °C above previous average temperatures [3]. Winters are predicted to become warmer (from 0.5 °C–0.8 °C), while the biggest changes will be during the months of June, July and August, with predicted changes of +1.4 °C in the north and +1.1 °C in southern areas. Precipitation is predicted to decrease by 10% in the west of the country and increase by 5% in the east. The autumn and winter seasons are expected to have the highest reduction in precipitation. Further significant temperature increases are expected during the period 2071–2100, with a predicted average rise in temperature up to 4 °C and precipitation decrease up to 50%.

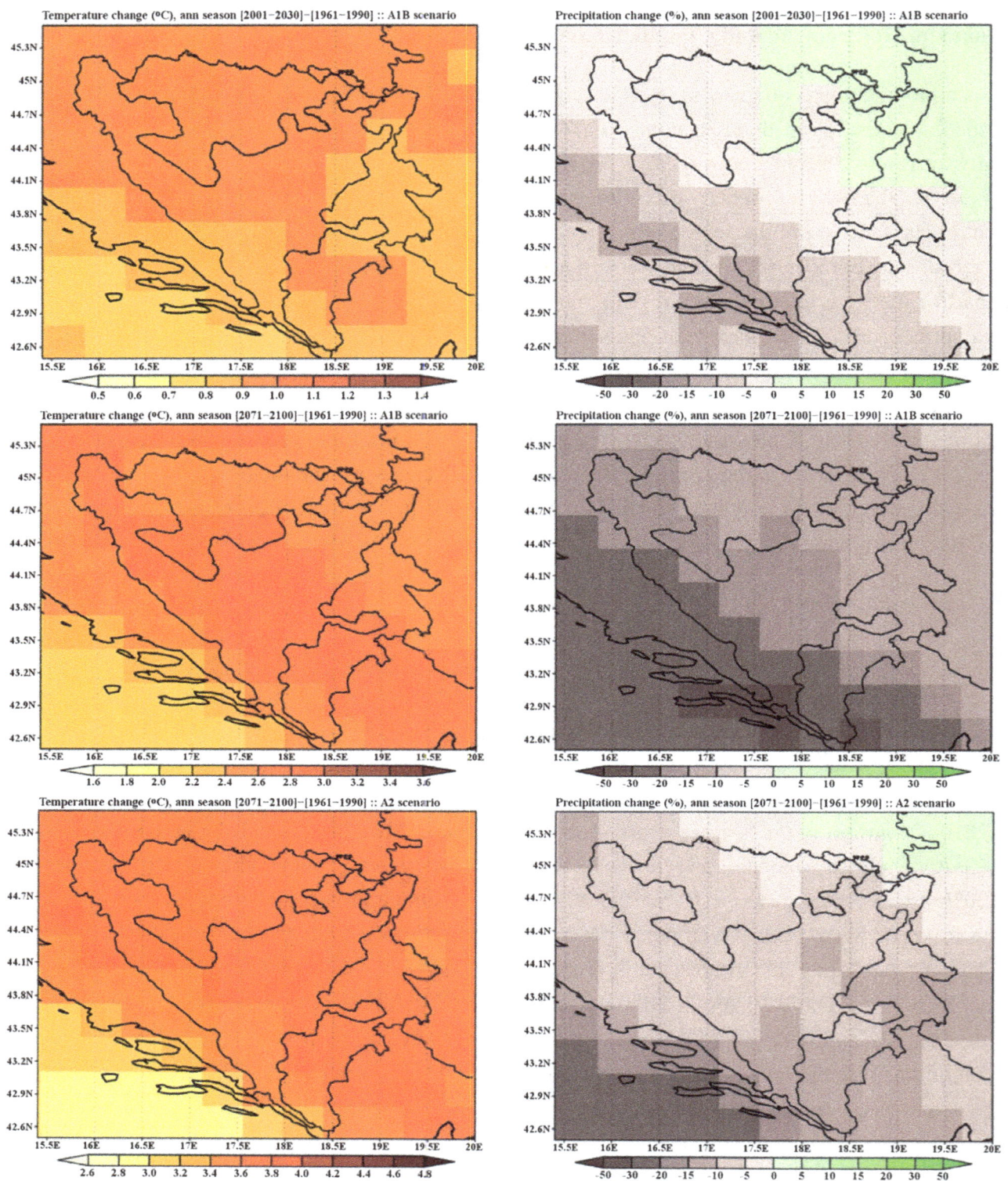

Figure 4. Temperature and precipitation projections for Bosnia and Herzegovina. Source: [3].

3.2. Recent Extreme Climate Events

Drought is a frequent adverse climatic event over the last decade in the Western Balkan and BH. Six drought periods were registered in past 14 years, resulting in enormous economic losses in agriculture. The extreme drought in 2012 was the culmination of a longer dry period, which resulted in a water supply crisis due to lowered water levels of rivers and groundwater. It is estimated that only in

2012, the drought periods caused losses of over USD 1 billion in agricultural production and yield reduction up to 70% [34]. The most affected was maize production, which is the main raw material in production of animal feed. Similar losses were found in production of barley, soybeans, alfalfa, clover, beans, meadows and pastures, which led to a lack of fodder. Lack of fodder influenced the reduction in the number of livestock and livestock production, production of milk and meat supply to the domestic market. At the same time, the effects of the drought have affected rises in food prices, and reduced the export of products. Current projections of drought impact on crop yields remain uncertain due to lack of research in this area. However, it is certain that this issue should not be ignored, taking into account the existing research. Research conducted in northeast Bosnia indicates the severity of the problems facing the country when it comes to the extension of drought periods and changes in precipitation distribution. Using climate, soil and crop data, the estimated average yield reduction in rain-fed agriculture for the six most common crops in the region for past five decades was 3.8%–20.6% and 9.3%–27.7% on loamy and heavy soil, respectively [35]. There is a big difference when the yield reduction from the last decade is compared to the rest of the research period, due to increased air temperature, precipitation, wind speed and lower relative air humidity. Most notably, the maize yield reduction in the past decade was 184% higher compared to the earlier period.

Flood is the other frequent major natural hazard related to weather and climate change in BH. In 2004, flooding affected over 300,000 people in 48 municipalities, destroyed 20,000 ha of farmland, washed away several bridges, and contaminated drinking water. In 2010, BH experienced the largest amount of precipitation recorded to that moment, which resulted in massive floods on the entire territory [36]. The flooding situation culminated with an extraordinary rainfall in May 2014, surpassing even the flood levels from 2010, affecting BH and the surrounding countries. The whole watershed of river Sava, the largest watershed in BH (67% of the total territory) was overflooded with the accumulated downstream flow of water, mud and debris, causing widespread floods in the plain. Breaches of embankments resulted in flash floods and the rivers carrying debris downstream created a path of destruction and desolation [37]. Early estimates indicate that 81 local government units suffered damages, losses, as well as social or environmental impacts to varying degrees. Around 90,000 people were temporarily displaced from their homes and more than 40,000 took extended refuge in public or private shelters. The floods in 2014 caused the damage equivalent to nearly 15% of GDP (cca 2.6 billion USD).

3.3. Impact of Climate Change on Agricultural Systems

There exist no detailed information concerning impact of climate change in BH. However, it is well documented that climate change and increased variability will lead to changes in land and water regimes, and thus have a direct impact on agriculture in the region [38] (Table 2). The impacts of climate change on agriculture are primarily reflected in changes in mean temperatures and precipitation, which subsequently lead to yield reduction and the emergence of new pathogens and diseases, crop failures long-term production declines [39]. Climate change will have different effects on agricultural systems in Europe, with likely increases in crop yields and ranges of grown crops in the north, while a significant decrease in yields are expected in the South [40]. The actual impacts of climate change depend greatly on the adaptive capacity of an affected system, region, or community to cope with the impacts and risks of climate change, which is again determined by its socioeconomic characteristics [41]. The response of crop yields to climate change varies widely, depending on agro climatic zones, species, cultivar, soil conditions, CO_2 level and other location factors [42]. Vulnerability of the agricultural sector in BH is characterised through appearance and frequency of droughts and floods, which can cause a significant yield loss or reduction.

It is expected that the duration of dry periods, the incidence of torrential flooding and intensity of soil erosion will increase during this century. In addition, a higher incidence of hail, storms and increased maximum wind speed may pose a threat to all forms of human activity [7]. This will significantly affect the water balance in soil and underground, as the increased intensity of rainfall and frequent episodes of rapid melting of snow increases the amount of water flowing over the surface and steep slopes of the mountains [3]. The result of this will be yield reductions due to reduced precipitation and increased evaporation, potentially reducing the productivity of livestock and increased incidence of pests and diseases of agricultural crops [43]. However, due to the extended vegetation period, growing seasons will be extended, with increasing potential for growing a wider range of crops.

3.4. Climate Institutions

As a signatory to UNFCCC and the Kyoto Protocol, BH is obliged to develop strategies for climate change mitigation and adaptation to changing climatic conditions, to cooperate in climate observations, research and technology transfers and to raise public awareness. It was quiet on the climate change front up until 2009, when the Initial National Communication of BH (INCBiH) under UNFCCC were published. The main objective of the INCBiH was to make an inventory of greenhouse gas emissions in line with UNFCCC reporting guidelines, which were available only for 1990 at the time. Besides the GHG inventory, the document also proposed some preliminary findings on climate change scenarios and their impact on different sectors [33]. The Second National Communication (SNCBiH) followed in 2012, which updated GHG inventory period (1991–2001), as well as vulnerability of main sectors and estimated potentials for mitigating climate change [3]. What is more important, the Climate Change Adaptation and Low-Emission Development Strategy for BH has been developed alongside SNCBiH. Although defined as the initial strategy, requiring more reliable data and additional development during the course of its implementation, it shows the political will towards building necessary capacities and policies towards low-emission and climate resilient development [44]. The strategy regards water resources and agriculture as the major priorities, which affect the other sectors to a greater or lesser extent and present some concrete activities and outputs, together with indicators, indicative budget and timeframe. Defining agriculture as a priority sector, along with the future integration of adaptation and mitigation measures in the agricultural strategic documents and the adequate funding, creates fertile ground for the realization of activities related to adaptation to climate change in agriculture proposed in the next section.

Table 2. Climate change impacts on agriculture in Western Balkan.

Nr.	Climate Change Variable	Impact	Country/Region	Source
1	Change in temperature and rainfall according to IPCC A1B and A2 scenarios	A1B (2001–2030) change in yields: winter wheat (−16% to 21%); maize (−6% to 71%); maize, irrigated (−5% to 6%); soybeans, irrigated (21%–67%) A2 (2071–2100): winter wheat (−10% to 6%); maize (−52% to −22%); maize, irrigated (−7% to 4%); soybeans, irrigated (−9% to 43%)	Serbia	[45]
2	Mean relative changes in water-limited crop yield simulated by the ClimateCrop model for the 2050s compared with 1961–1990 for 12 different climate models projections under the A1B scenario.	Projected changes in water-limited crop yield from 5% to −25%	Western Balkan	[46]
3	Rain-fed yield reduction in vulnerable areas calculated with FAO Crop Yield Response to Water Deficit using different IPCC SRES scenarios (2025–2100)	Yield reduction Vinegrape 46%–59% Tomato 72%–84% Winter wheat 8%–25% Alfalfa 58%–70% Apple 46%–59%	FYR Macedonia	[47]
	Cost of decreased production (million EUR—current prices)	Winter wheat 4.1 (2025)–8.6 (2100) Vinegrape 18.2 (2025)–23.4 (2100) Alfalfa 7.0 (2025)–8.5 (2100)		
4	Impact of a 2 °C global temperature rise on Mediterranean region according to HadCM3 simulation and CROPSYST crop simulation model	Significant yield decrease for all researched crops (legumes, C3 summer crops, tuber crops, cereals) except C4 summer crops	Mediterranean region (Serbia grid cell)	[48]

4. Policy Implications of Climate Change

The global challenges of adaptation to climate change in agriculture are many. The core challenge is to produce more food by using less resources, under changing production conditions and with net reduction in greenhouse gas emissions [49]. The extent of sustainable adaptation depends on the adaptive capacity, knowledge, skills, robustness of livelihoods and alternatives, resources and institutions accessible to enable undertaking effective adaptation [50]. Extreme weather events, such as increased intensity of droughts, frequency of heat waves, heavy precipitation events resulting in the floods and landslides, are becoming increasingly more frequent [51]. These events have significant impacts on both human lives and national economies and are expected to continue to increase in the future if we do not take sufficient actions. Extreme weather events have the potential to reverse development progress and entrench poverty, especially in developing countries like BH, characterized by limited social safety nets, lack of access to markets, capital, assets, or insurance mechanisms [52]. Below, we highlight relevant socio-political and technological processes that should be the basis for further strategic planning and development of necessary institutions in BH. These processes should address adequately the problem of climate change in agriculture.

4.1. Policy and Research Capacity

Significant difficulties for the agricultural sector in BH arise from the state's constitutional order, according to which all established levels of government—from national to municipal—have the authority for planning and management in agriculture. This sort of organization on the one hand does not allow for the establishment of functional coordinated networks of institutions, and on the other hand it leads to unnecessary and costly multiplications of institutions of the same or similar domains [8]. The unanimous attitude of most sectoral interest groups and non-governmental organizations is that BH needs a national Ministry of Agriculture, formation of which would represent a possible way to better coordinate agricultural policies and consistent articulation of the interests of the sector in international relations (especially in the process of joining with EU), as well as easier establishing the necessary information systems and registers [28]. The agricultural sector is already facing many challenges, which will be further exacerbated by increased frequency and severity of extreme weather events, eventually leading to increased vulnerability. Climate change is a global challenge whose environmental impacts knows no boundaries or borders. Only the unified stance and common policy will lead to effective coordination and harmonisation in terms of implementation of adaptation and mitigation measures in agriculture and all other vulnerable sectors.

Assessment of vulnerability to climate change is an important tool for the analysis and presentation of data at a national level, where the current and potential consequences of climate change are presented to stakeholders in a convenient way and serve as a base for further adaptation policy decisions [53]. The process of adaptation begins with an assessment of the different dimensions of vulnerability and the range of potential options for action, including their justification. A top-down approach is derived from global climate projections, which is further downscaled and applied to assess regional impacts of climate change, while the bottom-up approaches include the involvement of the population and stakeholders of the system in identifying climate-change stresses, impacts and adaptive strategies [54]. In terms of agriculture, assessment of agricultural vulnerability to climate change should lead to identification of particularly vulnerable regions and agricultural production systems, which should further lead to choice of specific adaptation measures and resource allocation for adaptation.

BH needs to have the political and scientific knowledge and public support to adapt to climate change. Investment in agricultural research and development has been declining, or stagnating at best, thus creating a knowledge gap between low and high-income countries [55]. It is crucial to take a long-term, strategic view and to conduct research in order to meet future climate challenges and to develop approaches to facilitate transfer of new knowledge and technologies into practical application. Extension services have played a key role in promoting agricultural productivity and dissemination of knowledge and their role in promoting adaptation measures to climate change will certainly have the same importance [56,57]. Therefore, it is necessary to strengthen the capacity of underdeveloped public agricultural extension services, enabling them to take leading role in terms of strengthening innovation process, building linkages between farmers and other agencies, and institutional development. Research institutes and agricultural universities as a country's leaders in knowledge generation need to create linkages and provide direct transfer of information by educating extension workers regarding the advantages and potential technologies and practices. This can be accomplished on different ways, such as demonstration fields, joint research projects, field trials, training seminars *etc.* [58].

4.2. Technological Development

Efforts to reduce vulnerability to climate change must include strengthening of adaptive capacity and resilience of rural communities. Such strategies will enable farmers to achieve food security and increased well-being under current climatic conditions and will directly contribute to increasing their ability to cope with future uncertainty. More productive and resilient agriculture requires a major shift in the way land, water, soil nutrients and genetic resources are managed to ensure that these resources are used more efficiently and sustainably [59]. Making this shift will require considerable changes in national and local governance, legislation, policies and financial mechanisms. The elements of this categorization include mechanical, biological, chemical, agronomic, biotechnological, and informational innovations [60]. Production that is more efficient and creates more opportunities and access to broader markets can boost smallholders' resilience and create sustainable livelihoods while helping to meet growing demand for food.

Development of new high yielding, input use-efficient, abiotic and biotic stress-resistant varieties with enhanced traits better suited to adapt to climate change is crucial for agricultural adaptation to climate change. Activities and research on plant genetic resources in BH started during eighties of last century, within a project called the "Gene Bank of Yugoslavia". Unfortunately, most of the documents from that period went missing or were destroyed during the last war. Activities such as *ex situ* and *in situ* inventory and conservation of plant genetic resources for agriculture have been restarted during the beginning of this century with the help of international donors [61]. After the inventory and identification of local genotypes, the next step should be linking the selection process with the other stakeholders by employing participatory techniques in order to improve the effectiveness and impact of agricultural research. Having farmers and other stakeholders involved in the development of varieties through participatory plant breeding [62,63] may lead quick and cost-effective production of new breeds and varieties of crops adapted to local needs.

Weather and climate information systems, including early warning systems, aim to reduce vulnerability and improve response capacities of those at risk by increasing their preparedness [64]. Prompt identification of a risk and communication can enable timely responses and assist in farm level adaptation. Farmers should use present and future climate-related information to plan and manage weather risks, maximize productivity, and minimize the environmental impacts of farming practices. Information from the network of agro-monitoring stations combined with remotely sensed information, enable the development of biophysical models used to estimate weather conditions, soil and nutrient status, crop water needs, soil erosion, pest and disease emergence, choice of crop variety best suited for local conditions, *etc.* [65]. Therefore, strengthening of technologies, human resources and development of monitoring, warning and forecasting networks related to hydrological, meteorological, climatic and environmental risks should be considered as one of the priorities.

Global climate change will continue to lead towards significant changes in annual precipitation patterns in south-eastern Europe. In such circumstances, irrigation can certainly be one of the key mechanisms of adaptation in agriculture. There exists no available data on irrigated areas or crops in BH. The total irrigated area according to unofficial data is only 0.4% of arable land, which is considerably less than in neighbouring countries, especially EU [66]. In this situation, irrigation is a measure that can reduce the problems of critical drought periods by improving and stabilizing yields. Due to the documented water deficits and the growing incidences of drought, irrigation should settle around 33% annual water needs for plants in south, 14% in north and 8% in central BH [67]. These needs will increase further according to future climate projections for BH, which include extension of vegetation period, increased frequency of extreme temperatures and longer frost-free periods, leading to additional increase in evapotranspiration and reduced soil moisture content [43,68]. The agricultural sector needs the support and cooperation of other sectors in order to operate in a sustainable manner and follow the principles of integrated water management. Public and private investments in irrigation would enable the expansion of irrigated areas and irrigation use as a supplement to rain-fed agriculture in order to stabilize and increase yields.

5. Conclusions

We have reviewed the agricultural sector in Bosnia and Herzegovina and the likely impacts of climate change upon it. We have also proposed policy options to optimize opportunities and mitigate consequences of climate change. We conclude that BH is a country rich in natural resources and biodiversity, and large parts of the country possess a favourable climate for agricultural production. However, despite such endowments, the agricultural sector is hampered by low productivity, extensive farming practices and technologies, carried out on small and fragmented farms. This is further exacerbated by a weak and inefficient agrarian policy and legislation, low budget allocations for agriculture, inadequate market access and a general lack of information and knowledge.

The legacy of the past socio-political system and the current complex political structure have significant consequences for the agricultural development. Agriculture is highly vulnerable to climate change. Higher temperatures and changes in precipitation are reducing crop yields and increasing the likelihood of short-term crop failures and long-term production declines. Bosnia and Herzegovina, as well as other Mediterranean countries in southern and south-eastern Europe, are expected to experience significant agricultural production losses. As a country that is a potential candidate for the EU-membership, BH needs to implement reforms in the agricultural and rural development sectors in order to reduce the significant policy gap compared to other European countries. In order to cope with the challenges of climate change and climate variability, it is imperative to raise the political awareness and increase recognition on all governmental levels about the impeding threats of climate change on the agricultural sector. The literature on climate change impacts and vulnerability in the agricultural sector stresses the importance of adaptation and urgency to implement adaptation and mitigation measures. The recently published Climate Change Adaptation and Low-Emission Development Strategy for BH shows that there is a political will to build necessary capacities and policies towards low-emission and climate resilient development. We addressed some of the relevant socio-political and technological processes that should be the basis for further strategic planning and institutions that are needed to create the capacity to address climate change in agriculture, agricultural development, knowledge transfer and implementation. Most of them require significant funds and time for their implementation. However, it is unlikely that the agricultural sector alone will be able to cope with the process of adaptation. Only the development of a favorable environment together with political, institutional, economic, social and other actors will lead to a successful agricultural adaptation to climate change.

Acknowledgments

This study is supported by the project "Agricultural Adaptation to Climate Change—Networking, Education, Research and Extension in the West Balkans", funded by HERD—Programme for Higher Education, Research and Development 2010–2014.

Special thanks to Melisa Ljusa from the Faculty of Agriculture and Food Sciences, University of Sarajevo, for providing valuable insights on the history of agricultural development in Bosnia and Herzegovina.

The authors would like to thank the anonymous reviewers for their helpful and constructive comments.

Conflicts of Interest

The authors declare no conflict of interest.

References

1. Brooks, N.; Adger, W.N.; Kelly, P.M. The determinants of vulnerability and adaptive capacity at the national level and the implications for adaptation. *Glob. Environ. Chang.* **2005**, *15*, 151–163.
2. Kreft, S.; Eckstein, D. *Global Climate Risk Index 2014*; Germanwatch: Bonn, Germany, 2013.
3. UNFCCC. Second National Communication of Bosnia and Herzegovina under the United Nation Framework Convention on Climate Change (SNCBIH); Available online: http://www.ba.undp.org/content/bosnia_and_herzegovina/en/home/library/environment_energy/sncbih-2013.html (accessed on 2 July 2014).

4. Smit, B.; Skinner, M.W. Adaptation options in agriculture to climate change: A typology. *Mitig. Adapt. Strateg. Glob. Chang.* **2002**, *7*, 85–114.

5. Eriksen, S.; Aldunce, P.; Bahinipati, C.S.; Martins, R.D.A.; Molefe, J.I.; Nhemachena, C.; O'Brien, K.; Olorunfemi, F.; Park, J.; Sygna, L.; *et al.* When not every response to climate change is a good one: Identifying principles for sustainable adaptation. *Clim. Dev.* **2011**, *3*, 7–20.

6. South East European Forum on Climate Change Adaptation (SEEFCCA). *Climate Vulnerability Assessment: Serbia*; SEEFCCA: Belgrade, Serbia, 2012.

7. Custovic, H.; Djikic, M.; Ljusa, M.; Zurovec, O. Effect of climate changes on agriculture of the western Balkan countries and adaptation policies. *Agric. For.* **2012**, *58*, 127–141.

8. Federal Ministry of Agriculture Water Management and Forestry. *Medium term development strategy of agricultural sector in Federation of Bosnia and Herzegovina for 2014–2018*; Federal Ministry of Agriculture Water Management and Forestry: Sarajevo, Bosnia and Herzegovina, 2013.

9. Kamberović, H. Osnovna obilježja razvoja društva u Bosni i Hercegovini od 1945 do 1953. *Časopis Suvrem. Povijest* **1998**, *30*, 359–376.

10. Golić, B. Bosanskohercegovačka ekonomija od ZAVNOBIH-a do Daytona. *Academy of Sciences and Arts of Bosnia and Herzegovina, Spec. Publ.* **2007**, *37*, 162–177.

11. Kamberović, H. Karakteristike društva u Bosni i Hercegovini neposredno nakon drugog svjetskog rata. *Academy of Sciences and Arts of Bosnia and Herzegovina, Spec. Publ.* **2007**, *37*, 214–227.

12. Katz, V. Društveni i ekonomski razvoj Bosne i Hercegovine (1945–1953); Institut za istoriju: Sarajevo, Bosnia and Herzegovina, **2001**.

13. Simon, D. Development revisited: Thinking about, practicing and teaching development after the Cold War. In *Development as Theory and Practice: Current Perspectives on Development and Development Cooperation*; Simon, D., Narman, A., Eds.; Routledge: London, UK, 1999, pp. 17–54.

14. Nurković, R. Distribution of industry in Bosnia and Herzegovina. *Glasnik ZRS Koper* **2008**, *8*, 67–69.

15. Mirjanić, S. Poljoprivredno Zemljište u Društvenom Planu Republike, *Savjetovanje o Temi Zemljište u prostornom planu SR BiH*; Poljoprivredni Fakultet: Sarajevo, Bosnia and Herzegovina, 1983.

16. Selak, V. Potražnja hrane i zemljišni prostor u SR BiH, *Savjetovanje o Temi Zemljište u Prostornom Planu SR BiH*; Poljoprivredni Fakultet: Sarajevo, Bosnia and Herzegovina, 1983.

17. Marinković, G. Stvaranje državne i društvene svojine na području Srbije i bivše Jugoslavije. *Zbornik Radova Građev. Fakulteta* **2012**, *21*, 135–147.

18. Mizik, T. A snapshot of western Balkan's agriculture from the perspective of EU accession. *Stud. Agric. Econ.* **2012**, *114*, 39–48.

19. Puljiz, V. *Eksodus poljoprivrednika*; Centar za sociologiju sela, grada i prostora Instituta za društvena istraživanja Sveučilišta u Zagrebu: Zagreb, Yugoslavia, 1977.

20. Žimbrek, T. *Agrarna politika-izabrana predavanja*; Agronomski fakultet Sveučilišta u Zagrebu: Zagreb, Croatia, 2008.

21. Ivankovic, M.; Bojnec, S.; Kolega, A.; Selak, V. Economic and social role of agriculture in the Federation of Bosnia and Herzegovina. *J. Gen. Soc. Issues* **2006**, *15*, 84–85.

22. Ministry of Foreign Trade and Economic Relations (MoFTER). *Agriculture Report for Bosnia and Herzegovina for 2012*; Ministry of Foreign Trade and Economic Relations (MoFTER): Sarajevo, Bosnia and Herzegovina, 2012.

23. European Bank for Reconstruction and Development (EBRD). *Transition Report 2013*; Available online: http://www.ebrd.com/news/publications/transition-report/transition-report-2013.html (accessed on 7 October 2014)

24. Agency for Statistics of Bosnia and Herzegovina (BHAS). *Bosnia and Herzegovina in figures—Statistical Bulletin*; Agency for Statistics of Bosnia and Herzegovina (BHAS): Sarajevo, Bosnia and Herzegovina, 2013.

25. Agency for Statistics of Bosnia and Herzegovina (BHAS). Structure of agricultural land in Bosnia and Herzegovina. Unpublished data, available on request; Sarajevo, Bosnia and Herzegovina, **2014**.

26. Agency for Statistics of Bosnia and Herzegovina (BHAS). *Harvested area, total production and yield of main crops, Statistical Bulletin*; Agency for Statistics of Bosnia and Herzegovina (BHAS): Sarajevo, Bosnia and Herzegovina, 2014.

27. Ministry of Foreign Trade and Economic Relations (MoFTER). *Competitiveness assessment of three agribusiness value-chains in Bosnia and Herzegovina*; Anteja ECG: Ljubljana, Slovenia, 2012.

28. Green Council. *Establishment of the BiH Ministry of agriculture, food and rural Development and other Structures as a way of achieving economic progress for BiH in the EU integration process*; Green Council: Sarajevo, Bosnia and Herzegovina, 2013.

29. FAOSTAT. Compare data tool—Production. Available online: http://faostat3.fao.org/compare/E (accessed on 8 December 2014).

30. Alibegovic-Grbic, S.; Civic, H.; Cengic, S.; Muratovic, S.; Dzomba, E. Effect of weather conditions, stage of plant growth and N application on yield and quality of grassland in Bosnia and Herzegovina. In *Land use systems in grassland dominated regions. Proceedings of the 20th General Meeting of the European Grassland Federation*, Luzern, Switzerland, 21–24 June 2004, pp. 897–899.

31. European Comission (EC). *Bosnia and Herzegovina country report*; Arcotrass GmbH: Germany, 2006.

32. World Bank. *Agricultural sector policy note for Bosnia and Herzegovina—Trade and integration policy notes*; World Bank: Washington, DC, USA, 2010.

33. UNFCCC. *Initial National Communication of Bosnia and Herzegovina under the United Nation Framework Convention on Climate Change (INCBIH)*; Ministry for Spatial Planning, Construction and Ecology of Republic of Srpska: Banja Luka, Bosnia and Herzegovina, 2009.

34. Hodzic, S.; Markovic, M.; Custovic, H. Drought conditions and management strategies in Bosnia and Herzegovina—Concise country report. Available online: http://www.ais.unwater.org/ais/pluginfile.php/548/mod_page/content/72/Bosnia_Herzegovina_CountryReport.pdf (accessed on 11 September 2014)

35. Zurovec, J.; Cadro, S. Climate changes, the need and importance of crop irrigation in northeastern Bosnia and Herzegovina. In Proceedings of the 21st Scientific-Expert Conference in Agriculture and Food Industry, Neum, Bosnia and Herzegovina, 29 September 2010—2 October 2 2010; pp. 705–716.

36. International Fund for Agricultural Development (IFAD). Environmental and climate change assessment—Bosnia and Herzegovina. Available online: http://www.ifad.org/climate/resources.htm (accessed on 7 September 2014)

37. European Comission (EC). Bosnia and Herzegovina recovery needs assessment, floods 14–19 May—Executive summary. Available online: http://ec.europa.eu/enlargement/pdf/press_corner/floods/rna-executive-summary.pdf (accessed on 11 September 2014).

38. Kurukulasuriya, P.; Rosenthal, S. Climate Change and Agriculture: A Review of Impacts and Adaptations. *Climate Change Series 91. Environment Department Papers*; World Bank: Washington, DC, USA, **2013**.

39. Nelson, G.C.; Rosegrant, M.W.; Koo, J.; Robertson, R.; Sulser, T.; Zhu, T.; Ringler, C.; Msangi, S.; Palazzo, A.; Batka, M. Climate Change: Impact on agriculture and costs of adaptation. International Food Policy Research Institute, Washington, DC, USA, 2009.

40. Alcamo, J.; Moreno, J.M.; Nováky, B.; Bindi, M.; Corobov, R.; Devoy, R.J.N.; Giannakopoulos, C.; Martin, E.; Olesen, J.E.; Shvidenko, A. Europe. In *Climate Change 2007: Impacts, Adaptation and Vulnerability. Contribution of Working Group II to the Fourth Assessment Report of the Intergovernmental Panel on Climate Change*; Parry, M.L., Canziani, O.F., Palutikof, J.P., van der Linden, P.J., Hanson, C.E., Eds.; Cambridge University Press: Cambridge, UK, 2007; pp. 541–580.

41. Smit, B.; Pilifosova, O. Adaptation to climate change in the context of sustainable development and equity. *Sustain. Dev.* **2003**, *8*, 9.

42. Thornton, P.K.; Jones, P.G.; Alagarswamy, G.; Andresen, J. Spatial variation of crop yield response to climate change in east Africa. *Glob. Environ. Chang.* **2009**, *19*, 54–65.

43. Bär, R.; Rouholahnejad, E.; Rahman, K.; Abbaspour, K.C.; Lehmann, A. Climate change and agricultural water resources: A vulnerability assessment of the Black Sea catchment. *Environ. Sci. Policy* **2015**, *46*, 57–69.

44. Council of Ministers of BiH. Climate change adaptation and low-emission development strategy for Bosnia and Herzegovina. Available online: http://www.ba.undp.org/content/bosnia_and_herzegovina/en/home/library/environment_energy/climate-change-adaptation-and-low-emission-development-strategy-/ (accessed on 9 August 2014).

45. Mihailović, D.T.; Lalić, B.; Drešković, N.; Mimić, G.; Djurdjević, V.; Jančić, M. Climate change effects on crop yields in Serbia and related shifts of Köppen climate zones under the SRES-A1B and SRES-A2. Available online: http://onlinelibrary.wiley.com/doi/10.1002/joc.4209/abstract (accessed on 11 September 2014).

46. Iglesias, A.; Quiroga, S.; Diz, A. Looking into the future of agriculture in a changing climate. *Eur. Rev. Agric. Econ.* **2011**, *38*, 427–447.

47. The Regional Environmental Center for Central and Eartern Europe (REC). The impacts of climate change on food production in the western Balkan region. Available online: http://documents.rec.org/topic-areas/Impacts-climage-change-food-production.pdf (accessed on 13 September 2014)

48. Giannakopoulos, C.; Bindi, M.; Moriondo, M.; LeSager, P.; Tin, T. Climate change impacts in the mediterranean resulting from a 2 °C global temperature rise. WWF: Gland, Switzerland, 2005.

49. Lybbert, T.J.; Sumner, D.A. Agricultural technologies for climate change in developing countries: Policy options for innovation and technology diffusion. *Food Policy* **2012**, *37*, 114–123.

50. IPCC. *Climate Change 2007: Synthesis Report. Contribution of Working Groups I, II and III to the Fourth Assessment Report of the Intergovernmental Panel on Climate Change, Core Writing Team*; Pachauri, R.K., Reisinger, A., Eds.; IPCC: Geneva, Switzerland, 2007; p. 104.

51. IPCC. Special Report—Managing the Risks of Extreme Events and Disasters to Advance Climate Change Adaptation; Cambridge University Press: New York, NY, USA, 2011.

52. Shepherd, A.; Mitchell, T.; Lewis, K.; Lenhardt, A.; Jones, L.; Scott, L. Muir-Wood, R. *Full Report: Geography of Disasters, Poverty and Climate Extremes in 2030*; ODI: London, UK, 2013.

53. Füssel, H.-M.; Klein, R.J.T. Climate change vulnerability assessments: An evolution of conceptual thinking. *Clim. Chang.* **2006**, *75*, 301–329.

54. Dessai, S.; Hulme, M. Does climate adaptation policy need probabilities? *Clim. Policy* **2004**, *4*, 107–128.

55. Beddington, J.; Asaduzzaman, M.; Clark, M.; Bremauntz, A.; Guillou, M.; Jahn, M.; Lin, E.; Mamo, T.; Negra, C.; Nobre, C.; *et al.* The role for scientists in tackling food insecurity and climate change. *Agric. Food Secur.* **2012**, *1*, 10.

56. Maponya, P.; Mpandeli, S. The role of extension services in climate change adaptation in Limpopo province, South Africa. *J. Agric. Ext. Rural. Dev.* **2013**, *5*, 137–142.

57. Bryan, E.; Deressa, T.T.; Gbetibouo, G.A.; Ringler, C. Adaptation to climate change in Ethiopia and South Africa: Options and constraints. *Environ. Sci. Policy* **2009**, *12*, 413–426.

58. FAO. *Management of Agricultural Research: A Training Manual. Module 8: Research-Extension Linkage*; FAO: Rome, Italy **1997**.

59. FAO. Climate-smart agriculture sourcebook. Available online: http://www.fao.org/climatechange/climatesmart/en/ (accessed on 12 November 2014).

60. Sunding, D.; Zilberman, D. The agricultural innovation process: Research and technology adoption in a changing agricultural sector. *Handb. Agric. Econ.* **2001**, *1*, 207–261.

61. FAO. *Country Report on the State of Plant Genetic Resources for Food and Agriculture—Bosnia and Herzegovina*; FAO: Rome, Italy, 2010.

62. Witcombe, J.R.; Joshi, K.D.; Gyawali, S.; Musa, A.M.; Johansen, C.; Virk, D.S.; Sthapit, B.R. Participatory plant breeding is better described as highly client-oriented plant breeding. I. Four indicators of client-orientation in plant breeding. *Exp. Agric.* **2005**, *41*, 299–319.

63. Mba, C.; Guimaraes, E.; Ghosh, K. Re-orienting crop improvement for the changing climatic conditions of the 21st century. *Agric. Food Secur.* **2012**, *1*:7.

64. Hogarth, J.R.; Campbell, D.; Wandel, J. Assessing human vulnerability to climate change from an evolutionary perspective. In *Reducing Disaster: Early Warning Systems for Climate Change*; Singh, A., Zommers, Z., Eds.; Springer: Netherlands 2014; pp. 63–87.

65. Weiss, A.; Van Crowder, L.; Bernardi, M. Communicating agrometeorological information to farming communities. *Agric. For. Meteorol.* **2000**, *103*, 185–196.

66. Siebert, S.; Döll, P.; Hoogeveen, J.; Faures, J.M.; Frenken, K.; Feick, S. Development and validation of the global map of irrigation areas. *Hydrol. Earth Syst. Sci.* **2005**, *9*, 535–547.

67. Vlahinic, M. Hydro accumulation, agriculture, and land and water management in Bosnia and Herzegovina. *Voda i mi* **2000**, *27*, 26–37.

68. Bazzaz, F.; Sombroek, W. Global climate change and agricultural production. Direct and indirect effects of changing hydrological, pedological and plant physiological processes; FAO and John Wiley & Sons: London, UK, 1996.

2

Achieving Water and Food Security in 2050: Outlook, Policies, and Investments

Dennis Wichelns

P.O. Box 2629, Bloomington, IN 47402, USA; E-Mail: dwichelns@csufresno.edu

Academic Editor: Stephen J. Herbert

Abstract: Food production in 2050 will be sufficient, globally, but many of the poor will remain food insecure. The primary cause of food insecurity will continue to be poverty, rather than inadequate food production. Thus, policies and investments that increase the incomes of the poor will remain the best ways to extend food security to all. Investments that promote growth in sustainable agriculture and provide non-farm employment opportunities in rural areas of lower income countries will be most helpful. There will be sufficient water, globally, to achieve food production goals and sustain rural and urban livelihoods, if we allocate and manage the resource wisely. Yet, water shortages will constrain agricultural production and limit incomes and livelihood opportunities in many areas. Policies and investments are needed to extend and ensure access to water for household use and agricultural production. Challenges requiring the attention of policy makers and investors include increasing urbanization and increasing demands for land and water resources. Policy makers must ensure that farmers retain access to the water they need for producing food and sustaining livelihoods, and they must create greater opportunities for women in agriculture. They must also motivate investments in new technologies that will enhance crop and livestock production, particularly for smallholders, and encourage the private sector to invest in activities that create employment opportunities in rural areas.

Keywords: agriculture; livelihoods; poverty; smallholders; sustainability; women

1. Introduction

The Millennium Development Goals (MDGs), and their targets pertaining to 2015, motivated notable advances in poverty reduction, and in the health and welfare of women and children in many lower income countries [1–3]. The international community is now engaged in the process of defining and agreeing upon a new set of global objectives pertaining more broadly to the notion of achieving sustainable economic development [4]. Two of the 17 proposed Sustainable Development Goals (SDGs) align closely with issues regarding water and food security [5]. In particular, SDG 2 calls for ending hunger, achieving food security, and improving nutrition, while promoting sustainable agriculture. SDG 6 calls for ensuring the availability and sustainable management of water and sanitation for all [6].

The objectives listed within SDG 2 describe both the demand and supply aspects of food security. In addition to calling for universal access to safe, nutritious, and sufficient food, the objectives call for doubling the agricultural productivity and incomes of small-scale food producers, with a particular focus on women, indigenous peoples, family farmers, pastoralists, and fishers. They note also the importance of ensuring secure and equal access to land, other productive resources and inputs, knowledge, financial services, markets and opportunities for value addition, and non-farm employment [6]. It is essential that smallholders and their households have access to the resources and inputs needed to engage in livelihoods that will enable them to purchase food, particularly at times of short supplies and high prices.

Also embedded within SDG 2 is the call for ensuring sustainable food production systems and implementing resilient agricultural practices that increase productivity and production, while maintaining ecosystems, and mitigating the potential impacts of climate change [6]. To this end, it is essential that the international research community continue to generate global public goods, such as state-of-the-art research and outreach regarding climate-resilient agriculture, new varieties of cultivated plants, and improvements in livestock health and performance.

Several of the objectives within SDG 6 pertain to water supply, sanitation, and wastewater recovery, and several reflect issues involving agriculture more directly. For example, some of the objectives describe the need to increase water use efficiency in all sectors, achieve sustainable withdrawals of freshwater resources, implement integrated water resources management, protect the quality of lakes, rivers, wetlands, and aquifers, and substantially reduce the number of people impacted by water scarcity [6]. Although not stated explicitly, the need to ensure access to water for use in food production and in support of other livelihood activities is implied within these objectives, as noted by the call for achieving sustainable freshwater withdrawals, protecting water sources, and alleviating the impacts of water scarcity.

My goals in this paper are to review the outlook for water and food security to 2050, and describe some of the policies and investments that will be helpful in achieving the objectives put forth regarding water and food in the Sustainable Development Goals. I focus largely on agriculture, given the sector's notable role in providing employment and livelihoods in rural areas of lower income countries, while recognizing the potential implications of increasing urbanization. I describe also the importance of considering gender, climate change, and the environment when selecting investments and designing policies to promote the wise and sustainable use of land and water resources.

2. The Outlook to 2050

2.1. Food Production

It is likely there will be sufficient food to support the global population in 2050, even though the annual rates of increase in the yields of several important food crops have declined in recent years [7,8].

Successful efforts to supply sufficient food will include closing yield gaps in lower income countries, in part by increasing the use of better seeds and fertilizer, improving resource management, and enhancing the technology of crop production through advances in genomics and phenomics [9–13]. In addition, public officials must choose policies and investments that encourage increases in agricultural productivity on large and small farms, and in irrigated and rainfed settings [14–16]. Yet, even with success in producing sufficient food, globally, food insecurity will remain a serious issue in some regions and countries where per capita food consumption will remain inadequate [7].

Global incomes are expected to rise substantially by 2050, yet areas of notable poverty will persist in some countries, particularly in sub-Saharan Africa. The per capita annual income in 2050 likely will remain below $1000 in 15 of the 98 lower income countries examined by Alexandratos and Bruinsma [7]. Average daily food consumption might remain below 2700 kcal per person in 16 of the 98 countries. Those 16 countries will be home to a population of 800 million. By comparison, an estimated 4.7 billion people (52% of global population) will live in countries with national daily averages of more than 3000 kcal per person in 2050, up from 1.9 billion (28%) in 2011 [7].

Estimates of the increase in food production required to ensure food security in 2050, from the global perspective, range from 60% to 100% above the production achieved in 2005 [17]. Those proportions are notably higher than the rate of increase in population to 2050, due largely to the increasing demands and changing preferences for food, that come with higher incomes. Household and per capita food consumption will increase in many countries, and many residents will consume more meat and vegetables. Those commodities, particularly beef, generally require more water and other productive inputs than grains, per calorie of food consumed [18]. The increasing demand for meat will place additional pressure on limited water resources in some regions. The projected increases in food demand, when realized, will reflect substantial improvement in food and nutritional security for those households with sufficient income to afford adequate food supplies.

As in the present, much of the food and nutritional insecurity that persists in 2050 will be found largely in poor households in countries with lower gross incomes and in areas where depleted or degraded natural resources no longer support viable livelihood activities for smallholders. The primary cause of food insecurity will be the persistent poverty that prevents households from gaining access to sufficient food and nutrition, particularly during periods of notable scarcity and high prices. Thus, the policies and investments most likely to enhance food security will be those that promote economic growth and increase incomes, particularly in rural areas, where many of the world's poor are engaged in agriculture. Substantial public and private-sector investments and policy interventions are needed between now and 2050, particularly in agriculture, to reduce poverty, increase incomes, and ensure food security for all.

The extent and severity of food insecurity in 2050 likely will correspond to the extent and depth of dollar-based poverty. In 2007, an estimated 47% of the world's poor earning less than $1.25 per day lived in South Asia, while 31% and 17% lived in sub-Saharan Africa and South Asia and the Pacific, respectively [19]. Just 2.3% of the world's poor earning less than $1.25 per day lived in Latin America and the Caribbean. Projecting future poverty levels and geographic distribution is imprecise, given the many factors that influence livelihoods and incomes. Edward and Sumner [20] examine several scenarios reflecting alternative assumptions regarding economic growth. They suggest that the number of poor earning less than $2.00 per day might increase or decline by 2030, yet they do not expect extreme poverty to be eradicated. In their most optimistic scenario, 300 million persons will live in extreme poverty in 2030.

The authors suggest also that even if the number of poor earning less than \$2.00 per day decreases, the number of moderately poor persons might increase. In their view, it is likely that 50% to 70% of the global population in 2030 might earn less than \$10 per day.

2.2. Water Resources

The volume of water withdrawn for irrigation, globally, will increase from 2.6 billion km^3 in 2005–2007 to an estimated 2.9 billion km^3 in 2050, with most of the net increase occurring in lower income countries [21,22]. The irrigation requirement (*i.e.*, the consumptive use portion of irrigation withdrawals) is estimated to increase from 1.27 billion km^3 to 1.34 billion km^3. Generally, there are sufficient freshwater resources to support this modest increase, although substantial water scarcity will persist in the Near East and North Africa, South Asia, and elsewhere. Water scarcity will intensify in areas where current rates of surface water and groundwater withdrawals are not sustainable, such as the North China Plain and portions of Central and South Asia [23–25].

Many analysts have suggested there will be sufficient water to produce the food needed in 2050 to support a global population of 9 to 10 billion, provided we allocate and manage water wisely, and we achieve notable gains in agricultural productivity [12,26–28]. Wise allocation and use involves understanding the role of water in crop and livestock production, and also in municipal, commercial, and industrial uses, and in the provision of ecosystem services. The demands for water will continue increasing with the size of the global population, with rising incomes, and with successful efforts to extend water supply and sanitation to all residents of urban and rural areas, particularly in lower income countries.

Meeting these increasing demands will require that policy makers provide effective leadership in communicating water scarcity conditions, allocating developed water supplies appropriately, encouraging wise use in all sectors, and conveying accurate perspectives regarding opportunities to enhance the broad spectrum of benefits obtained from water in productive and environmental uses. In countries with substantial numbers of smallholder farmers, policy makers must ensure access to land and water for agricultural households, whose livelihoods and food security are closely linked to the small amounts of land and water they utilize in producing crops and raising livestock.

Water scarcity will constrain agricultural production and livelihood activities in many additional areas in 2050, as the demands for water in agriculture and other sectors continue to increase. Land and water allocation between sectors will become an increasingly challenging political decision, with notable social and economic implications, as cities expand into agricultural areas, and as commerce and industry require additional water supplies. Smallholder farmers in peri-urban areas will be at risk of losing access to land and water to support their agricultural livelihoods. Persistent work is needed in policy and investment arenas, particularly in lower income countries, to extend and ensure access to water for household use and agricultural production.

3. Issues and Challenges

3.1. Urbanization

The rate of growth in global population is slowing, yet the population will continue increasing for many years, in both rural and urban areas. Projections suggest that the global population will reach 9 to 10 billion, before stabilizing and eventually declining. The declining rate of growth, in aggregate, will reduce to some degree the demand pressures on land and water resources [7]. Yet, local and regional resource issues will remain important and will require critical attention, particularly in countries where population growth remains strong, and where food insecurity persists. The proportion of the population living in urban areas in China has increased from 17% in 1978 to 50% in 2010, thus requiring the conversion of large areas of farmland during years of notable growth in population and in aggregate food demand [29].

Most of the net increase in global population between 2015 and 2050 will occur in urban areas of lower income countries. The increasing urbanization in many regions and the potential impacts of climate change on crop and livestock production add urgency to the question of whether or not food demands will be met in sustainable fashion. It is also essential that the food produced in 2050 is accessible and affordable to everyone, in the interest of achieving national and household food security in all countries.

Increasing urbanization will impact the volume and quality of water available for agriculture, particularly in peri-urban areas [30]. Substantial public and private investments in wastewater capture, treatment, and reuse will be needed to protect public health and to utilize both the water and the nutrients in effluent streams. As cities expand and urban populations increase, it will become increasingly important to capture the nitrogen, phosphorus, and other plant nutrients in wastewater, and to use those nutrients again in agriculture. Efforts to ensure that farmers in peri-urban areas retain access to water for irrigation will be needed, also, particularly when efforts to collect and treat wastewater are implemented.

Technological advances for capturing and treating wastewater in rural areas of lower income countries will improve water quality and enhance the safety and effectiveness of wastewater irrigation, particularly on smallholder farms [31]. Research regarding business models for generating income through the collection, treatment, and sale of wastewater products will encourage private companies to provide wastewater service in areas not served by public collection and treatment programmes [32–35]. Public agencies also can provide initial funding for entrepreneurs wishing to start new businesses in effluent collection, treatment, and processing, either as sole proprietors or in the form of public-private partnerships [36,37].

3.2. Agriculture and Rural Poverty

Even with increasing urbanization, much of the global population, and many of the poor, will live in rural areas and earn their living in agriculture in 2050. The rural population will decline by about 50% in China by 2050, while declining by just 9% in India, and increasing by 49% in Ethiopia (Table 1). Across Africa, the rural population will increase by 48%, while decreasing by 8% in South Asia and by 20% in Southeast Asia. Globally, the rural population will decline by just 7% by 2050 (Table 1). Thus, investments in agriculture in lower income countries will be remain critical in raising incomes of the poor and enabling them to achieve household food and nutritional security.

Table 1. Rural population projections for selected countries and regions, 2015 to 2050.

	2015	2025	2035	2050
	(million persons)			
Country				
Ethiopia	81	98	111	121
India	862	891	875	782
China	628	507	421	318
Region				
Africa	688	804	906	1015
South Asia	1182	1229	1213	1092
Southeast Asia	335	327	309	267
World	3367	3381	3333	3118

Source: Food and Agriculture Organization of the United Nations, FAOSTAT [38].

Smallholder crop and livestock production contributes directly to household and national food security by enhancing home consumption and providing a source of affordable food in local and regional markets [39,40]. Smallholder production provides households with the income needed to purchase the crop and livestock products they do not produce. Income also can be saved, as cash or in the form of durable assets, for purchasing food in years when crop production is impaired by inadequate rainfall. Savings are needed also when local food prices rise sharply, due to disruptions in local or international markets. Income from crop production also enables households to purchase meat and vegetables, thus enhancing household nutrition.

Smallholder agriculture is evolving along somewhat different trajectories in Africa and Asia, although the starting points also are quite different. Changes in the average farm size in both regions are determined largely by rural population density, as most rural residents engage in agriculture, and the amount of land available is essentially fixed [41]. Hazell [42] projects that the average annual growth rate in the rural population of Africa will slow from 2.8% during 1990 to 2010, to 1.35% from 2011 to 2030, and to just 0.63% from 2030 to 2050. By contrast, the projected average annual growth rates in the rural population of Asia are −0.35% for 2011 to 2030 and −0.83% for 2030 to 2050. These represent a substantial decline from the rate of 0.32% observed from 1990 to 2010. Thus, the average farm size in Asia likely will begin increasing, while the average farm size in Africa will continue to decline through 2050 [41].

In all areas, smallholders will become linked more closely with commercial traders and market chains, although the pace and degree of such interactions will vary notably across countries and regions. This will create both opportunities and challenges for smallholder farmers with limited experience interacting in formal markets. The increasing demand for agricultural output in both domestic and international markets should provide the impetus for public and private sector efforts to increase the productivity of crop and livestock production. Substantial investments are needed, particularly in Africa, where perpetually low rates of fertilizer application and inappropriate soil management practices have resulted in nutrient mining of farm soils for many years [43,44]. Sanchez and Swaminathan [45] have characterized the "crisis in soils" in Africa as a "quiet catastrophe" that has evolved over decades.

The impacts of climate change might be particularly severe on smallholders, given their limited adaptation opportunities. Small households with limited finance and little or no access to irrigation might be forced to seek new livelihood activities if the changes in temperature and rainfall preclude them from continuing to grow crops and raise livestock. Policy makers must evaluate both the global and local implications of climate change, while considering also the differing impacts on large farms and smallholders, as they select interventions that will ensure food security for smallholders engaged in crop, livestock, and fish production.

3.3. Water Use in Agriculture

Agriculture will continue to be the largest user of developed water resources in most countries, often accounting for 70% or more of water withdrawals from rivers, reservoirs, and aquifers. Increasing demands for water in cities and industries, and for environmental flows, will reduce the volume of water available for agriculture in many areas. Yet, globally, the volume of water transpired in crop and livestock production must increase between now and 2050, if we are to increase food production. Farmers in many regions must adapt to having less water available for irrigation, while facing increasing demands for their products. Innovations in technology and investments in education and training with regard to managing water in both irrigated and rainfed settings are needed to achieve sustainable agricultural production.

Several authors have suggested that agriculture must "produce more food with less water" in future, given the increasing demands for water in competing sectors [28]. While compelling at first read, this phrase is not sufficiently precise, as it does not distinguish between the water diverted and applied to farm fields, and the water transpired in the process of generating crop yields. Much of the water applied in irrigation runs off the ends of farm fields or percolates into shallow groundwater, where it is available for use again in irrigation or for some other purpose. Only the water consumed by crops in the process of transpiration, and the water that evaporates from plant and soil surfaces, are "lost" from the system at this point in the hydrologic cycle. Opportunities for saving water through investments in technology are limited by the extent to which water is lost in each setting.

The distinction between water diverted and water transpired is important when considering water requirements for food production. The relationship between crop yield or biomass and the amount of water transpired is largely linear for a given cultivar and production setting [46–48]. Thus, in a given setting, absent a technological advance, higher yields can be generated only by transpiring more water. Similarly, more water will be transpired in agriculture as planted areas are expanded in pursuit of higher aggregate production. Advances in crop production technology, including genetic enhancement, can modify the yield-transpiration relationship, such that more output is produced per unit of water transpired. Yet, absent major advances in technology, the amount of water transpired in agriculture will increase between now and 2050.

The water required to support additional transpiration in 2050 can come from several sources, including new development of surface water and groundwater resources for use in agriculture, and better efforts to capture and re-apply surface runoff and utilize shallow groundwater directly in crop production [49–51]. Farmers also can reduce evaporation and improve distribution uniformity by replacing surface irrigation methods with drip systems and micro-sprinklers, where feasible [52], and they can minimize transpiration by non-beneficial plants by removing vegetation from irrigation canals. Such efforts to optimize water use are consistent with the notion of sustainable agricultural intensification, in which higher yields are achieved with given resources, while reducing negative impacts on the environment and enhancing natural capital [52].

Competing demands in other sectors and public demands for environmental amenities will limit the amount of new development of surface water and groundwater for agriculture in many regions. Yet, many farmers can improve water management in ways that reduce non-beneficial evaporation and increase the portion of applied water that is transpired beneficially by crops. Farmers also can increase the amount of crop yield obtained per unit of water transpired by assuring that other essential inputs are available in adequate supply. Agricultural productivity generally is higher when high-quality seeds, soil moisture, plant nutrients, and farm chemicals are applied in sufficient amounts and at the appropriate times during the season.

3.4. Climate Change

The potential impacts of climate change influence the outlook regarding future gains in agricultural productivity. Some production areas will become warmer and drier, while others will receive more annual rainfall, although the timing of the additional rainfall might not be optimal from a seasonal crop production perspective [53]. Some regions likely will experience reductions in agricultural output, particularly in arid areas where water supplies already are limited. Other areas might experience beneficial changes in cropping patterns and increases in crop yields, with warmer temperatures and longer growing seasons [54–56]. Higher concentrations of CO_2 will increase the yields of C_3 crops (e.g., wheat, rice, barley, sugarbeets, and cotton) in some areas, while higher ozone concentrations will have an offsetting, negative impact in others [57]. In large countries such as China and India, the impacts of climate change and the appropriate policy responses and investments could vary substantially across production regions [56,58–60].

Livestock systems also are subject to potentially substantial impacts of climate change, with notable implications for food security and welfare, particularly in lower income countries [61–64]. Sustained high temperatures can impair livestock health and productivity, directly, while water shortages and higher ozone levels in the atmosphere can reduce the yields of livestock feed [65–68]. The potential impacts of climate change might be substantial in the livestock sector, as grazing and mixed rain-fed systems account for 70% of all ruminants and two-thirds of the milk and meat they produce, worldwide [65].

The net effects of climate change on crop and livestock production will influence efforts to achieve national food security in some countries. Poor residents of lower income countries are particularly vulnerable to climate-change induced impairment of food security, given their limited ability to modify production and consumption activities [69]. Rainfed production, which accounts for 80% of global cropland and 60% of global food output, might be notably affected by climate change, particularly in arid and semi-arid areas [70]. Yet, efforts to mitigate or adapt to climate change should not preclude interventions that increase crop yields and improve farm incomes in the near term. Successful efforts to increase fertilizer use in Africa and to reduce groundwater overdraft on the Indo-Gangetic Plain are needed urgently, yet they might increase the variance in farm output in a climate change scenario [71].

Climate change also can impact the availability and quality of both surface water and groundwater, with implications for agricultural production and for associated ecosystems. The increasing variability in rainfall can influence the flow of water in surface systems and the rates of recharge and discharge from aquifers [72,73]. Currently, an estimated 38% of the global irrigated area depends on groundwater [74].

Further research is needed to describe more fully the potential effects of climate change on groundwater dependent ecosystems, yet the impacts are thought to be larger in arid regions, on shallow aquifers, and on ecosystems already stressed in advance of climate change [72,75]. Improvements in the representation of groundwater systems in river basin models that normally highlight surface water, will enhance understanding of climate change impacts, given the importance of groundwater system characteristics in determining the number and duration of droughts [76–78].

Further study of interactions involving groundwater withdrawals, irrigation, and climate change will provide insight for policy makers considering adaptation strategies. Ferguson and Maxwell [79] show that the impacts of irrigation on groundwater storage and stream discharge in a semi-arid basin in the southern United States are quite similar to the simulated impacts of a 2.5 °C rise in temperature. The implications of this research, as described by the authors, are twofold: (1) Many semi-arid basins in which groundwater supports irrigation might already be experiencing some of the potential impacts of climate change; and (2) The actual impacts of climate change might be exacerbated by the additional stress placed on aquifers that support irrigation. Thus, policy makers might need to regulate groundwater pumping in semi-arid irrigated basins.

The increased frequency of major weather events and unexpected changes in weather patterns, brought about by climate change, might also cause more frequent occurrence of crop failures in key production regions, causing short-term reductions in food supplies and consequent price increases, as happened in 2008 and 2011. Given this likelihood, some degree of coordination among countries in establishing regional grain reserves that could be released in times of production shortfalls might limit the harmful effects of higher food prices on poor households during periods of regional crop failure. The cost of maintaining such a reserve might be shared by both importing and exporting countries.

National governments and donors must continue investing in measures that enhance adaptation to climate change at both regional and household levels, such as water storage structures, conjunctive use of groundwater and surface water, wastewater capture and reuse, and research that generates more resilient production systems for smallholders. Policy makers must also protect and sustain the upland areas and mountainous regions where much of the world's water supply originates.

3.5. Protecting Water Resources

In several key production regions, water resources are over-exploited or degraded in ways that are not sustainable. In large areas of South and East Asia, including northwestern India and the North China Plain, groundwater withdrawals exceed the rates of natural recharge, and aquifers are in decline [23,80,81]. Millions of households depend on water for production in such regions, and yet the over-exploitation cannot continue indefinitely [82]. Policy interventions are needed urgently, to achieve the necessary reductions in water withdrawals in a planned and gradual manner, while assisting households in pursuing alternative livelihood activities. Investments in groundwater treatment and pollution reduction also are needed, as many residents depend on groundwater for drinking and other household uses [83].

Improvements in irrigation efficiency in northwestern India and on the North China Plain will not be sufficient to eliminate the overdraft of groundwater, if the aggregate consumptive use of water continues to exceed the natural rate of recharge [23]. Reductions in consumptive use can be achieved by growing crops that require less water per season and by reducing the area planted, yet such efforts will reduce household net income and impact the regional economy. Irrigation supplies can be augmented somewhat by capturing and utilizing rainfall that, otherwise, would evaporate, yet the potential net gains from such activity are limited [23]). In areas where substantial rainfall occurs only in short seasons, efforts to capture more of the water in aquifers, for use later in the year might enhance irrigation and other water-dependent activities during the dry season [84].

Groundwater use in agriculture and other sectors has increased substantially since the middle 20th century, and in many areas, annual groundwater withdrawals exceed the rate of natural recharge. Global groundwater withdrawals in humid to semi-arid areas have increased from an estimated 312 km^3 per year in 1960 to an estimated 734 km^3 per year in 2006. As a result, groundwater depletions have increased from an estimated 126 km^3 per year in 1960 to an estimated 283 km^3 per year in 2006 [85]. Most of the increased withdrawals and depletion are attributed to the increasing use of groundwater for irrigation, in response to rising demands for agricultural output.

Technological advances during the 1950s through the 1980s, including high-capacity pumps and affordable, small-scale pumps and tubewells, also facilitated the rapid increase in groundwater pumping across large areas of North America, South Asia, and northern China [86–90]. Grogan et al. [91] find that groundwater mining accounts for 20% to 49% of gross irrigation water demand in a large portion of China, assuming all demand is met. Given this estimate, the authors suggest that 15% to 27% of China's current crop production is made possible by mining groundwater. Subsidized energy prices have contributed to the increase in groundwater pumping in South Asia [92]. Policy makers must consider issues pertaining to water, energy, and food production in a comprehensive manner, to determine the optimal plan for regulating groundwater pumping, while minimizing unintended impacts.

Optimizing the conjunctive use of surface water and groundwater will enhance the sustainability of irrigated agriculture in many regions, particularly in areas where excessive withdrawals cause costly increases in groundwater pumping depths, or the return flows from agriculture degrade water quality in receiving streams [87,93,94]. Siderius et al. [95] demonstrate the economic viability of conjunctively managing groundwater and rainfall in a tank irrigation system in Andhra Pradesh, India. The higher yields obtained of rice, groundnuts, and sugarcane generate sufficient revenue to offset the cost of rehabilitating the tank system at the start of the six-year experiment, while providing substantial net income per hectare of irrigated land.

Shah [96] reviews India's Groundwater Recharge Master Plan, designed to raise groundwater levels in the post-monsoon season to three meters below ground level. The programme will involve annual "managed artificial recharge" of 36.4 km^3 of water, using an estimated four million spreading-type recharge structures. While commending the intent of the ambitious recharge programme, the author recommends greater focus on the most depleted basins, while also utilizing the 11 million private dug wells already constructed by Indian farmers. In addition, Shah [96] recommends revising energy tariffs to encourage farm-level support for the groundwater recharge programme.

As in many areas of India, farmers in the Indus River Basin of Pakistan practice a de facto form of conjunctive use, as they utilize many small wells for irrigation, in combination with canal water deliveries [97]. Farm-level benefits vary with location along each delivery canal, as farmers located further from a turnout rely more on groundwater than do farmers located more closely. Access to groundwater provides many farmers with a higher degree of security regarding their water supply and enables them to optimize the timing of irrigation events. Thus, crop yields and cropping intensities generally are enhanced in groundwater zones. However, some degree of coordination will be required to prevent salinization of farm soils [97], which can impair agricultural productivity and degrade water quality in rivers and shallow aquifers.

The ability to monitor and analyze changes in groundwater levels has been notably enhanced in recent years with the development of satellite-based gravitational techniques for measuring changes in aquifer storage. Many researchers have used data gathered by the Gravity Recovery and Climate Experiment (GRACE), a multi-year research programme funded jointly by the United States and Germany, to characterize changes in aquifer storage in Africa, Asia, the Middle East, Australia, and in North and South America [98–106]. The characterizations, made possible by satellite coverage, are particularly helpful to water resource analysts and policy makers in regions where conventional data describing changes in aquifer storage are not readily available.

4. Policies and Investments

4.1. Technology for Sustainable Agriculture

Much has been learned in the last 50 years, regarding the role of technology in improving water management, increasing crop yields, and enhancing farm incomes. Many farmers in arid and semi-arid regions have adopted drip and sprinkler irrigation systems, while many also laser level their fields, and many deliver fertilizer via their on-farm irrigation systems in a process known as fertigation [107–109]. Optimizing the use of water and nutrients in crop production, planting hybrid varieties of some crops, using higher quality seeds, and implementing new methods of pest control, have contributed to the large and sustained increases in crop yields observed in many countries since the 1960s and 1970s [110–113].

Genetic improvements in crop varieties also have increased household incomes and enhanced food security in some areas. The adoption of Bt cotton production by smallholders in central and southern India has enabled households to consume larger amounts of more nutritious foods, thus generally improving their dietary status [114]. Other countries also are exploring the potential for increasing agricultural output with genetically improved crops. The National Technical Committee on Crop Biotechnology in the Ministry of Agriculture of Bangladesh has approved the import of Golden Rice, fruit-and shoot-borer resistant Bt eggplant, late blight resistant potato, insect resistant Bt chickpea, and ring spot virus resistant papaya for contained trials [115].

In Sri Lanka, micro-propagation via tissue culture of an improved banana variety has enabled smallholder rice farmers to diversify their cropping pattern and earn additional income [116]. Kabunga et al. [117] observed similar results in a survey of 385 diversified smallholder farming households in the Central and Eastern Provinces of Kenya. Farmers adopting tissue culture technology for vegetative propagation of bananas increased their farm and household incomes by 116% and 86%, respectively, due largely to higher net yields and beneficial adjustments in the mix of inputs. Food security was improved through higher incomes and larger amounts of bananas for home consumption.

Agriculture can benefit also from advances in technology that do not involve genetic enhancement. Advances in biotechnology can improve the detection and control of plant diseases, while biofertilizers and biopesticides can enhance plant nutrition and pest control [118]. Similar advances are available for use in livestock production and in aquaculture. Molecular-based serological techniques have notably improved animal health in lower income countries, while molecular-based pathogen detection systems are used to detect viruses in all countries producing commercial shrimp [118].

Continuous investment in research and development of technologies that will enhance smallholder crop, livestock, and fish production is essential. Improvements in crop and livestock genetics, and in production techniques that enable farmers to increase production with limited land and water resources are needed by smallholders, along with supporting investments in education, training, and outreach. Private sector investments and public-private partnerships will enhance the pace at which new technologies are developed and implemented.

4.2. Investing in Rainfed and Irrigated Production

The remarkable increases in agricultural productivity achieved since the 1960s have enabled farmers in many countries to produce sufficient food to support the world's population as it has increased from about 3 billion in 1960 to more than 7 billion in 2015. Much of the gain in aggregate output has been achieved through the expansion of planted area, while much has come also from higher yields. Much of the additional food production required by 2050 must come from further increases in crop and livestock yields, given the high costs and environmental impacts of continuing to expand agricultural area. Further investments in both rainfed and irrigated areas are needed to ensure that crop and livestock yields will continue increasing at a pace that is sufficient to feed a global population of 9 to 10 billion in 2050.

Irrigated agriculture accounts for about 20% of cultivated area, worldwide, while generating an estimated 40% of crop production [38,119,120]. Yields are notably higher with irrigation, in part because farmers apply larger amounts of fertilizer and farm chemicals when they have some control over the timing and amount of soil moisture in their fields [121]. Much of the world's food supply in 2050 will come from irrigated farms, yet much will also come from farms that rely fully on rainfall and those that supplement rainfall with partial irrigation. In many countries, achieving national food security in 2050 will require investments and interventions in both irrigated and rainfed areas.

Substantial research has been conducted in recent years on methods for improving water management in rainfed areas, such as rainwater harvesting, plant nutrient strategies, cropping systems, mulching, tillage, and other soil and water conservation practices [122,123]. Efforts to extend improved methods of farming in rainfed conditions will improve incomes and enhance livelihoods in areas where poverty is correlated with low crop yields and inadequate use of fertilizer and modern seeds [11,124].

Advances in agricultural technology will enable many farmers to increase production in rainfed and irrigated settings. Yet, technology, alone, will not be sufficient to completely offset limitations regarding land, water, and other natural resources. Accurate water accounting and water balance studies will be needed in many areas to identify the most appropriate interventions for increasing agricultural productivity with limited water supplies. In addition, public officials must support efforts to ensure that farmers have affordable access to complementary inputs, such as high-quality seeds, plant nutrients, and farm chemicals.

4.3. Closing the Yield Gaps

Continuous public investments are needed in the development of new technology and in technical assistance to support smallholder crop, livestock, and aquaculture production. Many rural households will remain engaged in agriculture in 2050, and their production will contribute to local and regional food supplies, while also enhancing household incomes. Closing the large gaps that exist between smallholder yields and those obtained on larger farms and experiment stations will serve both to increase food supply and enhance effective demand for food at household and community levels. New crop varieties, better methods of producing current varieties, and enhanced outreach by crop and livestock extension specialists are needed.

Evidence in the literature regarding the challenge of closing yield gaps is mixed [125,126]. The annual rate of increase in crop yields is slowing in key production areas, causing concern that future gains might not keep pace with rate of increase in global food demand [127,128]. Some authors suggest that improvements in soil and water management, facilitated in part by more affordable access to farm inputs in lower income countries, will be helpful in closing the existing yield gaps across a large portion of the world's agricultural landscape [129,130]. The potential yield gaps likely are larger in rainfed settings than in irrigated areas, yet the challenges of increasing yields in rainfed areas also are substantial [131,132]. In either setting, the desired increases in yield will take time, and progress will not be uniform, as outcomes will vary with soil and water conditions and with access to fertilizer [133,134]. Li *et al.* [135] report that wheat yields on the North China Plain have increased by about 115 kg per ha per year, since 1981, thus substantially closing the farm-level yield gap. However, they report also that wheat yields are no longer increasing in some areas.

Other authors suggest that advances in plant genetics, agronomy, biotechnology, and animal science will provide the improvements needed in crop and livestock technology to achieve further increases in yields [136–144]. However, some authors question whether the needed advances can be developed, tested, and implemented broadly within the time available between now and 2050 [145]. Substantial public investments in crop and livestock science are needed to move research programmes forward, particularly those that largely will benefit smallholders [146]. However, even with adequate financial support, substantial time will be required for producing new cultivars with traits that meet both global and farm-level objectives [147].

Livestock production and marketing are essential livelihood components for more than one billion poor persons in Asia and Africa [148]. Many of these persons are smallholders, for whom livestock represent a source of food and income, while serving also as a means for accumulating wealth. The increasing global demand for livestock products will create opportunities for smallholders to generate higher incomes, provided they have access to output markets and to the inputs and capital needed to expand their operations in sustainable fashion, while maintaining risk at acceptable levels [149,150].

4.4. Seeking Sustainability

While the potential for increasing crop and livestock yields and improving rural livelihoods through investments in technology, education, and outreach is substantial, continuous effort is needed also in developing agricultural practices and enterprises that manage natural resources wisely and do not impose excessive harm on the environment. The agriculture that thrives in 2050 and beyond must adapt to changes in weather patterns brought about by climate change, to changes in public preferences regarding environmental amenities, and to changes in food demands due to increasing incomes and changing demographics. This is no small order for a sector that involves much of the world's developed land and water resources, and still engages a large portion of the global population.

Some authors have characterized the quest for this new agricultural paradigm as that of seeking "sustainable intensification" or pursuing "climate-smart agriculture." Both phrases, as they are described in the literature, reflect the need to ensure the long-term viability of agriculture, with due consideration for natural resources, the environment, and climate change [151]. The first phrase also acknowledges the need to enhance productivity in future, by increasing the use of farm inputs in ways that have less impact on the environment than in current or earlier production settings. Both phrases express ideas, rather than providing a clear statement of the investments and changes in practices that a programme of sustainable intensification or climate-smart agriculture would require. Thus, the phrases are helpful in generating discussion, but they do not, by themselves, provide guidelines for achieving production outcomes that are sustainable or climate smart.

In a recent review of the sustainable intensification debate, Godfray [152] suggests that the perspective has stimulated helpful discussion of alternative agricultural models, yet further work in defining the production practices that are considered acceptable within the paradigm would provide greater clarity. Uphoff [153] urges caution in selecting specific practices before considering the notions of sustainability and intensification within a systems thinking perspective, and accounting more explicitly for the inherent inability to predict which practices and production models might actually be sustainable.

Garnett *et al.* [154] describe four premises supporting the sustainable intensification perspective: (1) Agricultural production must increase to achieve global food security goals; (2) Higher yields are essential to minimize environmental harm due to land expansion; (3) Protecting the environment is as important as increasing agricultural productivity; and (4) Achieving sustainable intensification is a goal, rather than a well-defined plan of action. The authors suggest also that sustainable intensification embeds goals pertaining to food and nutritional security, and sustainable economic development. Hanspach *et al.* [155] propose inverting that relationship by placing initial emphasis on achieving sustainable development, given that food insecurity is due largely to inadequate incomes, rather than the lack of food production. Vanlauwe *et al.* [156] suggest that sustainable intensification might not be pertinent in much of sub-Saharan Africa, where the average farm size is too small to intensify, and where many farmers must place the issues of risk and immediacy ahead of longer-term concerns. "Ultimately, the profitability of intensification will determine whether or not smallholders engage—its sustainability will not necessarily be *their* immediate concern [156]."

Scholarly discussion of sustainable intensification and climate-smart agriculture, in addition to reports from development practitioners, will add clarity to both perspectives, over time [151,157,158]. Successful efforts to ensure food and nutritional security in future, in sustainable fashion, must reflect the inherent complexities regarding agriculture, the environment, and livelihoods, and the notable heterogeneity in farm structures and opportunities in rainfed, irrigated, large-scale, and smallholder settings [159]. The characteristics of sustainable, climate resilient agriculture likely will be different in each setting. The time and investments required to achieve sustainable intensification, where appropriate, also will vary with initial conditions and with expected case-specific outcomes, as viewed by farmers, as they are impacted most directly when implementing the recommended practices.

4.5. Capturing and Reusing Plant Nutrients

The large yields of grains and other crops achieved in many countries, are made possible, in part, by the application of large amounts of nitrogen and other plant nutrients each season. Plants utilize much of the applied nutrients in the process of carbon assimilation, yet some portion of the nutrients enter the atmosphere, runoff in streams, or seep into groundwater. The portion taken up by plants is conveyed to processing plants and dining rooms, and eventually to the wastewater stream leaving households, villages, and cities. As urbanization intensifies in many areas, and as the direct and indirect costs of nutrient use in agriculture increase, over time, the need to recycle the water and plant nutrients in municipal wastewater will become more evident and more urgent.

The cost of producing nitrogen fertilizer depends largely on the price of energy, as the process is energy intensive. Although energy prices have declined sharply in 2014 and 2015, it is likely that energy prices will resume their long-term upward trend in the not-too-distant future. Phosphorus is an essential plant nutrient that is produced largely by mining phosphate rock [160]. The supply of phosphate rock is ultimately limited and large portions of the remaining supply are located in just a few countries [161–164]. Thus, there is some uncertainty regarding the security of newly mined supplies of phosphorus in future. Recycling the phosphorus in wastewater will extend the useful life of existing phosphate rock reserves, by reducing the demand for that source of phosphorus. There are mixed views in the literature, regarding if or when the world might exhaust its supply of phosphate rock [165]. An enhanced wastewater recycling programme, in which phosphorus, nitrogen, and other elements are obtained and reused, likely would be a wise hedging strategy from a global perspective.

There is also a sense of circularity or ecosystem closure in the notion of returning plant nutrients to farmland in the countryside, after food has been consumed in the city. The recovered nutrients can be used again to produce more food, and the cycle can be repeated. In addition, efforts to extend and intensify the capture and reuse of wastewater will reduce the negative environmental impacts of unregulated wastewater discharge into rivers and streams. In areas where the economics of wastewater recovery and reuse are positive, and local governments encourage private firms to engage in the activity for profit, wastewater management can become a widely acknowledged business enterprise that generates sustainable benefits for households, communities, and farmers [35]. The health risks to farmers, households, and consumers can be managed through appropriate policy interventions [166,167].

4.6. Investing in Sustainable Aquaculture

Fisheries and aquaculture are major sources of protein for much of the world's population, and the sectors support many livelihoods in both formal and informal economies [168]. An estimated 3 billion persons obtain about 20% of their intake of animal protein from the output of a capture fishery or an aquaculture operation [169]. An additional 1.3 billion persons obtain 15% of their protein intake from fish. These proportions represent averages across many countries. For individual countries, the shares can be much higher. In Gambia, Sierra Leone, and Ghana, the share of dietary protein from fish is higher than 60% [169]. The share ranges from 50% to 60% in Cambodia, Bangladesh, Indonesia, and Sri Lanka, where capture fisheries have long been important, and where aquaculture has developed rapidly since the 1990s [169].

Aquaculture currently generates more than 50% of the fish and shellfish products consumed worldwide [170,171]. More than 60% of global aquaculture production comes from China, while an additional 26% of production comes from other countries in southern and eastern Asia [171]. The Americas and Europe each account for about 4% of global aquaculture production, while Africa accounts for about 2% of the global sum. Although production in Africa is presently a small portion of global output, the rate of growth in African production has been quite high in recent years. African production has increased from about 81,000 tons in 1990 to 1.4 million tons in 2012, thus increasing by a factor of 18 within 22 years [171]. By comparison, China's production in 2012 (41 million tons) is about 6 times higher than its production in 1990 (6.7 million tons).

Across Africa, aquaculture employs about 920,000 persons and accounts for 0.15 percent of Gross Domestic Product [172]. These are small portions of the employment and income generated by both fisheries and aquaculture in Africa. The full sector employs about 12 million persons and generates annual income of about $24 billion, or 1.26% of African gross domestic income [172]. Yet, for the persons involved in small-scale aquaculture, often in conjunction with small-scale farming, the additional production and income enhance household food and nutritional security [173]. The increasing demand for fish and fish products in Africa presents a substantial opportunity for further expansion of small-scale, commercial aquaculture.

Fisheries and aquaculture provide livelihoods for many smallholders, often in conjunction with other activities, such as rice production, in which farmers utilize land and water for both fish and crops. In some areas, aquaculture competes with agriculture for water supply, and it impacts agriculture by degrading land and water quality. These and other environmental issues, including the use of fishmeal and fish oil as feed materials, and the off-site impacts of effluent from aquaculture operations will require policy intervention to ensure that aquaculture can continue contributing to global food and nutrition demands in a sustainable fashion.

4.7. Investing in Risk Management

Investments in agriculture and water, and policies designed to encourage wise use of resources, must recognize also the inherent risk and uncertainty in farming, particularly in smallholder settings, and also the potential impacts of climate change. Smallholders often are prevented from adopting new technologies or utilizing the appropriate amounts of farm inputs because they cannot risk losing their investment in expensive seeds or irrigation water if a dry spell or pest infestation destroys their crop. Programmes of crop insurance and access to affordable credit can assist in such situations, but they do not fully eliminate the farm-level risk.

Crop yields are determined in large part by the amounts of seeds or plants applied to each hectare, and the amounts of water, fertilizer, and chemicals used each season. Yet, yields also are influenced by weather, pests, and the timing by which inputs are applied. Farmers can manage the effects of weather and pests, to some degree, and they can choose the timing by which they apply key inputs, yet much of the resulting influence on crop yields is uncertain. The yield obtained in one season by applying 20 kg of seed, 100 kg of nitrogen, and 600 mm of irrigation water on a hectare of grain can be quite different from the yield achieved with the same inputs in a subsequent season, due largely to influences that farmers cannot control.

The nature of risk and uncertainty, and the degree of farm-level risk aversion vary across farms, with differences in farmer perspectives, household savings, access to crop insurance, crop choices, weather patterns, and market conditions. Perhaps the largest distinction is between farmers in higher income countries, with substantial savings accounts and crop insurance, and smallholder farmers in lower income countries with very limited savings and no access to insurance. The latter farmers often will limit their use of costly inputs, such as high-quality seeds and plant nutrients, as inadequate rainfall or a serious pest infestation can cause them to lose their entire expenditure. Smallholders can manage risk to some degree by diversifying their crop choices, but opportunities are limited in areas with very much or too little rainfall [174,175]. Interventions that assist farmers in accommodating risk, such as index-based weather insurance, can be helpful in improving household income and welfare in such settings [176].

Investments and programmes that enhance agricultural risk management, particularly for smallholders, will be critical in enabling farm households to adopt new technologies, diversify their activities, and sustain food security during periods of high input prices, low crop yields, and major weather events. In addition to insurance products, investments are needed also in infrastructure that enhances the availability and transport of farm inputs and crop and livestock products, and reduces the transaction costs of marketing farm produce. Such investments will increase the values that farmers generate with limited water resources, while also enhancing household food and nutritional security.

4.8. Investing in Water, Sanitation, and Health

Many poor households have inadequate access to clean water and sanitation. As a result, many women and children spend substantial time and effort fetching water for household use, and family members often suffer from ill health, caused by unclean or unsanitary living conditions. Such illness, and the time spent fetching and preparing water for use, reduce educational opportunities and limit labor productivity. Securing access to an affordable, safe water source can greatly enhance a household's likelihood of escaping poverty, by enabling family members to devote more time and effort to educational and productive activities.

Many rural households also lack secure title to the land and water they use to produce crops and raise livestock, as part of their essential livelihood activities. Many smallholders operate in rainfed settings, in which the crop water supply is inherently uncertain. Small reservoirs are helpful in capturing and storing rainwater for use in households or on crops, as needed, but not all farm households can afford such an investment, due partly to the cost of installation and partly to the opportunity cost of setting land aside from crop production. Efforts to assist farmers in constructing small reservoirs and training farmers to optimize rainwater harvesting would be helpful in many areas. Where water is available from an irrigation scheme or a wastewater treatment facility, many smallholders might benefit from assistance in securing a permanent or long-term right to receive some portion of the available water. Long-term security in land and water will motivate smallholders to invest in improving their crop, livestock, and aquaculture operations, as funds allow, over time.

In many areas of low-income countries, an investment in water can be viewed also as an investment in poverty reduction. The need for investments in water supply, water treatment, irrigation, drainage, flood control, and rainwater harvesting is quite large in many countries. Investors in the water sector can generate substantial improvements in livelihoods and greatly enhance the welfare of households, and communities across much of Africa and Asia. Yet, investments must be carefully planned, and they must account for many of the interactions and externalities that are inherent in water development projects. In many settings, the development of an irrigation scheme or construction of a rainwater harvesting structure in one location will enhance water supply for one set of users, while impairing the water supply for others. The impairment might be direct, in terms of the volume or flow of water available in an aquifer or stream, or it might be indirect, in the form of reduced flows to an estuary that supports an indigenous fishery or provides plant materials that are harvested each season by residents who produce crafts for sale in local markets. The best investments in water resources, from a poverty reduction perspective, will be those that enhance the volume and quality of water available for household use and production, while minimizing impacts on other water users and the environment.

Investments in drinking water supply, water quality, sanitation, and health care, with a particular focus on women and children, are essential to ensure that residents of urban and rural areas can fully utilize available food and nutrition, while not suffering from chronic diseases and other impairments that reduce household welfare and limit educational and productive opportunities. Good health is essential to successful utilization, and successful growth and development are essential to productivity, income generation, and food security. This virtuous cycle revolves around assured access to affordable clean water, sanitation, and health facilities.

4.9. Investing in Rural Economies

Many residents of rural areas earn income in non-farm activities. This is particularly true of landless households and women. Rural households with little or no land earn from 30% to 90% of their income in non-farm employment, while women account for one-quarter of the workforce in the rural non-farm economy lower income countries [177]. Households with insufficient land to rise above the poverty line in farming typically account for about half of rural families [178]. Most of the rural, non-farm households in lower income countries are poor, and in high density rural areas, most of the poor are landless or have too little land to support them in agriculture [178]. For those households, non-farm employment and viable opportunities for producing and selling non-tradable goods and services are essential.

Many farm households rely on income from non-farm employment to supplement farm income, and as a source of finance for farm inputs [177]. An estimated 65% of smallholder farmers in Latin America and the Caribbean rely increasingly on non-farm source of incomes to sustain their livelihoods [179]. The non-farm economy employs family members not needed to work full-time in smallholder farming operations. As the rural population density increases, so too does the importance of non-farm employment in providing livelihood opportunities for households with surplus labor, either seasonally or year-round. Quantifying the impact of non-farm employment on poverty reduction is challenging, due to inadequate data and the difficulty of identifying causality in environments where many macroeconomic variables change with time. Yet, it appears non-farm employment has accounted for substantial poverty reduction in some countries [177].

Further work is needed to determine with accuracy the impacts of non-farm employment on poverty reduction, and the impacts of agricultural growth on the non-farm economy. However, it is likely that the success of commercially oriented smallholder farmers will lead to greater expenditures on non-tradable goods and services in the rural non-farm sector, thus enhancing economic activity and providing new employment opportunities. This will reduce poverty and improve food security in rural areas [178].

Several authors in recent years have provided empirical evidence of the impacts of non-farm employment on household income and food security. In a survey of 220 farm households in Nigeria, Babatunde and Qaim [180] found that off-farm employment contributes to higher incomes, thus enabling greater consumption of calories and micronutrieints. Opportunities to earn off-farm income also significantly improved child height-for-age statistics in the villages in which the authors conducted their survey.

Kumanayake *et al.* [181] show that as rural households in Sri Lanka shifted from farm to non-farm sources of income, between 1990 and 2006, they became less poor. Education played a role in the ability of farm households to gain employment in the non-farm sector in Sri Lanka. Investments in education in rural areas can lead to greater participation in the non-farm economy, with consequent improvement in household welfare.

Wossen and Berger [182], combined large-scale survey data with information collected from 292 randomly selected households in northern Ghana, in a simulation of the potential impacts of access to off-farm employment opportunities and improved access to financial credit on poverty and food security. The authors determined that households with access to credit and off-farm employment can increase their incomes substantially and enhance their food security, particularly when subject to climate and price variability.

Imai *et al.* [183], using aggregate data collected in national household surveys in India and Vietnam, identified significant reductions in poverty and in vulnerability to shocks, for rural households engaged in non-farm employment. In addition, the authors report that employment in skilled jobs, such as in sales or professional activities, has a more notable impact on poverty and vulnerability than does employment in unskilled jobs, such as those involving manual labor. Thus, in India and Vietnam, while any form of non-farm employment is helpful in reducing household poverty and vulnerability, employment in skilled jobs is most desirable.

Policies and investments that enhance opportunities for off-farm employment in rural areas will increase incomes, reduce poverty, and enhance food security, particularly in areas where land and water resources are inadequate to support higher population densities. Higher incomes are essential to achieving food security, and in many rural areas, those higher incomes must come from new opportunities in off-farm employment.

4.10. Investing in Women

Much of the farming in Asia and Africa is conducted by women [38]. Yet, women often do not share the same status as men, regarding such issues as land tenure, water rights, access to credit, participation in outreach programmes, and representation in water user associations [184–186]. It is essential that the status of women be improved, so they may achieve the same degree of access as men, to the tenure, credit, and other inputs needed to produce and market crops successfully.

Efforts are needed also to encourage and support the role of women in agricultural research, extension, and teaching, and as representatives in farmer groups and marketing cooperatives. As agriculture intensifies and as new marketing opportunities arise for smallholders in lower income countries, interventions and investments must enhance the critical role of women in production and marketing activities. Policies and institutions must also acknowledge the role of women in the allocation of household income, with implications for food and nutritional security, and educational opportunities for children [148,186]. Engaging and empowering women in deliberative processes pertaining to climate change will enhance the resulting policies and interventions [187,188].

Several studies provide empirical evidence of gender aspects of crop production and marketing. Ndiritu *et al.* [189], in a survey involving 578 farm households in Kenya, find that women manage smaller plots than men, and they are less likely to adopt some sustainable intensification practices, such as manure application and minimum tillage. The authors find no gender differences in the adoption of other soil and water conservation practices, such as maize-legume intercropping, maize-legume rotations, improved seed varieties, and the use of chemical fertilizer. Such findings, while not fully explained, suggest that further work is needed to understand gender differences in the adoption of selected intensification practices.

The commercialization of smallholder agriculture provides opportunities for farmers to earn and retain higher revenues, as they gain access to a wider array of markets for their produce. One way in which smallholders can advance their participation in new markets is by forming cooperatives or farmer groups that interact in markets on behalf of the membership. Forming and joining farmer groups can modify crop choices and the distribution of farm income within households, if the representation and status of men and women in such groups are different. Fischer and Qaim [190] examine this issue, using data pertaining to banana production in the highlands of central Kenya.

Tissue culture propagation of bananas, in combination with a new mix of productive inputs, has enabled farmers in Kenya to achieve higher yields, thus providing the opportunity to expand their sales in commercial markets. Many smallholders have joined farmer groups that interact with potential buyers, and sell large lots of bananas at collectively negotiated prices. Membership in the groups is individual, and both men and women may join, yet the elected leadership generally is male dominated [190].

Using data from a survey of 444 member and non-member farm households, Fischer and Qaim [190] test hypotheses regarding the impacts of farmer groups on crop production and revenue, women's control of farm revenues, and household nutrition. The authors find that farmer groups tend to increase male control of banana production and revenues. This does not influence the number of calories consumed in the household, but it does have a negative marginal impact on dietary quality, perhaps due to differences in male and female spending preferences. Most notably, female membership in the groups can have a positive impact on the share of income controlled by women.

There are notable gender differences, also, in the formation and productivity of rural non-farm enterprises. In a study of survey data collected by the World Bank in Bangladesh, Ethiopia, Indonesia, and Sri Lanka, Rijkers and Costa [191] find that women are less likely than men to start a non-farm enterprise (with the exception of Ethiopia), women's enterprises tend to be small and home based, and firms operated by women are less productive, as measured by sales per worker (with the exception of Indonesia). Male managers generally are better educated than female managers, yet the authors do not find evidence that differences in human capital account for gender differences in firm performance. The authors also find no support for the hypothesis that gender productivity differences are due differential gender impacts in the local investment climate [191]. Further work is needed to understand fully the gender aspects of the rural non-farm economy in lower income countries.

Although women are responsible for much of the farming in Asia and Africa, many of the institutional settings that influence agriculture are not supportive of women's role in the sector. More appropriate institutions, supportive policies, and strategic investments are needed to enhance the role and success of women in agriculture, particularly in production, but also in research, education, and outreach. Policies that enhance women's security of land tenure, water rights, access to credit, and representation in water user associations and farmer cooperatives are essential. So, too, are programmes that encourage women to enter careers in agricultural research, extension, and teaching.

4.11. Water Policy, Institutions, and Incentives

With increasing competition for water in agriculture and other sectors, national and provincial governments will need to communicate water scarcity conditions effectively, and ensure that farmers, firms, and consumers use water wisely. They must also allocate water with the right mix of concerns for equity and efficiency. Economists often promote pricing as an effective mechanism for communicating scarcity conditions. Yet, pricing water often is difficult to implement, for political or cultural reasons, and water tariffs can be difficult to modify, once in place [192–194]. Nonetheless, it is helpful to consider water prices as a policy option, alongside other interventions, such as water allocations, withdrawal limits, pumping restrictions, rotational deliveries, and cropping pattern restrictions.

In areas where implementing higher prices is not yet politically feasible, public officials might consider implementing water allocations. Such an approach can be just as effective in communicating scarcity conditions, as a programme involving higher water prices. When the volume of water available in a river basin or irrigation district is limited, the aggregate volume can be divided among water users by assigning to each a pro-rated portion of that volume. When farmers know their water supply is limited, they have an incentive to optimize the values they obtain with the amount of water they receive.

Incentive programmes can encourage water users to improve water management practices in both irrigated and rainfed settings. In agriculture, public support for investments in land leveling and the purchase of drip or micro-sprinkler systems can be helpful, where appropriate, although such investments might not result in aggregate water savings. Industries also respond positively to subsidies for investments in water-saving processes and in wastewater capture and reuse. Such programmes are helpful, also, when implementing higher water prices. The higher prices modify the incremental price of water as desired, while the subsidies can limit the increase in the total cost of adjusting to higher water prices, from the water user's perspective.

5. Conclusions

Water resources and food production will be sufficient to support a global population of 9 to 10 billion in 2050, but many of the poor will struggle to retain access to water for household use and production, and many will remain food insecure. The primary cause of food insecurity will continue to be poverty, rather than inadequate food production. Thus, policy makers need to implement programs and encourage investments that will increase the incomes of the poor. Public and private investments that promote growth in sustainable agriculture and provide non-farm employment opportunities in rural areas of lower income countries will be most helpful.

Policy makers must ensure that farmers retain access to the water they need for producing food and sustaining livelihoods, particularly as increasing urbanization places additional pressure on land and water resources. They must create greater opportunities for women in agriculture, and they must motivate the development and adoption of new technologies that will enhance crop and livestock production, particularly for smallholders. Private sector investments also are essential to speed the pace of creating viable employment opportunities in rural areas. Public officials must communicate scarcity conditions by implementing the right mix of pricing policies, resource allocations, and incentives. Finally, all policies and investments must also account for the likely impacts of climate change on agriculture and on the livelihoods of smallholders.

Acknowledgments

I appreciate the helpful comments and suggestions received from scholars at the Food and Agriculture Organization of the United Nations and several other institutes engaged in research, outreach, and development activities. I am grateful also for the helpful comments of three anonymous reviewers.

Conflicts of Interest

The author declares no conflict of interest.

References

1. Cabero-Roura, L.; Rushwan, H. An update on maternal mortality in low-resource countries. *Int. J. Gynecol. Obstet.* **2014**, *125*, 175–180.
2. Cohen, R.L.; Alfonso, Y.N.; Adam, T.; Kuruvilla, S.; Schweitzer, J.; Bishai, D. Country progress towards the Millennium Development Goals: Adjusting for socioeconomic factors reveals greater progress and new challenges. *Glob. Health* **2014**, *10*, 1–19.
3. Lomazzi, M.; Borisch, B.; Laaser, U. The Millennium Development Goals: Experiences, achievements, and what's next. *Glob. Health Action* **2014**, *7* (Suppl. 1), 1–7.
4. Dora, C.; Haines, A.; Balbus, J.; Fletcher, E.; Adair-Rohani, H.; Alabaster, G.; Hossain, R.; de Onis, M.; Branca, F.; Neira, M. Indicators linking health and sustainability in the post-2015 development agenda. *Lancet* **2015**, *385*, 380–391.
5. Maurice, J. New goals in sight to reduce poverty and hunger. *Lancet* **2013**, *382*, 383–384.
6. United Nations. Report of the Open Working Group of the General Assembly on Sustainable Development Goals. Issued as document A/68/970, United Nations, 2014. Available online: http://undocs.org/A/68/970 (accessed on 12 April 2015).
7. Alexandratos, N.; Bruinsma, J. *World Agriculture towards 2030/2050: The 2012 Revision*; ESA Working paper No. 12–03; FAO: Rome, Italy, 2012.
8. Ray, D.K.; Mueller, N.D.; West, P.C.; Foley, J.A. Yield trends are insufficient to double global crop production by 2050. *PLoS ONE* **2013**, *8*, e66428, doi:10.1371/journal.pone.0066428.
9. Furbank, R.T.; Tester, M. Phenomics—Technologies to relieve the phenotyping bottleneck. *Trends Plant Sci.* **2011**, *16*, 635–644.

10. Afari-Sefa, V.; Tenkouano, A.; Ojiewo, C.O.; Keatinge, J.D.H.; Hughes, J.D.A. Vegetable breeding in Africa: Constraints, complexity and contributions toward achieving food and nutritional security. *Food Secur.* **2012**, *4*, 115–127.

11. Dzanku, F.M.; Jirström, M.; Marstorp, H. Yield gap-based poverty gaps in rural sub-Saharan Africa. *World Dev.* **2015**, *67*, 336–362.

12. Grafton, R.Q.; Williams, J.; Jiang, Q. Food and water gaps to 2050: Preliminary results from the global food and water system (GFWS) platform. *Food Secur.* **2015**, *7*, 209–220.

13. Rivers, J.; Warthmann, N.; Pogson, B.J.; Borevitz, J.O. Genomic breeding for food, environment and livelihoods. *Food Secur.* **2015**, *7*, 375–382.

14. Koning, N.B.J.; van Ittersum, M.K.; Becx, G.A.; van Boekel, M.A.J.S.; Brandenburg, W.A.; van den Broek, J.A.; Goudriaan, J.; van Hofwegen, G.; Jongeneel, R.A.; Schiere, J.B.; *et al.* Long-term global availability of food: Continued abundance or new scarcity. *NJAS Wagening. J. Life Sci.* **2008**, *55*, 229–292.

15. Koning, N.; van Ittersum, M.K. Will the world have enough to eat? *Curr. Opin. Environ. Sustain.* **2009**, *1*, 77–82.

16. McKenzie, F.C.; Williams, J. Sustainable food production: Constraints, challenges and choices by 2050. *Food Secur.* **2015**, *7*, 221–233.

17. Bruinsma, J. *The Resource Outlook to 2050: By How Much do Land, Water and Crop Yields Need to Increase by 2050? FAO Expert Meeting on How to Feed the World in 2050*; Food and Agriculture Organization: Rome, Italy, 2009.

18. Eshel, G.; Shepon, A.; Makov, T.; Milo, R. Land, irrigation water, greenhouse gas, and reactive nitrogen burdens of meat, eggs, and dairy production in the United States. *Proc. Natl. Acad. Sci. USA* **2014**, *111*, 11996–12001.

19. Sumner, A. Where Do The Poor Live? *World Dev.* **2012**, *40*, 865–877.

20. Edward, P.; Sumner, A. Estimating the scale and geography of global poverty now and in the future: How much difference do method and assumptions make? *World Dev.* **2014**, *58*, 67–82.

21. Bruinsma, J. The resources outlook: By how much do land, water and crop yields need to increase by 2050? In *Looking Ahead in World Food and Agriculture: Perspectives to 2050*; Conforti, P., Ed.; Food and Agriculture Organization: Rome, Italy, 2011.

22. FAO. *The State of the World's Land and Water Resources for Food and Agriculture (SOLAW): Managing Systems at Risk*; FAO: Rome, Italy; Earthscan: London, UK, 2011.

23. Zheng, C.; Liu, J.; Cao, G.; Kendy, E.; Wang, H.; Jia, Y. Can China cope with its water crisis? —Perspectives from the North China Plain. *Ground Water* **2010**, *48*, 350–354.

24. Biemans, H.; Speelman, L.H.; Ludwig, F.; Moors, E.J.; Wiltshire, A.J.; Kumar, P.; Gerten, D.; Kabat, P. Future water resources for food production in five South Asian river basins and potential for adaptation—A modeling study. *Sci. Total Environ.* **2013**, *468*, S117–S131.

25. Karthe, D.; Chalov, S.; Borchardt, D. Water resources and their management in central Asia in the early twenty first century: Status, challenges and future prospects. *Environ. Earth Sci.* **2014**, *73*, 487–499.

26. De Fraiture, C.; Molden, D.; Wichelns, D. Investing in water for food, ecosystems, and livelihoods: An overview of the comprehensive assessment of water management in agriculture. *Agric. Water Manag.* **2010**, *97*, 495–501.

27. De Fraiture, C.; Wichelns, D. Satisfying future water demands for agriculture. *Agric. Water Manag.* **2010**, *97*, 502–511.

28. Springer, N.P.; Duchin, F. Feeding nine billion people sustainably: Conserving land and water through shifting diets and changes in technologies. *Environ. Sci. Technol.* **2014**, *48*, 4444–4451.

29. Liu, L.; Xu, X.; Chen, X. Assessing the impact of urban expansion on potential crop yield in China during 1990–2010. *Food Secur.* **2015**, *7*, 33–43.

30. Qadir, M.; Wichelns, D.; Raschid-Sally, L.; McCornick, P.G.; Drechsel, P.; Bahri, A.; Minhas, P.S. The challenges of wastewater irrigation in developing countries. *Agric. Water Manag.* **2010**, *97*, 561–568.

31. Kim, M.; Lee, H.; Kim, M.; Kang, D.; Kim, D.; Kim, Y.; Lee, S. Wastewater retreatment and reuse system for agricultural irrigation in rural villages. *Water Sci. Technol.* **2014**, *70*, 1961–1968.

32. Murray, A.; Cofie, O.; Drechsel, P. Efficiency indicators for waste-based business models: Fostering private-sector participation in wastewater and faecal-sludge management. *Water Int.* **2011**, *36*, 505–521.

33. Wichelns, D.; Drechsel, P. Meeting the challenge of wastewater irrigation: Economics, finance, business opportunities and methodological constraints. *Water Int.* **2011**, *36*, 415–419.

34. Scott, C.A.; Raschid-Sally, L. The global commodification of wastewater. *Water Int.* **2012**, *37*, 147–155.

35. Otoo, M.; Drechsel, P.; Hanjra, M.A. Business models and economic approaches for nutrient recovery from wastewater and fecal sludge. In *Wastewater: Economic Asset in an Urbanizing World*; Drechsel, P., Qadir, M., Wichelns, D., Eds.; Springer: Berlin/Heidelberg, Germany, 2015; Chapter 13, pp. 247–270.

36. Murray, A.; Mekala, G.D.; Chen, X. Evolving policies and the roles of public and private stakeholders in wastewater and faecal-sludge management in India, China and Ghana. *Water Int.* **2011**, *36*, 491–504.

37. Amerasinghe, P.; Bhardwaj, R.M.; Scott, C.; Jella, K.; Marshall, F. Urban wastewater and agricultural reuse challenges in India. *IWMI Res. Rep.* **2013**, *147*, 1–28.

38. FAO. *FAOSTAT*; FAO: Rome, Italy, 2015.

39. Tscharntke, T.; Clough, Y.; Wanger, T.C.; Jackson, L.; Motzke, I.; Perfecto, I.; Vandermeer, J.; Whitbread, A. Global food security, biodiversity conservation and the future of agricultural intensification. *Biol. Conserv.* **2012**, *151*, 53–59.

40. High Level Panel of Experts (HLPE). *Investing in Smallholder Agriculture for Food Security. A Report by the High Level Panel of Experts on Food Security and Nutrition of the Committee on World Food Security*; High Level Panel of Experts (HLPE): Rome, Italy, 2013.

41. Masters, W.A.; Djurfeldt, A.A.; de Haan, C.; Hazell, P.; Jayne, T.; Jirström, M.; Reardon, T. Urbanization and farm size in Asia and Africa: Implications for food security and agricultural research. *Glob. Food Secur.* **2013**, *2*, 156–165.

42. Hazell, P. Comparative Study of Trends in Urbanization and Changes in Farm Size in Africa and Asia: Implications for Agricultural research. A Foresight Study of the Independent Science and Partnership Council. Available online: http://ispc.cgiar.org/publications/search?field_publication_date_value_1[value][year]=2013&combine=Changes+in+Farm+Size (accessed on 3 February 2015).

43. Bekunda, M.; Sanginga, N.; Woomer, P.L. Restoring soil fertility in sub-saharan Africa. *Adv. Agron.* **2010**, *108*, 183–236.

44. Kamau, M.; Smale, M.; Mutua, M. Farmer demand for soil fertility management practices in Kenya's grain basket. *Food Secur.* **2014**, *6*, 793–806.

45. Sanchez, P.A.; Swaminathan, M.S. Hunger in Africa: The link between unhealthy people and unhealthy soils. *Lancet* **2005**, *365*, 442–444.

46. Zwart, S.J.; Bastiaanssen, W.G.M. Review of measured crop water productivity values for irrigated wheat, rice, cotton and maize. *Agric. Water Manag.* **2004**, *69*, 115–133.

47. Tolk, J.A.; Howell, T.A. Field water supply: Yield relationships of grain sorghum grown in three USA Southern Great Plains soils. *Agric. Water Manag.* **2008**, *95*, 1303–1313.

48. Steduto, P.; Hsiao, T.C.; Raes, D.; Fereres, E. AquaCrop—The FAO crop model to simulate yield response to water: I. Concepts and underlying principles. *Agron. J.* **2009**, *101*, 426–437.

49. Clemmens, A.J.; Allen, R.G.; Burt, C.M. Technical concepts related to conservation of irrigation and rainwater in agricultural systems. *Water Resour. Res.* **2008**, *44*, W00E03, doi:10.1029/2007WR006095.

50. Ayars, J.E.; Shouse, P.; Lesch, S.M. *In situ* use of groundwater by alfalfa. *Agric. Water Manag.* **2009**, *96*, 1579–1586.

51. Satchithanantham, S.; Krahn, V.; Sri Ranjan, R.; Sager, S. Shallow groundwater uptake and irrigation water redistribution within the potato root zone. *Agric. Water Manag.* **2014**, *132*, 101–110.

52. FAO. *Save and Grow: A Policy Maker's Guide to the Sustainable Intensification of Smallholder Crop Production*; FAO: Rome, Italy, 2011.

53. Roudier, P.; Sultan, B.; Quirion, P.; Berg, A. The impact of future climate change on West African crop yields: What does the recent literature say? *Glob. Environ. Chang.* **2011**, *21*, 1073–1083.

54. Kang, Y.; Khan, S.; Ma, X. Climate change impacts on crop yield, crop water productivity and food security—A review. *Prog. Nat. Sci.* **2009**, *19*, 1665–1674.

55. Gerardeaux, E.; Giner, M.; Ramanantsoanirina, A.; Dusserre, J. Positive effects of climate change on rice in Madagascar. *Agron. Sustain. Dev.* **2012**, *32*, 619–627.

56. Zhou, L.; Turvey, C.G. Climate change, adaptation and China's grain production. *China Econ. Rev.* **2014**, *28*, 72–89.

57. Jaggard, K.W.; Qi, A.; Ober, S. Possible changes to arable crop yields by 2050. *Philos. Trans. R. Soc. B* **2010**, *365*, 2835–2851.

58. Chauhan, B.S.; Prabhjyot, K.; Mahajan, G.; Randhawa, R.K.; Singh, H.; Kang, M.S. Global warming and its possible impact on agriculture in India. *Adv. Agron.* **2014**, *123*, 65–121.

59. Wei, T.; Cherry, T.L.; Glomrød, S.; Zhang, T. Climate change impacts on crop yield: Evidence from China. *Sci. Total Environ.* **2014**, *499*, 133–140.

60. Xiong, W.; Holman, I.; Lin, E.; Conway, D.; Jiang, J.; Xu, Y.; Li, Y. Climate change, water availability and future cereal production in China. *Agric. Ecosyst. Environ.* **2010**, *135*, 58–69.

61. Thornton, P.K.; van de Steeg, J.; Notenbaert, A.; Herrero, M. The impacts of climate change on livestock and livestock systems in developing countries: A review of what we know and what we need to know. *Agric. Syst.* **2009**, *101*, 113–127.

62. Herrero, M.; Thornton, P.K. Livestock and global change: Emerging issues for sustainable food systems. *Proc. Natl. Acad. Sci. USA* **2013**, *110*, 20878–20881.

63. Godber, O.F.; Wall, R. Livestock and food security: Vulnerability to population growth and climate change. *Glob. Chang. Biol.* **2014**, *20*, 3092–3102.

64. Headey, D.; Taffesse, A.S.; You, L. Diversification and development in pastoralist Ethiopia. *World Dev.* **2014**, *56*, 200–213.

65. Nardone, A.; Ronchi, B.; Lacetera, N.; Ranieri, M.S.; Bernabucci, U. Effects of climate changes on animal production and sustainability of livestock systems. *Livest. Sci.* **2010**, *130*, 57–69.

66. Nielsen, A.; Steinheim, G.; Mysterud, A. Do different sheep breeds show equal responses to climate fluctuations? *Basic Appl. Ecol.* **2013**, *14*, 137–145.

67. Megersa, B.; Markemann, A.; Angassa, A.; Ogutu, J.O.; Piepho, H.-P.; Valle Záráte, A. Impacts of climate change and variability on cattle production in southern Ethiopia: Perceptions and empirical evidence. *Agric. Syst.* **2014**, *130*, 23–34.

68. Morignat, E.; Perrin, J.-B.; Gay, E.; Vinard, J.; Calavas, D.; Hénaux, V. Assessment of the impact of the 2003 and 2006 heat waves on cattle mortality in France. *PLoS ONE* **2014**, *9*, e93176, doi:10.1371/journal.pone.0093176.

69. High Level Panel of Experts (HLPE). *Food Security and Climate Change. A Report by the High Level Panel of Experts on Food Security and Nutrition of the Committee on World Food Security*; High Level Panel of Experts (HLPE): Rome, Italy, 2012.

70. Turral, H.; Burke, J.; Faurès, J.M. *Climate Change, Water and Food Security. FAO Water Reports 36*; Food and Agriculture Organization: Rome, Italy, 2011.

71. Lobell, D.B. Climate change adaptation in crop production: Beware of illusions. *Glob. Food Secur.* **2014**, *3*, 72–76.

72. Kløve, B.; Ala-Aho, P.; Bertrand, G.; Gurdak, J.J.; Kupfersberger, H.; Kværner, J.; Muotka, T.; Mykrä, H.; Preda, E.; Rossi, P.; *et al.* Climate change impacts on groundwater and dependent ecosystems. *J. Hydrol.* **2014**, *518*, 250–266.

73. Kurylyk, B.L.; MacQuarrie, K.T.B.; McKenzie, J.M. Climate change impacts on groundwater and soil temperatures in cold and temperate regions: Implications, mathematical theory, and emerging simulation tools. *Earth Sci. Rev.* **2014**, *138*, 313–334.

74. Siebert, S.; Henrich, V.; Frenken, K.; Burke, J. *Update of the Global Map of Irrigation Areas to Version 5*; FAO: Rome, Italy, 2013.

75. Menberg, K.; Blum, P.; Kurylyk, B.L.; Bayer, P. Observed groundwater temperature response to recent climate change. *Hydrol. Earth Syst. Sci.* **2014**, *18*, 4453–4466.

76. Van Lanen, H.A.J.; Wanders, N.; Tallaksen, L.M.; van Loon, A.F. Hydrological drought across the world: Impact of climate and physical catchment structure. *Hydrol. Earth Syst. Sci.* **2013**, *17*, 1715–1732.

77. Van Loon, A.F.; Tijdeman, E.; Wanders, N.; van-Lanen, H.A.J.; Teuling, A.J.; Uijlenhoet, R. How climate seasonality modifies drought duration and deficit. *J. Geophys. Res.* **2014**, *119*, 4640–4656.

78. Wanders, N.; van Lanen, H.A.J. Future discharge drought across climate regions around the world modelled with a synthetic hydrological modelling approach forced by three general circulation models. *Nat. Hazards Earth Syst. Sci.* **2015**, *15*, 487–504.

79. Ferguson, I.M.; Maxwell, R.M. Human impacts on terrestrial hydrology: Climate change *versus* pumping and irrigation. *Environ. Res. Lett.* **2012**, *7*, 044022, doi:10.1088/1748-9326/7/4/044022.

80. Tiwari, V.M.; Wahr, J.; Swenson, S. Dwindling groundwater resources in northern India, from satellite gravity observations. *Geophys. Res. Lett.* **2009**, *36*, L18401, doi:10.1029/2009GL039401.

81. Glazer, A.N.; Likens, G.E. The water table: The shifting foundation of life on land. *Ambio* **2012**, *41*, 657–669.

82. Shah, T. The groundwater economy of south Asia: An assessment of size, significance and socio-ecological impacts. In *The Agricultural Groundwater Revolution: Opportunities and Threats to Development*; Giordano, M., Villholth, K., Eds.; CAB International: Wallingford, UK, 2007.

83. Qiu, J. China to spend billions cleaning up groundwater. *Science* **2011**, *334*, 745, doi:10.1126/science.334.6057.745.

84. Khan, M.R.; Voss, C.I.; Yu, W.; Michael, H.A. Water resources management in the Ganges basin: A comparison of three strategies for conjunctive use of groundwater and surface water. *Water Resour. Manag.* **2014**, *28*, 1235–1250.

85. Wada, Y.; van Beek, L.P.H.; van Kempen, C.M.; Reckman, J.W.T.M.; Vasak, S.; Bierkens, M.F.P. Global depletion of groundwater resources. *Geophys. Res. Lett.* **2010**, *37*, L20402, doi:10.1029/2010GL044571.

86. Qureshi, A.S.; McCornick, P.G.; Qadir, M.; Aslam, Z. Managing salinity and waterlogging in the Indus Basin of Pakistan. *Agric. Water Manag.* **2008**, *95*, 1–10.

87. Shah, T. Climate change and groundwater: India's opportunities for mitigation and adaptation. *Environ. Res. Lett.* **2009**, *4*, 1–13.

88. Zhang, L.; Wang, J.; Huang, J.; Huang, Q.; Rozelle, S. Access to groundwater and agricultural production in China. *Agric. Water Manag.* **2010**, *97*, 1609–1616.

89. Green, T.R.; Taniguchi, M.; Kooi, H.; Gurdak, J.J.; Allen, D.M.; Hiscock, K.M.; Treidel, H.; Aureli, A. Beneath the surface of global change: Impacts of climate change on groundwater. *J. Hydrol.* **2011**, *405*, 532–560.

90. Shi, J.; Wang, Z.; Zhang, Z.; Fei, Y.; Li, Y.; Zhang, F.; Chen, J.; Qian, Y. Assessment of deep groundwater over-exploitation in the North China Plain. *Geosci. Front.* **2011**, *2*, 593–598.

91. Grogan, D.S.; Zhang, F.; Prusevich, A.; Lammers, R.B.; Wisser, D.; Glidden, S.; Li, C.; Frolking, S. Quantifying the link between crop production and mined groundwater irrigation in China. *Sci. Total Environ.* **2015**, *511*, 161–175.

92. Shah, T.; Giordano, M.; Mukherji, A. Political economy of the energy-groundwater nexus in India: Exploring issues and assessing policy options. *Hydrogeol. J.* **2012**, *20*, 995–1006.

93. Shah, T. Towards a managed aquifer recharge strategy for Gujarat, India: An economist's dialogue with hydro-geologists. *J. Hydrol.* **2014**, *518*, 94–107.

94. Singh, A. Conjunctive use of water resources for sustainable irrigated agriculture. *J. Hydrol.* **2014**, *519*, 1688–1697.

95. Siderius, C.; Boonstra, H.; Munaswamy, V.; Ramana, C.; Kabat, P.; van Ierland, E.; Hellegers, P. Climate-smart tank irrigation: A multi-year analysis of improved conjunctive water use under high rainfall variability. *Agric. Water Manag.* **2015**, *148*, 52–62.

96. Shah, T. India's master plan for groundwater recharge: An assessment and some suggestions for revision. *Econ. Political Wkly.* **2008**, *43*, 41–49.

97. Kazmi, S.I.; Ertsen, M.W.; Asi, M.R. The impact of conjunctive use of canal and tube well water in Lagar irrigated area, Pakistan. *Phys. Chem. Earth* **2012**, *47*, 86–98.

98. Frappart, F.; Seoane, L.; Ramillien, G. Validation of GRACE-derived terrestrial water storage from a regional approach over South America. *Remote Sens. Environ.* **2013**, *137*, 69–83.

99. Jin, S.; Feng, G. Large-scale variations of global groundwater from satellite gravimetry and hydrological models, 2002–2012. *Glob. Planet. Chang.* **2013**, *106*, 20–30.

100. Lenk, O. Satellite based estimates of terrestrial water storage variations in Turkey. *J. Geodyn.* **2013**, *67*, 106–110.

101. Famiglietti, J.S.; Rodell, M. Water in the balance. *Science* **2013**, *340*, 1300–1301.

102. Voss, K.A.; Famiglietti, J.S.; Lo, M.; de Linage, C.; Rodell, M.; Swenson, S.C. Groundwater depletion in the Middle East from GRACE with implications for transboundary water management in the Tigris-Euphrates-Western Iran region. *Water Resour. Res.* **2013**, *49*, 904–914.

103. Ahmed, M.; Sultan, M.; Wahr, J.; Yan, E. The use of GRACE data to monitor natural and anthropogenic induced variations in water availability across Africa. *Earth Sci. Rev.* **2014**, *136*, 289–300.

104. Castle, S.L.; Thomas, B.F.; Reager, J.T.; Rodell, M.; Swenson, S.C.; Famiglietti, J.S. Groundwater depletion during drought threatens future water security of the Colorado River Basin. *Geophys. Res. Lett.* **2014**, *41*, 5904–5911.

105. Chen, J.; Li, J.; Zhang, Z.; Ni, S. Long-term groundwater variations in Northwest India from satellite gravity measurements. *Glob. Planet. Chang.* **2014**, *116*, 130–138.

106. Wang, H.; Guan, H.; Gutiérrez-Jurado, H.A.; Simmons, C.T. Examination of water budget using satellite products over Australia. *J. Hydrol.* **2014**, *511*, 546–554.

107. Castellanos, M.T.; Tarquis, A.M.; Ribas, F.; Cabello, M.J.; Arce, A.; Cartagena, M.C. Nitrogen fertigation: An integrated agronomic and environmental study. *Agric. Water Manag.* **2013**, *120*, 46–55.

108. Chai, Q.; Gan, Y.; Turner, N.C.; Zhang, R.-Z.; Yang, C.; Niu, Y.; Siddique, K.H.M. Water-saving innovations in Chinese agriculture. *Adv. Agron.* **2014**, *126*, 149–201.

109. Gheysari, M.; Loescher, H.W.; Sadeghi, S.H.; Mirlatif, S.M.; Zareian, M.J.; Hoogenboom, G. Water-yield relations and water use efficiency of maize under nitrogen fertigation for semiarid environments: Experiment and synthesis. *Adv. Agron.* **2015**, *130*, 175–229.

110. Biazin, B.; Sterk, G.; Temesgen, M.; Abdulkedir, A.; Stroosnijder, L. Rainwater harvesting and management in rainfed agricultural systems in sub-Saharan Africa: A review. *Phys. Chem. Earth* **2012**, *47*, 139–151.

111. Wright, B.D. Grand missions of agricultural innovation. *Research Policy* **2012**, *41*, 1716–1728.

112. Stevenson, J.R.; Villoria, N.; Byerlee, D.; Kelley, T.; Maredia, M. Green Revolution research saved an estimated 18 to 27 million hectares from being brought into agricultural production. *Proc. Natl. Acad. Sci. USA* **2013**, *110*, 8363–8368.

113. Alston, J.M.; Pardey, P.G. Agriculture in the global economy. *J. Econ. Perspect.* **2014**, *28*, 121–146.

114. Qaim, M.; Kouser, S. Genetically modified crops and food security. *PLoS ONE* **2013**, *8*, 1–7.

115. Fahmi, A. Benefits of new tools in biotechnology to developing countries in South Asia: A perspective from UNESCO. *J. Biotechnol.* **2011**, *156*, 364–369.

116. Lagoda, P.J.L. Use of tissue culture and mutation induction to improve banana production for smallholders in Sri Lanka. In *Biotechnologies at Work for Smallholders: Case Studies from Developing Countries in Crops, Livestock and Fish*; Ruane, J., Dargie, J.D., Mba, C., Boettcher, P., Makkar, H.P.S., Bartley, D.M., Sonnino, A., Eds.; Food and Agriculture Organization: Rome, Italy, 2013.

117. Kabunga, N.S.; Dubois, T.; Qaim, M. Impact of tissue culture banana technology on farm household income and food security in Kenya. *Food Policy* **2014**, *45*, 25–34.

118. Ruane, J.; Sonnino, A. Agricultural biotechnologies in developing countries and their possible contribution to food security. *J. Biotechnol.* **2011**, *156*, 356–363.

119. Turral, H.; Svendsen, M.; Faurès, J.M. Investing in irrigation: Reviewing the past and looking to the future. *Agric. Water Manag.* **2010**, *97*, 551–560.

120. FAO. *AQUASTAT, FAO's Global Water Information System*; FAO: Rome, Italy, 2015.

121. Monjardino, M.; McBeath, T.M.; Brennan, L.; Llewellyn, R.S. Are farmers in low-rainfall cropping regions under-fertilising with nitrogen? A risk analysis. *Agric. Syst.* **2013**, *116*, 37–51.

122. Karpouzoglou, T.; Barron, J. A global and regional perspective of rainwater harvesting insub-Saharan Africa's rainfed farming systems. *Phys. Chem. Earth* **2014**, *72*, 43–53.

123. Kurothe, R.S.; Kumar, G.; Singh, R.; Tiwari, S.P.; Vishwakarma, A.K.; Sena, D.R.; Pande, V.C. Effect of tillage and cropping systems on runoff, soil loss and crop yields under semiarid rainfed agriculture in India. *Soil Tillage Res.* **2014**, *140*, 126–134.

124. Affholder, F.; Poeydebat, C.; Corbeels, M.; Scopel, E.; Tittonell, P. The yield gap of major food crops in family agriculture in the tropics: Assessment and analysis through field surveys and modelling. *Field Crops Res.* **2013**, *143*, 106–118.

125. Waddington, S.R.; Li, X.; Dixon, J.; Hyman, G.; de Vicente, M.C. Getting the focus right: Production constraints for six major food crops in Asian and African farming systems. *Food Secur.* **2010**, *2*, 27–48.

126. Sumberg, J. Mind the (yield) gap(s). *Food Policy* **2012**, *4*, 509–518.

127. Grassini, P.; Eskridge, K.M.; Cassman, K.G. Distinguishing between yield advances and yield plateaus in historical crop production trends. *Nat. Commun.* **2013**, *4*, 2918, doi:10.1038/ncomms3918.

128. Jat, M.L.; Bijay-Singh; Gerard, B. Nutrient management and use efficiency in wheat systems of South Asia. *Adv. Agron.* **2014**, *125*, 171–259.

129. Spiertz, H. Avenues to meet food security. The role of agronomy on solving complexity in food production and resource use. *Eur. J. Agron.* **2012**, *43*, 1–8.

130. Pasuquin, J.M.; Pampolino, M.F.; Witt, C.; Dobermann, A.; Oberthür, T.; Fisher, M.J.; Inubushi, K. Closing yield gaps in maize production in Southeast Asia through site-specific nutrient management. *Field Crops Res.* **2014**, *156*, 219–230.

131. Lobell, D.B.; Cassman, K.G.; Field, C.B. Crop yield gaps: Their importance, magnitudes, and causes. *Annu. Rev. Environ. Resour.* **2009**, *34*, 179–204.

132. Kassie, B.T.; van Ittersum, M.K.; Hengsdijk, H.; Asseng, S.; Wolf, J.; Rötter, R.P. Climate-induced yield variability and yield gaps of maize (*Zea mays* L.) in the Central Rift Valley of Ethiopia. *Field Crops Res.* **2014**, *160*, 41–53.

133. Bryan, B.A.; King, D.; Zhao, G. Influence of management and environment on Australian wheat: Information for sustainable intensification and closing yield gaps. *Environ. Res. Lett.* **2014**, *9*, 044005, doi:10.1088/1748-9326/9/4/044005.

134. Connor, D.J.; Mínguez, M.I. Evolution not revolution of farming systems will best feed and green the world. *Glob. Food Secur.* **2012**, *1*, 106–113.

135. Li, K.; Yang, X.; Liu, Z.; Zhang, T.; Lu, S.; Liu, Y. Low yield gap of winter wheat in the North China Plain. *Eur. J. Agron.* **2014**, *59*, 1–12.

136. Bruce, T.J.A. Tackling the threat to food security caused by crop pests in the new millennium. *Food Secur.* **2010**, *2*, 133–141.

137. Powell, N.; Ji, X.; Ravash, R.; Edlington, J.; Dolferus, R. Yield stability for cereals in a changing climate. *Funct. Plant Biol.* **2012**, *39*, 539–552.

138. Blum, A. Drought resistance is it really a complex trait? *Funct. Plant Biol.* **2011**, *38*, 753–757.

139. Blum, A. Heterosis, stress, and the environment: A possible road map towards the general improvement of crop yield. *J. Exp. Bot.* **2013**, *64*, 4829–4837.

140. Cabello, J.V.; Lodeyro, A.F.; Zurbriggen, M.D. Novel perspectives for the engineering of abiotic stress tolerance in plants. *Curr. Opin. Biotechnol.* **2014**, *26*, 62–70.

141. Dolferus, R. To grow or not to grow: A stressful decision for plants. *Plant Sci.* **2014**, *229*, 247–261.

142. Rothschild, M.F.; Plastow, G.S. Applications of genomics to improve livestock in the developing world. *Livest. Sci.* **2014**, *166*, 76–83.

143. Vadez, V.; Palta, J.; Berger, J. Developing drought tolerant crops: Hopes and challenges in an exciting journey. *Funct. Plant Biol.* **2014**, *41*, v–vi.

144. Langridge, P.; Reynolds, M.P. Genomic tools to assist breeding for drought tolerance. *Curr. Opin. Biotechnol.* **2015**, *32*, 130–135.

145. Hall, A.J.; Richards, R.A. Prognosis for genetic improvement of yield potential and water-limited yield of major grain crops. *Field Crops Res.* **2013**, *143*, 18–33.

146. Anthony, V.M.; Ferroni, M. Agricultural biotechnology and smallholder farmers in developing countries. *Curr. Opin. Biotechnol.* **2012**, *23*, 278–285.

147. Spiertz, H. Agricultural sciences in transition from 1800 to 2020: Exploring knowledge and creating impact. *Eur. J. Agron.* **2014**, *59*, 96–106.

148. McDermott, J.; Aït-Aïssa, M.; Morel, J.; Rapando, N. Agriculture and household nutrition security: Development practice and research needs. *Food Secur.* **2013**, *5*, 667–678.

149. Herrero, M.; Thornton, P.K.; Gerber, P.; Reid, R.S. Livestock, livelihoods and the environment: Understanding the trade-offs. *Curr. Opin. Environ. Sustain.* **2009**, *1*, 111–120.

150. Tiwari, R.; Dileep Kumar, H.; Dutt, T.; Singh, B.P.; Pachaiyappan, K.; Dhama, K. Future challenges of food security and sustainable livestock production in India in the changing climatic scenario. *Asian J. Anim. Vet. Adv.* **2014**, *9*, 367–384.

151. Campbell, B.M.; Thornton, P.; Zougmoré, R.; van Asten, P.; Lipper, L. Sustainable intensification: What is its role in climate smart agriculture? *Curr. Opin. Environ. Sustain.* **2014**, *8*, 39–43.

152. Godfray, H.C.J. The debate over sustainable intensification. *Food Secur.* **2015**, *7*, 199–208.

153. Uphoff, N. Systems thinking on intensification and sustainability: Systems boundaries, processes and dimensions. *Curr. Opin. Environ. Sustain.* **2014**, *8*, 89–100.

154. Garnett, T.; Appleby, M.C.; Balmford, A.; Bateman, I.J.; Benton, T.G.; Bloomer, P.; Burlingame, B.; Dawkins, M.; Dolan, L.; Fraser, D.; *et al.* Sustainable intensification in agriculture: Premises and policies. *Science* **2013**, *341*, 33–34.

155. Hanspach, J.; Abson, D.J.; Loos, J.; Tichit, M.; Chappell, M.J.; Fischer, J. Develop, then intensify. *Science* **2013**, *341*, 713, doi:10.1126/science.341.6147.713-a.

156. Vanlauwe, B.; Coyne, D.; Gockowski, J.; Hauser, S.; Huising, J.; Masso, C.; Nziguheba, G.; Schut, M.; van Asten, P. Sustainable intensification and the African smallholder farmer. *Curr. Opin. Environ. Sustain.* **2014**, *8*, 15–22.

157. Kuyper, T.W.; Struik, P.C. Epilogue: Global Food Security, Rhetoric, and the Sustainable Intensification Debate. *Curr. Opin. Environ. Sustain.* **2014**, *8*, 71–79.

158. Struik, P.C.; Kuyper, T.W.; Brussaard, L.; Leeuwis, C. Deconstructing and unpacking scientific controversies in intensification and sustainability: Why the tensions in concepts and values? *Curr. Opin. Environ. Sustain.* **2014**, *8*, 80–88.

159. Habel, J.C.; Teucher, M.; Hornetz, B.; Jaetzold, R.; Kimatu, J.N.; Kasili, S.; Mairura, Z.; Mulwa, R.K.; Eggermont, H.; Weisser, W.W.; *et al.* Real-world complexity of food security and biodiversity conservation. *Biodivers. Conserv.* **2015**, *24*, doi:10.1007/s10531-015-0866-z.

160. Johnston, A.E.; Poulton, P.R.; Fixen, P.E.; Curtin, D. Phosphorus: Its efficient use in agriculture. *Adv. Agron.* **2014**, *123*, 177–228.

161. Lott, J.N.A.; Kolasa, J.; Batten, G.D.; Campbell, L.C. The critical role of phosphorus in world production of cereal grains and legume seeds. *Food Secur.* **2011**, *3*, 451–462.

162. Ryan, J.; Ibrikci, H.; Delgado, A.; Torrent, J.; Sommer, R.; Rashid, A. Significance of phosphorus for agriculture and the environment in the West Asia and North Africa region. *Adv. Agron.* **2012**, *114*, 91–153.

163. Cordell, D.; White, S. Life's bottleneck: Sustaining the world's phosphorus for a food secure future. *Annu. Rev. Environ. Resour.* **2014**, *39*, 161–188.

164. Cordell, D.; White, S. Tracking phosphorus security: Indicators of phosphorus vulnerability in the global food system. *Food Secur.* **2015**, *7*, 337–350.

165. Ziadi, N.; Whalen, J.K.; Messiga, A.J.; Morel, C. Assessment and modeling of soil available phosphorus in sustainable cropping systems. *Adv. Agron.* **2013**, *122*, 85–126.

166. Hanjra, M.A.; Blackwell, J.; Carr, G.; Zhang, F.; Jackson, T.M. Wastewater irrigation and environmental health: Implications for water governance and public policy. *Int. J. Hyg. Environ. Health* **2012**, *215*, 255–269.

167. Keraita, B.; Medlicott, K.; Drechsel, P.; Mateo-Sagasta Dávila, J. Health risks and cost-effective health risk management in wastewater use systems. In *Wastewater: Economic Asset in an Urbanizing World*; Drechsel, P., Qadir, M., Wichelns, D., Eds.; Springer: Berlin/Heidelberg, Germany, 2015; Chapter 3, pp. 39–54.

168. Béné, C.; Barange, M.; Subasinghe, R.; Pinstrup-Andersen, P.; Merino, G.; Hemre, G.-I.; Williams, M. Feeding 9 billion by 2050—Putting fish back on the menu. *Food Secur.* **2015**, *7*, 261–274.

169. High Level Panel of Experts (HLPE). *Sustainable Fisheries and Aquaculture for Food Security and Nutrition. A Report by the High Level Panel of Experts on Food Security and Nutrition of the Committee on World Food Security*; High Level Panel of Experts (HLPE): Rome, Italy, 2014.

170. Naylor, R.L.; Hardy, R.W.; Bureau, D.P.; Chiu, A.; Elliott, M.; Farrelle, A.P.; Forstere, I.; Gatlin, D.M.; Goldburg, R.J.; Hua, K.; *et al.* Feeding aquaculture in an era of finite resources. *Proc. Natl. Acad. Sci. USA* **2009**, *106*, 15103–15110.

171. FAO. *The State of World Fisheries and Aquaculture 2014. Opportunities and Challenges*; FAO: Rome, Italy, 2014.

172. De Graaf, G.; Garibaldi, L. *The Value of African Fisheries. FAO Fisheries and Aquaculture Circular. No. 1093*; FAO: Rome, Italy, 2014; p. 76.

173. Beveridge, M.C.M.; Thilsted, S.H.; Phillips, M.J.; Metian, M.; Troell, M.; Hall, S.J. Meeting the food and nutrition needs of the poor: The role of fish and the opportunities and challenges emerging from the rise of aquaculture. *J. Fish Biol.* **2013**, *83*, 1067–1084.

174. Bezabih, M.; di Falco, S. Rainfall variability and food crop portfolio choice: Evidence from Ethiopia. *Food Secur.* **2012**, *4*, 557–567.

175. Kandulu, J.M.; Bryan, B.A.; King, D.; Connor, J.D. Mitigating economic risk from climate variability in rain-fed agriculture through enterprise mix diversification. *Ecol. Econ.* **2012**, *79*, 105–112.

176. Hazell, P.B.R.; Hess, U. Drought insurance for agricultural development and food security in dryland areas. *Food Secur.* **2010**, *2*, 395–405.

177. Haggblade, S.; Hazell, P.; Reardon, T. The rural non-farm economy: Prospects for growth and poverty reduction. *World Dev.* **2010**, *38*, 1429–1441.

178. Mellor, J.W. High rural population density Africa—What are the growth requirements and who participates? *Food Policy* **2014**, *48*, 66–75.

179. Berdegué, J.A.; Ricardo Fuentealba, R. Latin America: The state of smallholders in agriculture. In Proceedings of the IFAD Conference on New Directions for Smallholder Agriculture, Rome, Italy, 24–25 January 2011; p. 38.

180. Babatunde, R.O.; Qaim, M. Impact of off-farm income on food security and nutrition in Nigeria. *Food Policy* **2010**, *35*, 303–311.

181. Kumanayake, N.S.; Estudillo, J.P.; Otsuka, K. Changing sources of household income, poverty, and sectoral inequality in Sri Lanka, 1990–2006. *Dev. Econ.* **2014**, *52*, 26–51.

182. Wossen, T.; Berger, T. Climate variability, food security and poverty: Agent-based assessment of policy options for farm households in Northern Ghana. *Environ. Sci. Policy* **2015**, *47*, 95–107.

183. Imai, K.S.; Gaiha, R.; Thapa, G. Does non-farm sector employment reduce rural poverty and vulnerability? Evidence from Vietnam and India. *J. Asian Econ.* **2015**, *36*, 47–61.

184. FAO. *The State of Food and Agriculture 2010–11. Women in Agriculture: Closing the Gender Gap for Development*; FAO: Rome, Italy, 2011.

185. Ibnouf, F.O. Challenges and possibilities for achieving household food security in the Western Sudan region: The role of female farmers. *Food Secur.* **2011**, *3*, 215–231.

186. Mohapatra, S. The pillars of Africa's agriculture. *Rice Today* **2011**, *10*, 22–23.

187. Arora-Jonsson, S. Virtue and vulnerability: Discourses on women, gender and climate change. *Glob. Environ. Chang.* **2011**, *21*, 744–751.

188. Figueiredo, P.; Perkins, P.E. Women and water management in times of climate change: Participatory and inclusive processes. *J. Clean. Prod.* **2013**, *60*, 188–194.

189. Ndiritu, S.W.; Kassie, M.; Shiferaw, B. Are there systematic gender differences in the adoption of sustainable agricultural intensification practices? Evidence from Kenya. *Food Policy* **2014**, *49*, 117–127.

190. Fischer, E.; Qaim, M. Gender, agricultural commercialization, and collective action in Kenya. *Food Secur.* **2012**, *4*, 441–453.

191. Rijkers, B.; Costa, R. Gender and rural non-farm entrepreneurship. *World Dev.* **2012**, *40*, 2411–2426.

192. Ruijs, A.; Zimmermann, A.; van den Berg, M. Demand and distributional effects of water pricing policies. *Ecol. Econ.* **2008**, *66*, 506–516.

193. Dono, G.; Giraldo, L.; Severini, S. Pricing of irrigation water under alternative charging methods: Possible shortcomings of a volumetric approach. *Agric. Water Manag.* **2010**, *97*, 1795–1805.

194. Cooper, B.; Crase, L.; Pawsey, N. Best practice pricing principles and the politics of water pricing. *Agric. Water Manag.* **2014**, *145*, 92–97.

3

Effect of Ozone Treatment on Inactivation of *Escherichia coli* and *Listeria* sp. on Spinach

Shreya Wani, Jagpreet K. Maker, Joseph R. Thompson, Jeremy Barnes * and Ian Singleton *

School of Biology, Newcastle University, Newcastle upon Tyne, NE1 7RU, UK;
E-Mails: shrenrique@gmail.com (S.W.); jagmaker@gmail.com (J.K.M.);
j.thompson5@newcastle.ac.uk (J.R.T.)

* Authors to whom correspondence should be addressed; E-Mails: jerry.barnes@newcastle.ac.uk (J.B.);
ian.singleton@ncl.ac.uk (I.S.)

Academic Editor: Pascal Delaquis

Abstract: The efficacy of "gaseous" ozone in reducing numbers and re-growth of food-borne pathogens, (*Escherichia coli* and *Listeria* spp.), on leafy salads was investigated using spinach. A preliminary *in vivo* study showed 1-log reduction in six strains of *E. coli* and two species of *Listeria* spp. on spinach exposed to 1 ppm ozone for 10 min. A range of ozone treatments were explored to deliver optimal bacterial inactivation while maintaining the visual appearance (color) of produce. Exposure to a higher ozone concentration for a shorter duration (10 ppm for 2 min) significantly reduced *E. coli* and *Listeria* spp. viable counts by 1-log and the pathogens did not re-grow following treatment (over a nine-day storage period). Impacts of 1 and 10 ppm ozone treatments were not significantly different. Approximately 10% of the pathogen population was resistant to ozone treatment. We hypothesized that cell age may be one of several factors responsible for variation in ozone resistance. *E. coli* cells from older colonies demonstrated higher ozone resistance in subsequent experiments. Overall, we speculate that gaseous ozone treatment constitutes the basis for an alternative customer-friendly method to reduce food pathogen contamination of leafy produce and is worth exploring on a pilot-scale in an industrial setting.

Keywords: gaseous ozone; *E. coli*; *Listeria* spp.

1. Introduction

In addition to reducing produce spoilage, an increased incidence in the outbreaks of microbial borne diseases associated with the consumption of raw leafy produce have added to the need to find alternative methods to reduce microbial loads [1]. All types of leafy produce have the potential to harbor pathogens, such as *Escherichia coli*, *Listeria monocytogenes*, *Shigella* spp., and *Salmonella* spp., which are ultimately responsible for the majority of foodborne outbreaks [2,3]. Contamination of fresh produce with pathogens can occur either pre-harvest and/or post-harvest. Pre-harvest sources of pathogens generally include organic fertilizers, irrigation water, and soil, whereas post-harvest sources mainly result from handling procedures including equipment, transport vehicles, and containers [4].

A recent investigation of retail leafy salads revealed contamination of a significant proportion with *E. coli* and *L. monocytogenes* [5]. *E. coli* is a Gram-negative, facultative anaerobic member of the *Enterobacteriaceae* family. It is commonly present in gastrointestinal tract of humans and animals including deer, cattle, and pigs [6]. Although most *E. coli* are harmless to humans, epidemiological research has documented that intake of leafy produce contaminated with *E. coli* O157:H7 and variants thereof, with a dose as low as 10 cells, can pose severe threat to human health [7]. *E. coli* that cause disease are categorized on the basis of pathogenic mechanisms, virulence, and clinical syndrome. For example, *E. coli* O157:H7 belongs to the enterohaemorrhagic (EHEC) group [8]. Infection with *E. coli* O157:H7 causes major outbreaks particularly associated with raw leafy produce [5]. Although leaf surfaces are not a suitable environment for *E. coli*, it can survive both harsh field and post-harvest storage conditions [5].

L. monocytogenes is a Gram-positive, facultative anaerobic, non-spore forming rod, which is capable of growing at low temperatures. This pathogen is widely present in soil, plant, and water surfaces [9]. It causes less than one percent of foodborne diseases, but it is responsible for causing listeriosis in human [10]. In healthy individuals, the main symptoms are fever and diarrhea, whereas in pregnant women, *L. monocytogenes* causes septicaemia, meningitis, abortion, or stillbirth [5,10]. *L. monocytogenes* is capable of growing at refrigeration temperatures and also surviving in food-processing sites [5].

Microbial contamination of fresh produce by pathogenic microbes not only poses significant risk to public health but also affects the industry financially by resulting in costly product recalls. For example, the recent Shiga toxin-producing *E. coli* O157 outbreak in watercress is estimated to have required the recall of 200,000 items in the United Kingdom [11]. Foodborne outbreaks are common in many countries. This could be due to the pathogens developing resistance to traditional sanitizing agents, thus posing a hazard to the safety of the food supply [12].

Ozone has been successfully used as a principal sanitizer for treating drinking and municipal waters for 100 years, but recently gained attention in the food and agriculture industry [13]. It is well known for its strong oxidizing capacity and has been recognized as a powerful antimicrobial agent, reacting with organic substances approximately 3000 times quicker than chlorine [14]. Ozone is capable of inactivating microorganisms including both Gram positive and Gram negative bacteria, bacterial spores, fungi, fungal spores, viruses, and protozoa [15]. In 1997, the United States Food and Drug Administration (US-FDA) in union with an expert panel granted ozone as GRAS (Generally Recognised as Safe) status [16], and in 2003, it received formal approval from the US-FDA as a "direct contact food sanitizing agent" [17]. One of the major advantages of ozone treatment is the fact the gas leaves no detectable residues in/on treated products, as ozone rapidly decomposes into oxygen unlike other sanitizers used in the food processing industry [13].

Given the importance of controlling pathogen contamination of leafy fresh produce, the present study aimed to determine the antimicrobial effects of ozone for the control of different strains of *E. coli* and *Listeria* spp. and to observe the regrowth of these pathogenic bacteria on ozone-treated produce during storage of produce for nine days at 4 °C. Previous studies have shown the impact of ozone on pathogens [14,18,19] but have not investigated re-growth after treatment. We also wanted to use ozone levels that did not damage produce, *i.e.*, we investigated commercially-relevant ozone levels for produce treatment. Previous research has shown that not all pathogens are killed by ozone treatment [20] and another objective of this work was to establish why this may occur. Accordingly, we examined the effect of cell age on *E. coli* resistance to ozone. Spinach was artificially contaminated by inoculating with *E. coli* or *Listeria* spp. before ozone treatment. Six different strains of non-pathogenic *E. coli* were used as a representative model for *E. coli* O157:H7, as there have been no reports suggesting significant differences in growth pattern and survival strategy between non-pathogenic *E. coli* and pathogenic *E. coli* O157:H7 [21]. In addition, *L. innocua* and *L. seeligeri* were used as surrogates for *L. monocytogenes* because these offer safe non-pathogenic alternatives for experimental purposes whilst exhibiting similar growth characteristics and behavior on leafy produce as *L. monocytogenes* [20,22].

2. Results and Discussion

2.1. Effect of Ozone Exposure on E. coli and Listeria sp. in Vitro

Colony numbers (CFU) of *E. coli* K12 and *L. innocua in vitro* were significantly reduced ($p < 0.05$) by all ozone treatments (Figure 1), even at the lowest level used (1 ppm for 10 min). Less than 1-log reduction was achieved when colonies on agar were exposed to 1 ppm ozone for 10 min, but more than 1-log reduction was achieved when both the strains of food pathogens were treated with ozone concentrations of 10 and 50 ppm. Similar results were observed by Alwi [18], when *E. coli* O157, *L. monocytogenes*, and *Salmonella typhimurium* were treated *in vitro* with 0.1, 0.3, 0.5, and 1.0 ppm ozone concentration for exposure times of 0.5, 3, 6, and 24 h, respectively. They also observed increases in ozone concentration, and exposure time increased the antibacterial activity.

Interestingly, the agar based *in vitro* assay on both Gram-positive and Gram-negative pathogens showed no significant difference in colony counts between 10 ppm and 50 ppm ozone concentration treatment. This is possibly due to cells being physically protected by others on the surface of the agar plates, *i.e.*, when the cells are spread on agar some cells may not be present as individuals but as groups that provide physical protection; thus, this could reduce the effectiveness of ozone treatment [18]. Alternatively, some cells may have an intrinsic resistance to ozone exposure perhaps due to their age and exposure to stress (see below Section 2.6). Fan and colleagues [20] reported that the maximum inactivation of *L. innocua* cells was observed in less than 2 h and inactivation reached a plateau after 4 h when treated with gaseous ozone *in vitro*.

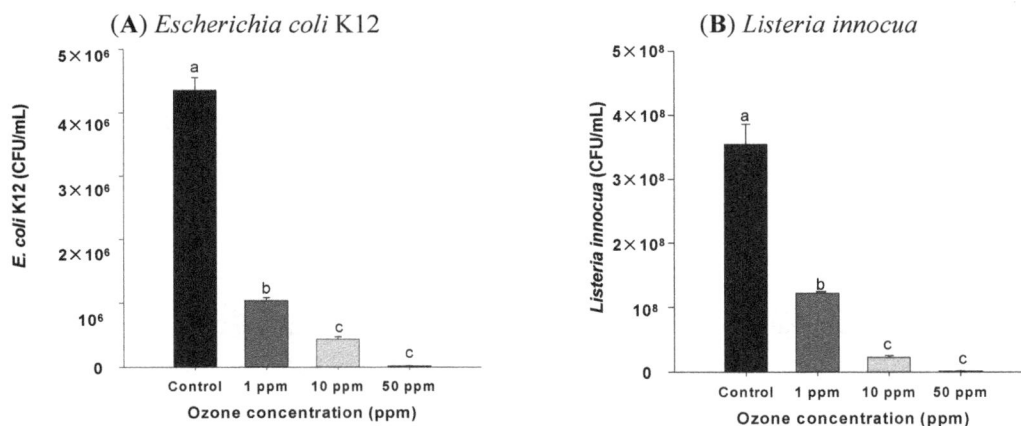

Figure 1. Impacts of ozone treatment on (**A**) *E. coli* K12 and (**B**) *L. innocua* (CFU/mL) grown on agar plates. The treatment chamber was ventilated with 1, 10, or 50 ppm ozone for 10 min. Controls were exposed to "clean air". Values represent the mean (± Standard Error) of measurements made on three independent plates per treatment. Bars with different letters are statistically significantly different ($p < 0.05$).

2.2. Optimization of the Concentration and Duration of Ozone Exposure Levels to Treat Spinach without Causing Visual Damage

The visual appearance and freshness of leafy produce has been the main judging criteria for quality distinction at purchase or consumption [23]. No visual ozone damage was observed when spinach was treated with 1 ppm gaseous ozone, but higher levels, e.g., 10 ppm for 10 min, caused significant visual blemishes and discoloration to spinach (Figure 2A). Similar results were previously observed on fresh produce like lettuce, spinach, and rocket leaves when treated with similar ozone concentrations [24]. Figure 2B illustrates ozone injury/visual damage on spinach when exposed to 10 ppm ozone concentration for 10 min. It is evident that the impact of ozone treatment on the quality of leafy produce is dependent on concentration; it may be beneficial up to a certain level to apply ozone, whereas after a critical level acceleration of browning responses will result in inferior quality.

No visual ozone damage was observed when spinach was exposed to higher concentrations such as 10, 15, and 20 ppm ozone for shorter durations (Table 1). Ozone treated produce visually looked as fresh and as attractive as untreated produce (control) after seven days of storage. Ozone injury/visible damage were observed on spinach when exposed to 25 ppm ozone concentration for all durations examined (30 s, 45 s, and 2 min).

Subsequent experiments examined the effect of varying ozone levels and exposure times on pathogens inoculated onto spinach surfaces.

Table 1. The maximum ozone exposure levels that can be applied on spinach without causing visible damage.

	Duration of the Exposure of Spinach			
Ozone concentration	10 ppm	15 ppm	20 ppm	25 ppm
Time	2 min	45 s	30 s	Damaged at 30 s, 45 s and 2 min

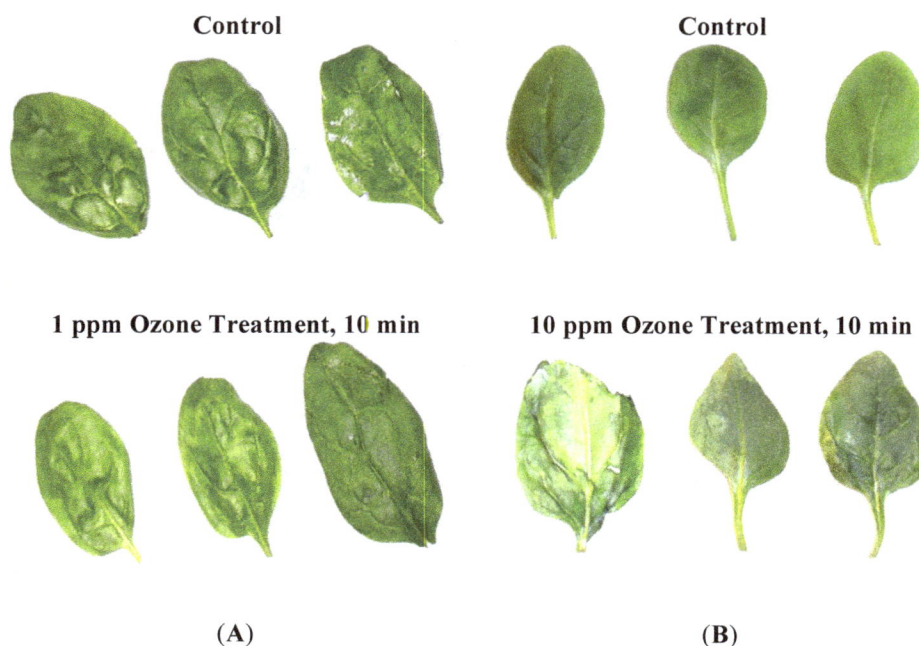

Figure 2. (A) Impact of ozone exposure levels on visual quality of spinach when treated at 1 ppm ozone concentration for 10 min and (B) Ozone injury/visual damage on spinach when exposed to 10 ppm ozone concentration for 10 min.

2.3. Effect of Ozone Exposure (1 ppm for 10 min) on Different Strains of E. coli Inoculated onto Spinach Leaf Surfaces

Colony numbers (CFU) of all six strains of *E. coli*, *i.e.*, *E. coli* O157:K88a, *E. coli* O25:H4, *E. coli* O128:K67, *E. coli* K12, *E. coli* O55:K59, and *E. coli* O104:H12 obtained from ozone exposed leaves were significantly reduced ($p < 0.05$) compared with non-ozone exposed controls (Figures 3 and 4). No *E. coli* colonies were isolated from non-inoculated spinach leaves. In the past, gaseous ozone treatment at 1 ppm for 5 min showed 3–5 log_{10} reduction of *E. coli* O157:H7 on spinach after 24 h of storage [25]. An experiment conducted in vacuum-cooling in combination with ozone gas (10 ppm for up to three days) showed 1.4 log_{10} reduction of *E. coli* O157:H7 on spinach [17]. Gaseous ozone treatment has also proved to be effective in reducing *E. coli* on many products like lettuce [14], parsley [17], mushrooms [19], blueberries [26], and dried figs [27]. Singh *et al.* [14] reported that the bactericidal effect of ozone against *E. coli* O157:H7 increased with exposure time and ozone concentration. For example, they observed 0.79–1.79 log_{10} CFU/g reduction of *E. coli* O157:H7 population on lettuce when exposed to ozone for 15 min. However, ozone treatment for 5 or 10 min did not decrease the *E. coli* O157:H7 population.

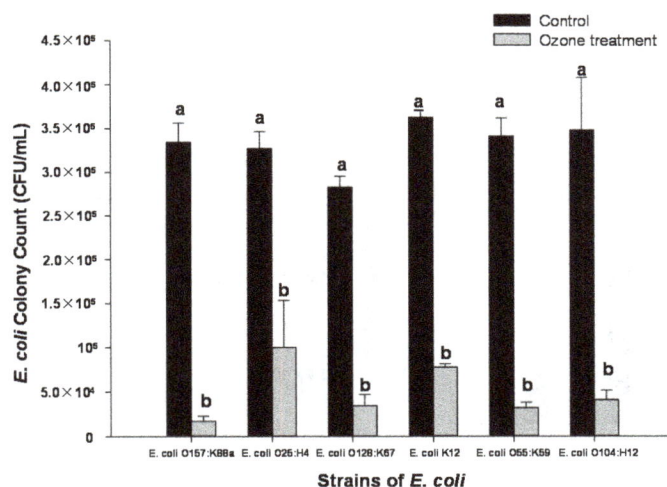

Figure 3. Impacts of ozone-enrichment on six strains of *E. coli* inoculated onto the surface of spinach leaves. Leaves were either treated with 1 ppm ozone concentration (grey bar) or untreated (black bar) for 10 min. Values represent means (±Standard Error) of measurements made on three independent spinach leaves per treatment. Bars with different letters are significantly different (*p* < 0.05).

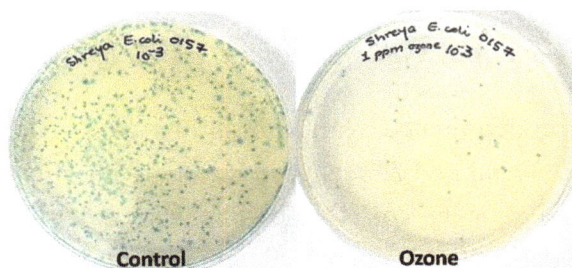

Figure 4. Colonies of *E. coli* O157:K88a on Tryptone Bile X-Glucuronide (TBX) agar recovered from leaves after being exposed to either "clean" air (control) or 1 ppm ozone concentration for 10 min.

2.4. Impact of Ozone Treatment on Listeria innocua and L. seeligeri Inoculated onto Spinach Leaves

In the present study, *L. innocua* and *L. seeligeri* were used as microbial surrogate of *L. monocytogenes*, as they are useful indicators of contamination and have also demonstrated behavior similar to *L. monocytogenes* on fresh produce [28]. Results from spinach artificially contaminated with *L. innocua* and *L. seeligeri* treated with 1 ppm ozone for an exposure time of 10 min showed a 1-log reduction in colony count compared with the untreated control (Figure 5). Karaca and his colleague [17] reported a reduction in *L. innocua* of 1.14 $\log_{10}$ CFU/g on flat-leaved parsley when treated with high ozone concentration of 950 ppm for 20 min. Similar results have been shown by previous research on mushrooms, alfalfa sprouts, alfalfa seeds, and lettuce [19]. The growth of *L. innocua* and *L. seeligeri* on spinach remained significantly reduced after Day 9 of storage (Figure 5). This may be due to the interactions between the natural background microflora of spinach and *L. innocua*, which can affect its growth and survival [22]. O'Berine and his colleague [22] reported that lactic acid bacteria and mixed population of natural microflora isolated from shredded lettuce reduced *L. innocua* growth in model media. Rodgers and colleagues [19] demonstrated complete inactivation of *L. monocytogenes* on lettuce during nine days of storage when treated with 3 ppm ozone for 3 min.

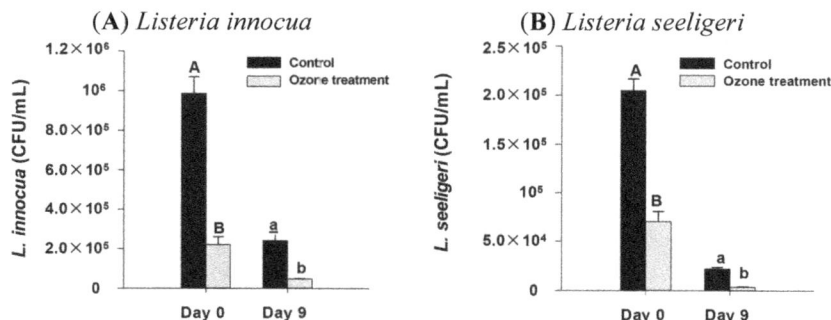

Figure 5. Impacts of ozone-enrichment on (**A**) *L. innocua* and (**B**) *L. seeligeri* inoculated onto the surface of spinach leaves. Leaves were either treated with 1 ppm ozone concentration (grey bar) or untreated (black bar) for 10 min. Colonies were enumerated either directly after the treatments, *i.e.*, Day 0 or after nine days of storage. Values represent means (±Standard Error) of measurements made on three independent spinach leaves per treatment. Bars with different letters are significantly different ($p < 0.05$).

2.5. Effect of Higher Ozone Treatment on E. coli and Listeria sp. Inoculated onto Spinach Leaf Surface

Results of spinach artificially contaminated with two strains of *E. coli* (*E. coli* O157:K88a and *E. coli* O25:H4) and *Listeria* (*L. innocua* and *L. seeligeri*) treated with 10 ppm of ozone concentration for 2 min are shown in Figure 6. For *E. coli* O157:K88a and *E. coli* O25:H4, ozone treatment significantly ($p < 0.05$) reduced counts by 1-log compared with the untreated control (Figure 6A). Ozone had less than 1-log effect on *L. innocua* and *L. seeligeri* (Figure 6B). Awli [18] achieved reduction of 2.89 and 3.06 $\log_{10}$ for *E. coli O157* and *L. monocytogenes*, respectively, on bell pepper when exposed to 9 ppm ozone for 6 h. Their work met the standards for an antimicrobial agent by attaining a minimum of 2 $\log_{10}$ reduction [18]. Similar reductions were observed from application of 5 ppm ozone for 3 min on whole tomato [29]. When results from this work (on leafy produce) are compared with other hardy produce, it appears that ozone treatment was less successful. This is most probably due to the delicate nature of leafy produce, which limits the use of increased ozone concentration and exposure time (results from Section 2.2). In addition, the results obtained from this treatment, *i.e.*, 10 ppm for 2 min were not significantly more effective in reducing bacterial viable counts in comparison to previous ozone treatment used in this study, *i.e.*, 1 ppm for 10 min (from Sections 2.3 and 2.4).

Ozone inactivates bacterial cells by the progressive oxidation of important cellular constituents [17], and suggestions for the principal target of ozonation include the bacterial cell surface. Bacterial cell death was observed as a consequence of a ruptured cell membrane and as a result of disintegration of cell wall to function as a barrier [17,18,20]. *E. coli*, a Gram-negative bacterium, is more susceptible to ozone treatment because it has a thin peptidoglycan lamella that is covered by an outer membrane made of polysaccharides and lipoproteins [30]. In contrast, some studies claimed that Gram-negative bacteria were more resistant to ozone treatment as compared with Gram-positive bacteria [31]. Results from this study show that ozone treatment was effective in both *E. coli* and *Listeria* spp. inactivation but

Listeria spp. were slightly more resistant. These results are in line with Yuk and colleagues [19], who showed that *E. coli* O157:H7 is more sensitive than *Listeria monocytogenes*.

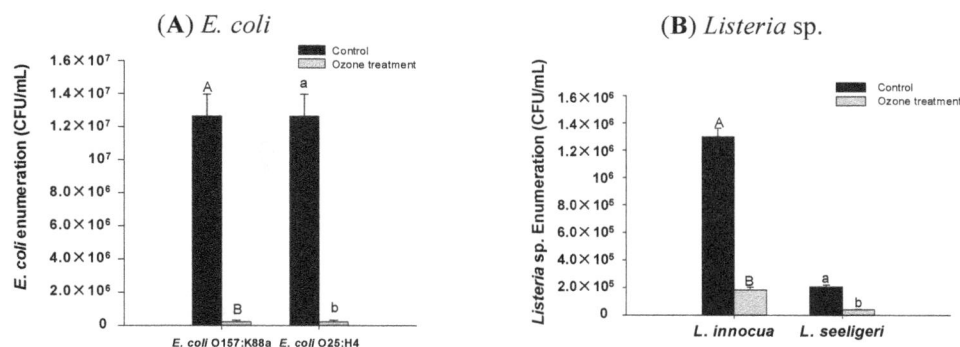

Figure 6. Impacts of increased levels of ozone exposure on two strains of (**A**) *E. coli* and (**B**) *Listeria* sp. inoculated onto the surface of spinach leaves. Leaves were either treated with 10 ppm ozone concentration (grey bar) or untreated (black bar) for 2 min. Values represent means (±Standard Error) of measurements made on three independent spinach leaves per treatment. Bars with different letters are significantly different ($p < 0.05$).

To investigate the after effects of the ozone treatment on pathogen growth, artificially contaminated spinach was stored at 7 °C for nine days. Figure 7 shows that populations of both *E. coli* (*E. coli* O157:K88a and *E. coli* O25:H4) and *Listeria* sp. (*L. innocua* and *L. seeligeri*) after nine days of storage did not regrow, as a significant reduction in number of colonies was observed in comparison with the untreated control. However, effect of higher ozone treatment on pathogen recovery did not show a significant difference in count as compared with treatment with lower ozone concentration.

2.6. Effect of Age on Ozone Resistance of E. coli O157:K88a in Vitro

Throughout the study, we observed that a certain proportion of cells survived ozone exposure and we were interested to make initial investigations into potential ozone resistance mechanisms. *E. coli* cells of increasing colony age were exposed to ozone (*in vitro*) and results demonstrated a clear increase in ozone resistance of *E. coli* O157:K88a with increasing colony age. For example, survival of *E. coli* O157:K88a was observed to be greater (approximately 15%) after five days of growth compared with the day 1 time point. Survival levels increased even further by Day 7 (Figure 8) suggesting that cells in older bacterial colonies are more ozone resistant than cells from younger colonies. This is possibly because the older *E. coli* cells may to be in their long-term stationary phase (fifth phase of bacterial growth cycle that survives on the nutrient released by the dead population of bacteria). These older cells can survive external stress unlike the younger cells (probably in first or second phase of bacterial growth cycle) and can remain viable for months or even years once they enter long-term stationary phase [32]. This stationary phase is dominated by the accumulation of the sigma factor RpoS [32]. The entire cellular physiology of *E. coli* is influenced by RpoS that directly or indirectly affects the expression of 10% of the *E. coli* genes. These genes are involved in morphological variations within the cell and responsible for increasing resistance during numerous stress conditions, e.g., oxidative stress, osmotic stress, heat shock, *etc.* [32].

(A) *E. coli*   **(B)** *Listeria* sp.

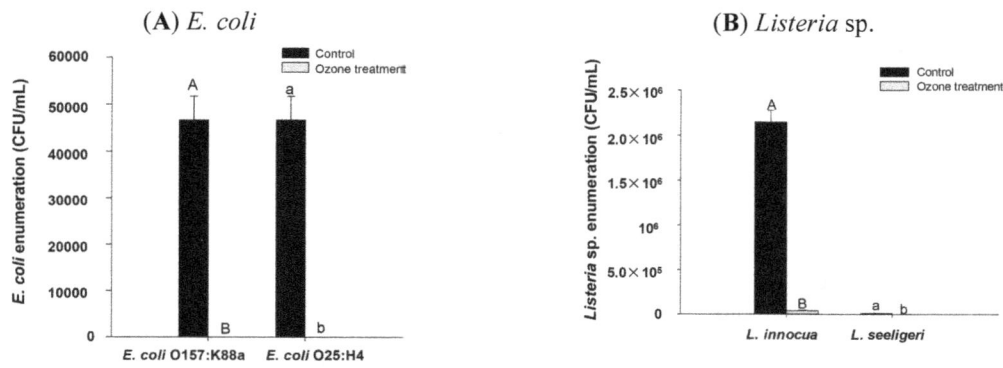

Figure 7. Impacts of ozone-enrichment on two strains of (**A**) *E. coli* and (**B**) *Listeria* sp. inoculated onto the surface of spinach leaves. Leaves were either treated with 10 ppm ozone concentration (grey bar) or untreated (black bar) for 2 min. Colonies were enumerated after nine days of storage. Values represent means (±Standard Error) of measurements made on three independent spinach leaves per treatment. Bars with different letters are significantly different ($p < 0.05$).

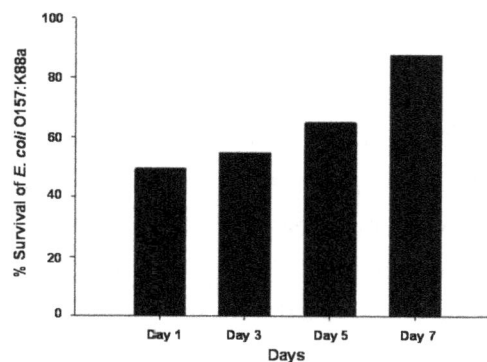

Figure 8. Survival of cells obtained from different colony ages of *E. coli* O157:K88a exposed to 10 ppm ozone concentration for 2 min. After ozone exposure, the culture plates were maintained at 37 °C for seven days.

3. Experimental Section

3.1. Ozone Fumigation System

A purpose designed ozone fumigation system (Figure 9) was housed in a fume hood and constructed of stainless steel (diameter 35 cm). An inlet pipe was used to add ozone generated by electric discharge from oxygen (model SGA01 Pacific Ozone Technology Inc., Brentwood, CA, USA), and the introduction of ozone was manually controlled via stainless steel needle valves/gap flow meters. Once the desired ozone concentration was achieved, Petri plates/produce to be exposed to ozone was placed at the bottom of the system and the fumigation system closed with the Pyrex cover (Figure 9). The ozone concentration in the system was recorded using a photometric analyzer (model 450, manufactured by Advanced Pollution Instrumentation Division, 9480 Carroll Park Drive, San Diego, CA, USA). The ozone monitor employed in these studies was serviced weekly and calibrated routinely against standards using a Dasibi 1008PC unit.

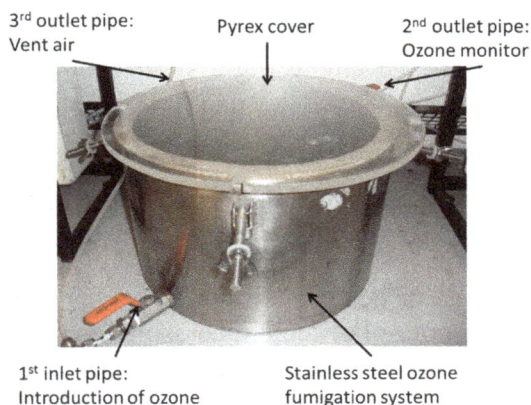

Figure 9. Ozone fumigation system.

3.2. Assessing the Impact of Ozone Treatment on Food Pathogens E. coli and L. innocua in Vitro

E. coli K12 and *L. innocua* were obtained from a culture collection maintained by Geneius Laboratories Ltd. (44 Colbourne Crescent, Nelson Park, Cramlington, UK). These cultures were sub-cultured by spread plating on Nutrient agar (NA) and Agar Listeria according to Ottaviani and Agosti (ALOA) agar plates, respectively. A single colony was isolated from each culture plate after incubation at 37 °C for 24 h and 30 °C for 48 h, respectively, and transferred to minimum recovery diluent (MRD). A standardized concentration of 10^4 cells per mL (100 μL) of each culture was spread onto sterile NA and ALOA agar plates, respectively. These plates were then either exposed to 1 ppm, 10 ppm, 50 ppm ozone concentration, or charcoal filtered "clean air" (controls) for 10 min at room temperature. After treatment, NA and ALOA agar plates were incubated at 37 °C for 24 h and 30 °C for 48 h, respectively. The number of colonies produced on control plates (non-ozone exposed) were compared with the numbers found on ozone-treated plates based on three replicate observations.

3.3. Optimization of Ozone Exposure Levels (Concentration and Duration) to Treat Leafy Salads without Causing Visual Damage to Produce

This experiment focused on optimizing the concentration and duration of ozone exposure to which fresh produce could be exposed without causing visible damage/deterioration. To determine the impact on visual quality of the produce, baby spinach was received from Vitacress Ltd. (Hampshire, UK) and then exposed to 1, 10, 25, 50 ppm ozone or "clean air" (controls) for varying periods of time (1–60 min). Following exposure to ozone, the produce was then packed in a sterile self-seal bag and maintained at 4 °C in dark conditions. Ozone injury was assessed visually by comparing ozone exposed produce with control (non-ozone exposed) produce every alternate day for seven days.

3.4. Ozone Resistance of Different Strains of E. coli: Inoculation of E. coli onto Spinach Leaves and Ozone Exposure Conditions

Six strains of *E. coli* (*E. coli* O157:K88a, *E. coli* O25:H4, *E. coli* O128:K67, *E. coli* K12, *E. coli* O55:K59, and *E. coli* O104:H12) were obtained from a culture collection maintained by Geneius Laboratories Ltd. (Nelson Park, Cramlington, UK). Cultures were stored at 4 °C on Luria-Bertani (LB) agar plates, and activated in LB broth at 37 °C. Baby-leaf spinach was purchased from a local retailer and aseptically cut into discs measuring 1.13 cm^2 using a sterile cork borer. A suspension of *E. coli* (overnight culture, 10^8–10^9 CFU/mL LB broth) was applied directly to the leaf disc in 300 µL aliquots, and then the inoculated leaves were stored overnight at 7 °C to mimic produce storage conditions and to allow attachment of *E. coli* to the leaf surface. Inoculated leaves were either exposed to 1 ppm ozone or charcoal filtered "clean air" for 10 min at room temperature. To determine the number of *E. coli* remaining (control and ozone exposed), the leaf discs were vigorously shaken in MRD for 2 min and then serially diluted using MRD, followed by pour plate technique using Tryptone Bile X-Glucuronide (TBX) agar plates. Plates were incubated at 44 °C for 24 h, and presumptive colonies were counted on the basis of three replicate observations.

3.5. Impact of Ozone Treatment on L. innocua and L. seeligeri Inoculated onto Spinach Leaves

Two strains of *Listeria* (*L. innocua* and *L. seeligeri*) were obtained from a culture collection maintained by Geneius Laboratories Ltd. Cultures and stored at 4 °C on ALOA agar plates. Spinach leaves were then aseptically cut into discs measuring 1.13 cm^2 using a sterile cork borer. A suspension of *Listeria* sp. (10^7–10^8 CFU/mL MRD) was applied directly to the leaf disc in 300 µL aliquots, and the inoculated leaves were maintained at 7 °C to mimic produce storage conditions for 2 h to allow attachment of *Listeria* sp. to the leaf surface. Inoculated leaves were either exposed to 1 ppm ozone or charcoal filtered "clean air" for 10 min at room temperature and survival rate enumerated (see below). For determining the survival and growth of *Listeria* sp. during storage, a proportion of the treated and untreated inoculated leaves were maintained at 7 °C for a further nine days. The number of colonies remaining (control and ozone exposed) on Day 0 and Day 9 was determined by vigorously shaking the leaf disc in MRD for 2 min after 1 h incubation at room temperature, and then serially diluting in MRD followed by standard spread technique on ALOA agar plates. Plates were incubated at 30 °C for 48 h, and colonies were counted.

3.6. Modified Ozone Fumigation System—Delivery of High Ozone Concentrations for Short Time Durations (Seconds)

A modified ozone fumigation system was engineered to improve the application of ozone to produce surfaces, and to reduce the time required to build up the desired ozone concentrations needed for produce treatment. The aim was to develop a system allowing application of higher ozone concentrations for shorter durations to achieve better bacterial kill without damaging the produce. This system was developed after discussions/meetings with industrial partners who ideally wanted to be able to expose produce to ozone quickly during their harvest and processing procedures. The modified ozone exposure apparatus was housed in a fume hood and constructed from 20 cm^2 Perspex. Produce was placed on a steel mesh in a 2 cm deep tray within the box and produce was then exposed to ozone once the desired concentration was achieved (Figure 10). An inlet pipe was used to add ozone generated by electric discharge from oxygen, with the introduction of ozone controlled manually. The ozone concentration was recorded by a photometric analyzer (model 450, manufactured by Advanced Pollution Instrumentation Inc., San Diego, CA, USA). The ozone monitor employed in these studies was serviced routinely.

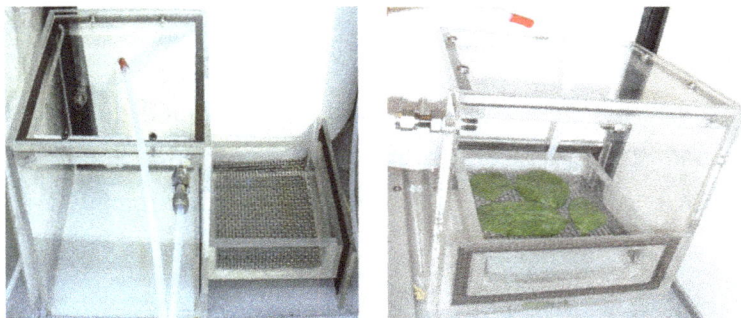

Figure 10. Modified ozone fumigation system.

3.7. Exploration of Higher Ozone Exposure Levels to Treat Spinach without Causing Visual Damage

This experiment aimed to determine the highest ozone concentration and exposure time that could be used on organic baby spinach without causing visible damage/deterioration to the produce. Produce was exposed to 10, 15, 20, 25 ppm ozone or "clean air" (controls) for varying periods of time ranging from 30 s to 2 min. Following exposure to ozone, produce was then packed in a sterile self-seal bag and maintained at 4 °C in the dark. Ozone injury was assessed visually by comparing ozone exposed produce with control (non-ozone exposed) produce every alternate day for seven days.

3.8. Impact of Higher/Increased Ozone Concentrations on Two Strains of E. coli and Listeria Inoculated onto Spinach Leaves

This experiment aimed to use the highest ozone exposure levels that did not cause produce damage (data obtained from Section 3.7—Result Section 2.2) to try and achieve higher reductions in pathogenic bacteria on the surface of baby spinach leaves. Two strains of *E. coli* (*E. coli* O157:K88a and *E. coli* O25:H4) and *Listeria* (*L. innocuous* and *L. seeligeri*) were inoculated onto spinach leaves as described in Sections 3.4 and 3.5, respectively. Inoculated leaves were either treated with 10 ppm ozone concentration or charcoal filtered "clean air" for 2 min. The number of *E. coli* and *Listeria* sp. remaining (control and ozone exposed) was determined as described above (Sections 3.4 and 3.5).

To determine the impact of highest ozone exposure levels on the survival and growth of *E. coli* (*E. coli* O157:K88a and *E. coli* O25:H4) and *Listeria* (*L. innocua* and *L. seeligeri*) during storage, the inoculated leaves were treated as mentioned in Section 3.5. After the treatment, inoculated and control leaves were maintained at 7 °C for nine days. The number of colonies remaining (control and ozone exposed) on day 9 was determined as mentioned in Section 3.5.

3.9. Age Effects on Ozone Resistance of E. coli in Vitro

To determine whether cell age affected the ozone resistance of the bacteria, a colony of *E. coli* O157:K88a obtained from a culture collection maintained by Geneius Laboratories Ltd. was sub-cultured onto NA plates and incubated at 37 °C for seven days. A single colony was isolated on the first, third, fifth, and seventh day of the incubation and transferred to MRD. A standardized concentration of 10^4 cells per mL (100 µL) of each cell age was spread onto sterile NA plates and these plates were then exposed to either 10 ppm ozone concentration or charcoal filtered "clean air" for 2 min. Colony count was determined after incubating NA plates at 37 °C for 24 h.

3.10. Statistical Analysis

Data were analyzed using SPSS (IBM SPSS Statistics 19 64Bit) and graphs were produced using Microsoft Office Excel 2010 and SigmaPlot 12.5. Normal data distribution was tested using a Normality test and significant differences between mean values were verified using LSD ($p < 0.05$) following one-way ANOVA.

4. Conclusions

Exposure to 1 ppm and 10 ppm gaseous ozone treatment for 10 and 2 min, respectively, significantly reduced *E. coli* and *Listeria* spp. populations on spinach. In addition, the pathogens did not re-grow after treatment, *i.e.*, over a nine-day storage period. Although ozone treatment only reduced bacterial loads by 1 log, there is still commercial potential as ozone is easy to produce on site and apply at levels which do not damage sensitive leafy produce. The findings from this study show that some bacteria in populations are resistant to ozone treatment and increasing cell (colony) age of *E. coli* was shown to be linked to enhanced ozone resistance. Further work is needed to better understand the exact mechanism of resistance, and this may lead to determining methods that can overcome resistance. Such applications could deliver immense potential benefits for commercial use and improving public health.

Acknowledgments

This study was supported by the ADHB/Horticultural Development Company (HDC) (FV 386) via the award of a PhD studentship to Shreya Wani. All *E. coli* and *Listeria* spp. strains were kindly provided by Geneius laboratories Ltd. (www.geneiuslabs.com). We thank Matthew Peake for technical help and Alan Craig for maintaining and calibrating the ozone fumigation.

Author Contributions

The work presented is a part of a PhD project funded by the UK-ADHB/HDC; the data were managed, designed, collected, and analyzed by Shreya Wani who took the lead on writing the manuscript. Ian Singleton and Jeremy Barnes supervised the work, assisted in experimental design and data interpretation, and won the grant award. Joseph Thompson was a summer intern and assisted with some experimental work in the laboratory. Jagpreet Maker was a post-graduate student who contributed to the editing and approval of the manuscript.

Conflicts of Interest

The authors declare no conflict of interest.

References

1.   Burnett, S.L.; Beuchat, L.R. Human pathogens associated with raw produce and unpasteurized juices, and difficulties in decontamination. *J. Ind. Microbiol. Biotechnol* **2000**, *25*, 281–287.

2. Abadias, M.; Usall, J.; Anguera, M.; Solsona, C.; Vinas, I. Microbiological quality of fresh, minimally-processed fruit and vegetables, and sprouts from retail establishments. *Int. J. Food Microbiol.* **2008**, *123*, 121–129.

3. Velusamy, V.; Arshak, K.; Korostynska, O.; Oliwa, K.; Adley, C. An overview of foodborne pathogen detection: In the perspective of biosensors. *Biotechnol. Adv.* **2010**, *28*, 232–254.

4. Olaimat, A.N.; Holley, R.A. Factors influencing the microbial safety of fresh produce: A review. *Food Microbiol.* **2012**, *32*, 1–19.

5. Engels, C.; Weiss, A.; Carle, R.; Schmidt, H.; Schieber, A.; Ganzle, M.G. Effect of gallotannin treatment on attachment, growth and survival of *Escherichia coli* O157:H7 and *Listeria monocytogenes* on spinach and lettuce. *Eur. Food Res. Technol.* **2012**, *234*, 1081–1090.

6. Griffin, P.M.; Tauxe, R.V. The epidemiology of infections caused by *Escherichia coli* O157:H7, other enterohemorrhagic *E.coli*, and the associated Hemolytic Uremic Syndrome. *Epidemiol. Rev.* **1991**, *13*, 60–99.

7. Tomas-Callejas, A.; Lopez-Velasco, G.; Camacho, A.B.; Artes, F.; Artes-Hernandez, F.; Suslow, T.V. Survival and distribution of *Escherichia coli* on diverse fresh-cut baby leafy greens under preharvest through postharvest conditions. *Int. J. Food Microbiol.* **2011**, *151*, 216–222.

8. Coia, J.E. Clinical, microbiological and epidemiological aspects of *Escherichia coli* O157 infection. *FEMS Immunol. Med. Microbiol.* **1998**, *20*, 1–9.

9. Farber, J.M.; Peterkin, P.I. *Listeria monocytogenes*, a Food-Borne pathogen. *Microbiol. Rev.* **1991**, *55*, 476–511.

10. Notermans, S.; Todd, E.C.D. Surveillance of listeriosis and its causative pathogen, *Listeria monocytogenes*. *Food Control* **2011**, *22*, 1484–1490.

11. Launders, N.; Byrne, L.; Adams, N.; Glen, K.; Jenkins, C.; Tubin-Delic, D.; Locking, M.; Williams, C.; Morgan, D.; Outbreak Control Team. Outbreak of Shiga toxin-producing *E.coli* O157 associated with consumption of watercress, United Kingdom, August to September 2013. *Euro Surveill.* **2013**, *18*, doi:10.2807/1560-7917.ES2013.18.44.20624.

12. Bower, C.K.; Daeschel, M.A. Resistance responses of microorganisms in food environments. *Int. J. Food Microbiol.* **1999**, *50*, 33–44.

13. Mahapatra, A.K.; Muthukumarappan, K.; Julson, J.L. Applications of ozone, bacteriocins and irradiation in food processing: A review. *Crit. Rev. Food Sci. Nutr.* **2005**, *45*, 447–461.

14. Singh, N.; Singh, R.K.; Bhunia, A.K.; Stroshine, R.L. Efficacy of chlorine dioxide, ozone, and thyme essential oil or a sequential washing in killing *Escherichia coli* O157:H7 on lettuce and baby carrots. *LWT Food Sci. Technol.* **2002**, *35*, 720–729.

15. Goncalves, A.A. Ozone—An emerging technology for the seafood industry. *Braz. Arch. Biol. Technol.* **2009**, *52*, 1527–1539.

16. Tzortzakis, N.; Borland, A.; Singleton, I.; Barnes, J. Impact of atmospheric ozone-enrichment on quality-related attributes of tomato fruit. *Postharvest Biol. Technol.* **2007**, *45*, 317–325.

17. Karaca, H.; Velioglu, Y.S. Effects of ozone treatments on microbial quality and some chemical properties of lettuce, spinach, and parsley. *Postharvest Biol. Technol.* **2014**, *88*, 46–53.

18. Alwi, N.A.; Ali, A. Reduction of *Escherichia coli* O157, *Listeria monocytogenes* and *Salmonella enterica* sv. Typhimurium populations on fresh-cut bell pepper using gaseous ozone. *Food Control* **2014**, *46*, 304–311.

19. Yuk, H.G.; Yoo, M.Y.; Yoon, J.W.; Marshall, D.L.; Oh, D.H. Effect of combined ozone and organic acid treatment for control of *Escherichia coli* O157:H7 and Listeria monocytogenes on enoki mushroom. *Food Control* **2007**, *18*, 548–553.

20. Fan, L.; Song, J.; McRae, K.B.; Walker, B.A.; Sharpe, D. Gaseous ozone treatment inactivates *Listeria innocua in vitro*. *J. Appl. Microbiol.* **2007**, *103*, 2657–2663.

21. Gleeson, E.; O'Beirne, D. Effects of process severity on survival & growth of *Escherichia coli* & *Listeria innocua* on minimally processed vegetables. *Food Control* **2005**, *16*, 677–685.

22. O'Beirne, D.; Francis, A.G. Effects of the indigenous microflora of minimally processed lettuce on the survival and growth of *Listeria innocua*. *Int. J. Food Sci. Technol.* **1998**, *33*, 477–488.

23. Rico, D.; Martín-Diana, A.B.; Barat, J.M.; Barry-Ryan, C. Extending and measuring the quality of fresh-cut fruit and vegetables: A review. *Trends Food Sci. Technol.* **2007**, *18*, 373–386.

24. Alexopoulos, A.; Plessas, S.; Ceciu, S.; Lazar, V.; Mantzourani, I.; Voidarou, C.; Stavropouloua, E.; Bezirtzogloua, E. Evaluation of ozone efficacy on the reduction of microbial population of fresh cut lettuce (*Lactuca sativa*) and green bell pepper (*Capsicum annuum*). *Food Control* **2013**, *30*, 491–496.

25. Klockow, P.A.; Keener, K.M. Safety and quality assessment of packaged spinach treated with a novel ozone-generation system. *LWT Food Sci. Technol.* **2009**, *42*, 1047–1053.

26. Bialka, K.L.; Demirci, A. Decontamination of *Escherichia coli* O157:H7 and *Salmonella enterica* on blueberries using ozone and pulsed UV-light. *J. Food Sci.* **2007**, *72*, 391–396.

27. Akbas, M.Y.; Ozdemir, M. Application of gaseous ozone to control populations of *Escherichia coli*, *Bacillus cereus* and *Bacillus cereus* spores in dried figs. *Food Microbiol.* **2008**, *25*, 386–391.

28. Scifò, G.O.; Randazzo, C.L.; Restuccia, C.; Fava, G.; Caggia, C. *Listeria innocua* growth in fresh cut mixed leafy salads packaged in modified atmosphere. *Food Control* **2009**, *20*, 611–617.

29. Bermúdez-Aguirre, D.; Barbosa-Cánovas, G.V. Disinfection of selected vegetables under nonthermal treatments: Chlorine, acid citric, ultraviolet light and ozone. *Food Control* **2013**, *29*, 82–90.

30. Zuma, F.; Lin, J.; Jonnalagadda, S.B. Ozone-initiated disinfection kinetics of *Escherichia coli* in water. *J. Environ. Sci. Health A Tox. Hazard. Subst. Environ. Eng.* **2009**, *44*, 48–56.

31. Vaz-Velho, M.; Silva, M.; Pessoa, J.; Gibbs, P. Inactivation by ozone of *Listeria innocua* on salmon-trout during cold-smoke processing. *Food Control* **2006**, *17*, 609–619.

32. Navarro Llorens, J.M.; Tormo, A.; Martinez-Garcia, E. Stationary phase in gram-negative bacteria. *FEMS Microbiol. Rev.* **2010**, *34*, 476–495.

4

Effectiveness of Organic Wastes as Fertilizers and Amendments in Salt-Affected Soils

Mariangela Diacono [1,*] and Francesco Montemurro [2]

[1] Consiglio per la Ricerca e l'analisi dell'economia Agraria, CRA-SCA, Research Unit for Cropping Systems in Dry Environments, Via Celso Ulpiani 5, 70125, Bari, Italy
[2] Consiglio per la Ricerca e l'analisi dell'economia Agraria, CRA-SCA, Research Unit for Cropping Systems in Dry Environments (Azienda Sperimentale Metaponto), SS 106 Jonica, km 448.2, 75010, Metaponto (MT), Italy; E-Mail: francesco.montemurro@entecra.it

* Author to whom correspondence should be addressed; E-Mail: mariangela.diacono@entecra.it

Academic Editor: Stephen R. Smith

Abstract: Excessive salt rate can adversely influence the physical, chemical, and biological properties of soils, mainly in arid and semi-arid world regions. Therefore, salt-affected soils must be reclaimed to maintain satisfactory fertility levels for increasing food production. Different approaches have been suggested to solve these issues. This short review focuses on selected studies that have identified organic materials (e.g., farmyard manures, different agro-industrial by-products, and composts) as effective tools to improve different soil properties (e.g., structural stability and permeability) in salt-affected soils. Organic fertilization is highly sustainable when compared to other options to date when taken into consideration as a solution to the highlighted issues. However, further experimental investigations are needed to validate this approach in a wider range of both saline and sodic soils, also combining waste recycling with other sustainable agronomic practices (crop rotations, cover crops use, *etc.*).

Keywords: salts; salinization; Mediterranean environment; agro-industrial by-products; organic fertilization

1. Introduction

The expected increase in the world's population (9.6 billion by 2050) needs food productivity to step up within a few decades [1,2]. Unfortunately, extensive areas of irrigated lands are unproductive, due to the accumulation of salts in the soil profile occupied by root systems. It is estimated that about 15% of the total land area of the world has been degraded by salinization and soil erosion, which are among the major causes of desertification [3]. Dajic [4] reported that the total world area affected by saline and sodic soils is 397×10^6 and 434×10^6 ha, respectively. On this matter, according to the Joint Research Centre Institute for Environment and Sustainability (European Commission), soil salinization affects an estimated one to three million hectares in the EU [5].

Soil salinization and drought stress mainly occur in the arid and semiarid regions of Mediterranean area, which are characterized by high evapotranspiration rates and low rainfall. In these areas, the leaching of salts is very low, therefore, salt accumulates in soil surface layers. Since high salts content may adversely influence soil properties and crop yields, food security could be limited as a consequence. Therefore, salt-affected soils must be reclaimed to maintain satisfactory levels of fertility for sustaining food production.

To date, different approaches have been suggested to solve these issues, such as soil leaching with water, chemical amendment, and phytoremediation [6,7]. On the other hand, the implementation of sustainable farming practices may prevent and, in some cases, reverse soil salinization conditions. For example, rational management of brackish water for land-irrigation should be employed and, in rain-fed agriculture, crops rotation can be promoted for improving the balance between rainfall and water use by crop. Biotechnological strategies and application of breeding and screening methodologies to enhance the tolerance of crops to salinity conditions, as well as organic fertilization for reclaiming saline and sodic soils, and increase their fertility, have also been assessed [8].

This paper provides a brief overview of the present knowledge regarding organic fertilization by different waste-recycling in soils that are under stress due to salinization conditions. The overview, focusing on recently-published data, aimed to investigate the main approaches and effects of this agronomic practice on some soil properties, thus, verifying the potential of organic amendments to restore soil quality.

2. Salinity: Causes and Effects

In order to identify proper strategies of organic fertilization of soils in salinized areas, it is essential to understand how salinity develops in the soil. Salinity can be defined as an accumulation of dissolved mineral salts in soil water, and an excess of sodium ions in the rizosphere [9]. The origin of soil salts can be natural (*i.e.*, primary salinization) or human-induced (*i.e.*, secondary salinization). The main natural source of salinity is the weathering of minerals in rocks, sediments, and soils. Other common sources of soil salts are the atmospheric deposition of oceanic salts and the intrusion of seawater into the groundwater of coastal areas, where the over-exploitation of water can considerably move down the normal water table [5,9]. Under high water table conditions, salts can move upward due to evaporation and evapotranspiration processes.

Secondary salinization can be the result of irrigation, which is, also, sometimes carried out with brackish water on saline soils. This use of unconventional water helps to face the current scarcity of water resources for farming, which is determined by the competition with different human and industrial uses. Attention must be paid, however, on the quality of the water used, as well as on the fact that seasonal/temporary salinization can be partially controlled by fulfilling appropriate leaching requirements [10]. In addition, repeated application of animal manures and sewage sludge to cropland may be considered as an anthropogenic source of salts. Therefore, appropriate wastes management strategies, such as controlled biodegradation processes (*i.e.*, composting), are crucial to minimize the potentially negative environmental impact of waste application prior to their use in agriculture [11].

The distribution of soluble salts in the soil profile is influenced by leaching and evaporation from soil surfaces. Some of the accumulated specific ions (such as Cl^-) can be directly toxic, depending on plant-specific tolerance, and may induce physiological disorders. Excessive amounts of salts can inhibit the uptake of mineral nutrients, cause premature senescence, and reduce the photosynthetic activity to a level that cannot sustain crop growth and yields [12]. Water deficit or osmotic effects are among the major factors that brings decline in cell division and reduction of plant growth, thus, limiting crop production [9].

In addition, excess of salts may adversely influence the biological, physical, and chemical properties of soil. Sodicity (*i.e.*, excess of Na^+ in the rhizosphere) is a secondary consequence of salinity, which is typical for clayey soils and affects their physical properties. In these soils, the exchangeable Na^+ is bound to the negative charges of clay, thus, causing deflocculation of clayey particles. As Lauchli and Epstein [13] highlighted, the high exchangeable Na^+ percentage can lead to swelling and dispersion of clays, as well as breaking of soil aggregates. As a consequence, both water infiltration and water-holding capacity could be reduced. Saline soils are easier to be reclaimed than sodic ones, because, generally, the former requires leaching of soluble salts, while the latter also requires a Ca^{2+} source to replace the excess Na^+.

Salinity also affects soil chemical properties, such as pH, cation exchange capacity (CEC), exchangeable sodium percentage (ESP), soil organic carbon, and alters the osmotic and matric potential of the soil solution [14]. Most salt-affected soils are deficient in several nutrients, thus, more fertilizer applications may be required. Micronutrient deficiency appears to be a side-effect of salinization and may derive both from soil alkalinization and ions competition [15]. Moreover, Garcia and Hernandez [16] showed that an increase in soil salinity inhibits several soil enzymatic activities, such as alkaline phosphatase and β-glucosidase, while Rietz and Haynes [17] indicated the effects of salinity, both on soil microbial biomass carbon and enzyme activities. In particular, the fungal part of the microbial biomass was strongly reduced in saline soils [18]. Therefore, effective tools to improve soil properties in salt-affected soils are crucial to guarantee an income for farmers particularly in arid world regions.

3. Organic Fertilization on Salt-Affected Soils: Organic Wastes Recycling

It is known that several organic materials, such as farmyard manures, agro-industrial by-products and composts can be used as amendments to enhance and sustain the overall soil fertility [19,20]. The same amendments could likely be considered for soil remediation in the salt-affected areas due to their high organic matter content. In fact, organic matter has several beneficial effects on agricultural fields, such as the slow release of nutrients, soil structure improvement, and the protection of soils against erosion [21].

Selected studies (from literature of the last 10–15 years) are summarized in Table 1, focusing on the effects of application of organic matter (*i.e.*, different organic waste materials) to salt stressed soils. In particular, such effects can be referred to chemical, biological, and physical soil properties, as it will be discussed in the next subsections. The reported findings offer powerful evidence on the potential of organic fertilizers in improving soil properties.

Table 1. Effects of various organic matter inputs under soil salinity conditions (essential data).

Organic Materials	Soil Salinity/Salt Levels	Effects	Reference
Cotton gin crushed compost and poultry manure	ESP 15.7 EC 9 mS·cm^{-1} pH 8	Improving soil structure, reducing (by 50%) the ESP and increasing different enzyme activities	[8]
Mixture of green waste compost, sedge peat and furfural residue	ESP 15.8 EC 3.69 mS·cm^{-1} pH 7.75	Decreasing bulk density, EC, and ESP and increasing total porosity and organic carbon. The combination of amendments had substantial potential for ameliorating saline soils, working better than each amendment alone	[14]
(i) Pig manure (ii) Pig manure+rice straw (iii) Rice straw (iv) control	Total salts 3.3 g·kg^{-1} pH 8.86	Urease activity increased by more than 150% in the mixed treatment, compared to the control. The incorporation of organic manure into the soil significantly increased soil alkaline phosphatase activity and soil respiration rate	[22]
Green manure mixed with farmyard manure	1%–2% salt EC 8.5–20.4 mS·cm^{-1} pH 4.58–4.79	The OM application in paddy fields could effectively alleviate the problem of soil salinity, also resulting in yield improvement	[23]
Cassava-industrial waste compost and vermicompost with or without earthworms	EC 4.26 mS·cm^{-1} pH 7.30	Compost and vermicompost amendments decreased electrical conductivity, improved CEC, soil organic carbon, total nitrogen and extractable phosphorus	[24]
Compost produced from by-products of the olive oil industry and poultry manure	EC 1.85 mS·cm^{-1} pH 7.7	Increasing soluble and exchangeable-K$^+$ (thus limiting the entry of Na$^+$ into the exchange complex) as well as CEC	[25]
Farmyard manure + saline water (EC 2.25 mS·cm^{-1})	EC 4.8–6.3 mS·cm^{-1}	Improvement of infiltration rate by about 89%, and decreasing soil sodicity by 41.3%. Decreasing soil bulk density, allowing an enhancement of soil porosity and aeration, and improving saline water leaching	[26]
Compost (animal wastes and plant residues)	ESP 34–37 EC 4.03–5.11 mS·cm^{-1} pH 8.62–8.75	Decreasing EC and sodium adsorption ratios of the saturation extracts of the soils. Organic amendments co-applied with chemical amendments seemed to have a high value for reducing soil pH, soil salinity, and soil sodicity	[27]
Municipal wastewater	EC 60 mS·cm^{-1} pH 7.48	Decreasing soil pH and bulk density, while increasing EC and OM content of soil	[28]

Table 1. *Cont.*

Organic Materials	Soil Salinity/Salt Levels	Effects	Reference
Farm yard manure	EC 3.7–5.0 mS·cm^{-1} pH 8.69–9.18	Gypsum + sulfuric acid + Farm yard manure decreased bulk density but increased the porosity, void ratio, water permeability and hydraulic conductivity	[29]
Municipal solid waste compost and sewage sludge	EC 75 mS·cm^{-1} pH 8.2	13.3 g·kg^{-1} of compost significantly improved soil physical-chemical properties, especially C and N contents. Enzyme activities were substantially promoted in presence of both amendments	[30]

Note: ESP, exchangeable sodium percentage; EC, Electrical conductivity; OM, organic matter; CEC, cation exchange capacity.

3.1. Effects of Organic Materials on Soil Chemical Properties

As Hu and Schmidhalter [31] highlighted, the uptake of phosphorous (P) by crops is reduced in dry-soil conditions and the availability of this macronutrient can be reduced in saline soils. Conversely, during the mineralization process, organic matter releases humic substances, which may convert soil phosphates into available forms, improving release from hardly soluble rock minerals due to high total acidity [32]. Additionally, under saline soils the available fraction of potassium (K) can increase through the increase of CEC linked to organic matter content. In particular, the application of poultry manure and compost to soil can increase both the CEC and the soluble and exchangeable-K$^+$, which is a competitor of Na$^+$ under sodicity conditions, thus, limiting the entry of Na$^+$ into the exchange complex [25]. Moreover, K$^+$ is likewise important to maintain the turgor pressure of plant under drought and salinity stress. In a recent study, a mixture of green waste compost, sedge peat, and furfural residue (1:1:1 by volume) significantly reduced Na$^+$ + K$^+$ content and improved CEC and the contents of available N, P, and K [14].

Hao and Chang [33] showed that the soluble ions and the adsorption ratios of Na$^+$ and K$^+$ increased with 25 years of high rates of cattle manure application, particularly under non-irrigated conditions. Another study suggested that, even in a region with abundant rainfall, there was potential risk for secondary soil salinization by successive applications of chicken and pigeon manure [34]. Therefore, proper selection of organic fertilizers as nutrient sources, timing, as well as method of their application to soil, can be considered equally important [19]. As regards to method of application, Khaled and Fawy [35] found that the effect of interaction between salt and soil humus application was statistically significant showing that, under salt stress, both soil and foliar application of humic substances in corn field increased the uptake of nutrients. In particular, soil application of humus increased the N uptake, whereas foliar application increased the uptake of other macro- and micronutrients. The authors indicated to not exceed 2 g of well-humified organic matter/kg in the soil to obtain benefit from humic substances under salt conditions.

3.2. Effects of Organic Materials on Soil Biological Properties

Biological soil properties are very reactive to small changes occurring in management practices, therefore, it is possible to use them for evaluating the effects of the application of organic matter on soil characteristics [19].

Salinization may greatly disturb a large variety of microbially mediated processes in soil. Sardinha *et al.* [36] demonstrated that in different sites affected by saline liquid residues, microbial biomass C, biomass N, and fungal ergosterol had the highest values at the low-saline site (content of soluble salts 2.1 mg·g^{-1} soil) and the lowest at the high-saline site (soluble salts 9.7 mg·g^{-1} soil).

Exogenous organic matter applications to cropland are known to improve soil biological functions, also showing positive effects in the salt-affected soils. Liang *et al.* [22] showed that, in soil derived from alluvial and marine deposits (with 3.3 g·kg^{-1} total salts), soil urease and alkaline phosphatase activity, and respiration rate were significantly stimulated by incorporation of organic manure. Similarly, Chandra *et al.* [37] pointed out that, at low concentration, salts had a stimulating effect on carbon mineralization, but they can become toxic to microorganisms with increasing concentrations. Soil salinity can alter the organic matter turnover process, and the response pattern of C and N mineralization to salinity stress could depend on the type of organic material incorporated into the soil [18]. In particular, rice straw, plus pig manure treatment had higher significant effects on enzymatic and microbial activity in salt-affected soil, than rice straw and manure alone [38]. This result confirms that incorporation of organic manure can be an effective low-input agro-technological approach to minimize toxicity conditions induced by salinization. In addition, it has been demonstrated that non-composted manure and compost application to a saline soil in dryland conditions can reduce ESP (by 50% than unamended soil), at the same time, significantly increasing different enzyme activities (e.g., urease, alkaline phosphatase, and dehydrogenase) [8]. Amendment incorporation under high soil salinity or sodicity may also provide a buffer of pH in saline and alkaline soils, influencing the activity of microorganisms [32].

Moreover, Rao and Pathak [39] found that organic matter (green manure) improved microbial activity at salinities of EC ≤ 26, showing an increase in urease activity of saline and alkali soils following the amendment addition.

3.3. Effects of Organic Materials on Soil Physical Properties

Organic matter promotes the stability of soil aggregates through the bonding or adhesion properties, both of waste products of bacteria (polysaccharides) and fungal and/or bacterial hyphae [19]. The improvement of aggregate stability can also be obtained by a reduction of soil sodicity. In fact, the Ca^{2+} contained in composts could decrease the proportion of Na$^+$ in the exchange complex and step up the leaching of exchanged Na$^+$ [40]. The flocculation of clay minerals is, thus, promoted, playing an important role in the control of erosion in saline soils. Oo *et al.* [24] reported that combinations of organic amendments resulted in substantial flocculation and in the formation of a large number of soil aggregates. As a consequence of aggregate stability, soil porosity, water infiltration, and water-holding capacity of soil are improved, thus minimizing the impact of drought. Sodium adsorption ratio of the soil decreased significantly when soil was treated with sulfuric acid, gypsum, farm yard manure, and their various combinations [29]. Moreover, another study found that the physical properties of the soil, such as structural stability, infiltration rate, and water holding capacity, were considerably improved by municipal solid waste application in a soil salinized by saline water irrigation, during tomato crop cultivation [41]. A direct correlation between organic matter additions and decrease of soil bulk density was also commonly found. This decrease can allow the enhancement of soil porosity and, consequently, the improvement of saline water leaching [8,26]. However, repeated and/or elevated application rates of animal manures or composts could not be sustainable in the case of their relatively high salt contents.

Recently, the above reported study by Wang *et al.* [14] found that a mixture of organic wastes decreased bulk density, EC, and ESP by 11%, 87%, and 71%, respectively, and increased total porosity and organic carbon by 25% and 96% respectively, than the control. These results suggest the effectiveness of combination of different amendments for reclaiming salt-affected soil.

4. Conclusions

In this short review we attempted to highlight some crucial aspects of organic fertilization in salt-affected soils. Basic recommendations for organic fertilization in non-saline conditions are also suitable for high saline soils, therefore, it is important to properly select organic materials, taking into account nutrients content, timing and method of application. As a matter of fact, organic fertilization in saline and sodic soils fulfils the sustainability of resources use, being able to recycle wastes locally stored, thus, contributing to solve the disposal problem of different agro-industrial sectors.

From the review of existing data it can be concluded that most of the well-known effects of organic materials on the chemical, biological, and physical properties of soil are of particular relevance under conditions of salinization, and the achievable effect size is relevant. Therefore, appropriate use of organic amendments must be considered an effective measure to restore soil quality in salt-affected soils.

However, further experimental investigations are needed to validate the application of different organic materials and to step up organic fertilization use in a wide range of saline and sodic soils. Moreover, the combination of waste recycling with different proper agronomic practices (e.g., crop rotations, cover crops use, *etc.*) should be promoted.

Conflicts of Interest

The authors declare no conflict of interest.

References

1. Pitman, M.G.; Läuchli, A. Global impact of salinity and agricultural ecosystems. In *Salinity: Environment-Plants-Molecules*; Läuchli, A., Lüttge, U., Eds.; Kluwer Academic Publishers: Dodrecht, The Netherlands, 2002; pp. 3–20.

2. United Nations, Department of Economic and Social Affairs, Population Division. *World Population Prospects: The 2012 Revision. Volume I: Comprehensive Tables ST/ESA/SER.A/336*; United Nations: New York, NY, USA, 2013.

3. Tóth, G.; Montanarella, L.; Rusco, E. *Updated Map of Salt Affected Soils in the European Union Threats to Soil Quality in Europe*; Official Publications of the European Communities: Luxembourg, Luxembourg, 2008; pp. 61–74.

4. Dajic, Z. Salt Stress. In *Physiology and Molecular Biology of Stress Tolerance*; Madhava Rao, K.V., Raghavendra, A.S., Janardhan Reddy, K., Eds.; Springer: Dordrecht, The Netherlands, 2006; pp. 41–99.

5. Joint Research Centre. Soil Themes > Soil Salinization. Available online: http://eusoils.jrc.ec. europa.eu/library/themes/Salinization/ (accessed on 8 February 2012).

6. Sharma, B.R.; Minhas, P.S. Strategies for managing saline/alkali waters for sustainable agricultural production in South Asia. *Agric. Water Manag.* **2005**, *78*, 136–151.

7. Ahmad, R.; Chang, M.H. Salinity control and environmental protection through halophytes. *J. Drain. Water Manag.* **2002**, *6*, 17–25.

8. Tejada, M.; Garcia, C.; Gonzalez, J.L.; Hernandez, M.T. Use of organic amendment as a strategy for saline soil remediation: Influence on the physical, chemical and biological properties of soil. *Soil Biol. Biochem.* **2006**, *38*, 1413–1421.

9. Tanji, K.K. Salinity in the soil environment. In *Salinity: Environment-Plants-Molecules*; Läuchli, A., Lüttge, U., Eds.; Kluwer Academic Publishers: Dordrecht, The Netherlands, 2002; pp. 21–51.

10. Maggio, A.; de Pascale, S.; Fagnano, M.; Barbieri, G. Saline agriculture in Mediterranean environments. *Ital. J. Agron.* **2011**, *6*, 36–43.

11. Montemurro, F.; Diacono, M.; Vitti, C.; Debiase, G. Biodegradation of olive husk mixed with other agricultural wastes. *Bioresour. Technol.* **2009**, *100*, 2969–2974.

12. Romero-Aranda, R.; Soria, T.; Cuartero, J. Tomato plant-water uptake and plant-water relationships under saline growth conditions. *Plant Sci.* **2001**, *160*, 265–272.

13. Lauchli, A.; Epstein, E. Plant response to salinity and sodic conditions. In *Agricultural Salinity Assessment and Management*; Manual and Reports on Engineering Practice 71; Tanji, K.K., Ed.; American Society of Civil Engineers: New York, NY, USA, 1990; pp. 113–137.

14. Wang, L.; Sun, X.; Li, S.; Zhang, T.; Zhang, W.; Zhai, P. Application of organic amendments to a coastal saline soil in North China: Effects on soil physical and chemical properties and tree growth. *PLoS ONE* **2014**, *9*, e89185, doi:10.1371/journal.pone.0089185.

15. Grattan, S.R.; Grieve, C.M. Salinity-mineral nutrient relations in horticultural crops. *Sci. Hort.* **1999**, *78*, 127–157.

16. Garcia, C.; Hernandez, T. Influence of salinity on the biological and biochemical activity of a calciothid soil. *Plant Soil* **1996**, *178*, 255–263.

17. Rietz, D.N.; Haynes, R.J. Effects of irrigation-induced salinity and sodicity on soil microbial activity. *Soil Biol. Biochem.* **2003**, *35*, 845–854.

18. Walpola, B.C.; Arunakumara, K.K.I.U. Effect of salt stress on decomposition of organic matter and nitrogen mineralization in animal manure amended soils. *J. Agric. Sci.* **2010**, *5*, 9–18.

19. Diacono, M.; Montemurro, F. Long-term effects of organic amendments on soil fertility: A review. *Agron. Sustain. Dev.* **2010**, *30*, 401–422.

20. Montemurro, F.; Vitti, C.; Diacono, M.; Canali, S.; Tittarelli, F.; Ferri, D. A three-year field anaerobic digestates application: Effects on fodder crops performance and soil properties. *Fresenius Environ. Bull.* **2010**, *19*, 2087–2093.

21. Roy, R.N.; Finck, A.; Blair, G.J.; Tandon, H.L.S. *Plant Nutrition for FOOD Security. A Guide for Integrated Nutrient Management*; FAO Fertilizer and Plant Nutrition Bulletin 16; Food and Agriculture Organization of the United Nations: Rome, Italy, 2006; p. 347.

22. Liang, Y.C.; Yang, Y.F.; Yang, C.G.; Shen, Q.Q.; Zhou, J.M.; Yang, L.Z. Soil enzymatic activity and growth of rice and barley as influenced by organic matter in an anthropogenic soil. *Geoderma* **2003**, *115*, 149–160.

23. Cha-um, S.; Kirdmanee, C. Remediation of salt-affected soil by the addition of organic matter—An investigation into improving glutinous rice productivity. *Sci. Agric.* **2011**, *68*, 406–410.

24. Oo, A.N.; Iwai, C.B.; Saenjan, P. Soil properties and maize growth in saline and nonsaline soils using cassava-industrial waste compost and vermicompost with or without earthworms. *Land Degrad. Dev.* **2013**, *26*, 300–310, doi:10.1002/ldr.2208.

25. Walker, D.J.; Bernal, P.M. The effects of olive mill waste compost and poultry manure on the availability and plant uptake of nutrients in a highly saline soil. *Bioresour. Technol.* **2008**, *99*, 396–403.

26. Kahlown, M.A.; Azam, M. Effect of saline drainage effluent on soil health and crop yield. *Agric. Water Manag.* **2003**, *62*, 127–138.

27. Mahdy, A.M. Comparative effects of different soil amendments on amelioration of saline-sodic soils. *Soil Water Res.* **2011**, *6*, 205–216.

28. Mojiri, A. Effects of municipal wastewater on physical and chemical properties of saline soil. *J. Biol. Environ. Sci.* **2011**, *5*, 71–76.

29. Hussain, N.; Hassan, G.; Arshadullah, M.; Mujeeb, F. Evaluation of amendments for the improvement of physical properties of sodic soil. *Int. J. Agric. Biol.* **2001**, *3*, 319–322.

30. Lakhdar, A.; Scelza, R.; Scotti, R.; Rao, M.A.; Naceur, J.; Gianfreda, L.; Abdelly, C. The effect of compost and sewage sludge on soil biologic activities in salt affected soil. *R.C. Suelo Nutr. Veg.* **2010**, *10*, 40–47.

31. Hu, Y.; Schmidhalter, U. Drought and salinity: A comparison of their effects on mineral nutrition of plants. *J. Plant Nutr. Soil Sci.* **2005**, *168*, 541–549.

32. Lakhdar, A.; Rabhi, M.; Ghnaya, T.; Montemurro, F.; Jedidi, N.; Abdelly, C. Effectiveness of compost use in salt-affected soil. *J. Hazard. Mater.* **2009**, *171*, 29–37.

33. Hao, X.; Chang, C. Does long-term heavy cattle manure application increase salinity of a clay loam soil in semi-arid southern Alberta? *Agric. Ecosys. Environ.* **2003**, *94*, 89–103.

34. Yao, L.-X.; Li, G.-L.; Tu, S.-H.; Sulewski, G.; He, Z.-H. Salinity of animal manure and potential risk of secondary soil salinization through successive manure application. *Sci. Total Environ.* **2007**, *383*, 106–114.

35. Khaled, H.; Fawy, H.A. Effect of different levels of humic acids on the nutrient content, plant growth, and soil properties under conditions of salinity. *Soil Water Res.* **2011**, *6*, 21–29.

36. Sardinha, M.; Müller, T.; Schmeisky, H.; Joergensen, R.G. Microbial performance in soils along a salinity gradient under acidic conditions. *Appl. Soil Ecol.* **2003**, *23*, 237–244.

37. Chandra, S.; Joshi, H.C.; Pathak, H.; Jain, M.C.; Kalra, N. Effect of potassium salts and distillery effluent on carbon mineralization in soil. *Bioresour. Technol.* **2002**, *83*, 255–257.

38. Liang, Y.; Nikolic, M.; Peng, Y.; Chen, W.; Jiang, Y. Organic manure stimulates biological activity and barley growth in soil subject to secondary salinization. *Soil Biol. Biochem.* **2005**, *37*, 1185–1195.

39. Rao, D.L.N.; Pathak, H. Ameliorative influence of organic matter on biological activity of salt-affected soils. *Arid Soil Res. Rehabil.* **1996**, *10*, 311–319.

40. Qadir, M.; Oster, J.D. Crop and irrigation management strategies for saline-sodic soils and waters aimed at environmentally sustainable agriculture. *Sci. Total Environ.* **2004**, *323*, 1–19.

41. Lax, A.; Diaz, E.; Castillo, V.; Albaladejo, J. Reclamation of physical and chemical properties of a salinized soil by organic amendment. *Arid Soil Res. Rehabil.* **1994**, *8*, 9–17.

5

Provenancing Flower Bulbs by Analytical Fingerprinting: *Convallaria Majalis*

Saskia M. van Ruth [1,2,†,*] **and Ries de Visser** [3,†]

[1] RIKILT Wageningen UR, P.O. Box 230, 6700 EV Wageningen, The Netherlands

[2] Food Quality and Design Group, Wageningen University, P.O. Box 17, 6700 AA Wageningen, The Netherlands

[3] IsoLife B.V., P.O. Box 349, 6700 AH Wageningen, The Netherlands; E-Mail: ries.devisser@isolife.nl

[†] These authors contributed equally to this work.

[*] Author to whom correspondence should be addressed; E-Mail: saskia.vanruth@wur.nl

Academic Editor: Takayuki Shibamoto

Abstract: The origin of agricultural products is gaining in appreciation while often hard to determine for various reasons. Geographical origin may be resolved using a combination of chemical and physical analytical technologies. In the present case of Lily of the Valley (*Convallaria majalis*) rhizomes, we investigated an exploratory set of material from The Netherlands, three other European (EU) countries and China. We show that the geographical origin is correlated to patterns of stable isotope ratios (isotope fingerprints) and volatile organic carbon (VOC) compounds (chemical fingerprints). These fingerprints allowed clear distinction using exploratory and supervised statistics. Isotope ratio mass spectrometry of $^{12}C/^{13}C$, $^{14}N/^{15}N$ and $^{16}O/^{18}O$ isotopes separated materials from Europe and China successfully. The VOC patterns measured by Proton Transfer Reaction Mass Spectrometry (PTR-MS) allowed distinction of three groups: material from The Netherlands, the other EU countries and China. This knowledge is expected to help developing a systematic and efficient analytical tool for authenticating the origin of flower bulbs.

Keywords: authenticity; fingerprint; isotope ratio mass spectrometry (IRMS); Lily of the Valley; origin; PTR-MS; stable isotopes

1. Introduction

Globalization of markets gives rise to a growing need for analytical tools capable of identifying origins of local products in a reliable and efficient way. Various bio-molecular, chemical and physical technologies have been tried in attempts to link agricultural products to their site of production or origin, also called authentication [1]. Analytical techniques for authenticating local products have been described for a wide range of agricultural and other products, like beef [2], Trappist beers [3], and pharmaceuticals [4]. However, much research is still needed towards the development of a comprehensive system of authentication based on scientific analytical methods. Little knowledge is currently available on the major issue of how to select the appropriate (combination of) analytical techniques that will be the most likely approach towards successful authentication of a particular product. Lilies are commonly kept ornamental flowering plants that are used in holiday celebrations, weddings, and funerals, and in various floral arrangements. Two thirds of the worldwide flower bulb production area is located in The Netherlands. The land area for flower bulb production in The Netherlands is *ca.* twenty-four thousand hectares in The Netherlands, including *ca.* twelve thousand hectares of tulips, and five thousand hectares of lilies. The Dutch have been known for their flower bulb production and export over the last 500 years [5]. By 1636, the tulip bulb became the fourth leading export product of The Netherlands—after gin, herring and cheese. The price of tulips skyrocketed because of speculation in tulip futures among people who never saw the bulbs. Many men made and lost fortunes overnight.

The *Liliaceae*, or lily family, is composed of 280 to 300 genera made up of 4000 to 4600 different species. The numbers vary because botanists differ in how to classify this diversity based on flowering type, ovary position, and distribution. There are ornamental plants within the group (lilies, tulips, hyacinths, daffodils, and amaryllis); food plants (onions, garlic, asparagus, leeks, shallots, and chives); and a variety of toxic species in the family, some of which are quite deadly [6].

Another floral plant associated with lilies is *Convallaria majalis*, commonly known as Lily of the Valley. Although part of the family *Asparagaceae*, in earlier classification systems the species were often treated as belonging to the family *Liliaceae*. It is a poisonous woodland flowering plant native throughout the cool temperate Northern Hemisphere in Asia, Europe and in the southern Appalachian Mountains in the United States. *C. majalis* is a herbaceous perennial plant that forms extensive colonies by spreading underground stems. These are called rhizomes. In botany, the term 'bulb' designates underground plant stems surrounded by modified leaves called scales which store nutrients, while in horticulture and gardening the term refers to any bulbous plant organ or underground stem, be it a corm, tuber, rhizome or true bulb. The *C. majalis* stems grow to 15–30 cm tall. Flowering stems have two leaves and 5–15 flowers on the stem apex. The flowers are usually white tepals, shaped like small bells, and sweetly scented (Figure 1). Flowering is in late spring (April/May) in the Northern Hemisphere.

C. majalis contains potent cardenolide glycosides, which are often toxic, and specifically heart-arresting [7]. In fact most of the literature on *Convallaria* deals with its toxic components and their biogenetic synthesis [8].

By nature, agricultural products have a land-based, and therefore geographical origin. Historically, application habits were shaped by socio-cultural factors and available local natural resources [9].

Such links between agricultural produce and territory have disappeared over time by various means. However, the last ten years consumers have a renewed interest in agricultural products strongly identified with a place of origin. The EU has recognized and supported the potential of differentiating quality products on a regional basis [10]. The EU regulation allows the application of the following geographical indications to a food product: Protected Designation of Origin (PDO), Protected Geographical Indication (PGI) and Traditional Specialties Guaranteed (TSG). PDO is the term used to describe foodstuffs that are produced, processed and prepared in a given geographical area using recognized methods. Examples are Roquefort (France), Traditional Balsamic Vinegar of Modena (Italy), and Farmers cheese from Leiden (The Netherlands). The use of geographical indications allows producers to obtain market recognition and often a premium price, not only for food products but also for flowers and bulbs. False use of geographical indications by unauthorized parties is detrimental to consumers and legitimate producers. From this point of view, the development of new and increasingly sophisticated techniques for determining the geographical origin of agricultural products is highly desirable for consumers, agricultural farmers, retailers and authorities. It is an analytically challenging problem that receives much attention in Europe. Reports on analytical methods for determining the geographical origin of agricultural products have been increasing since the 1980s [11]. Food and feed received considerable attention, but to the authors knowledge no authentication approaches have been reported for the provenance of flower bulbs in the scientific literature. Fingerprinting techniques combined with chemometrics are state-of-the-art analytical techniques in product authentication. These fingerprinting techniques aim to find a specific pattern for the authentic product (*i.e.*, for each geographical origin), which might allow discrimination between different products (*i.e.*, those from different geographical origins).

Stable isotope composition (or 'fingerprint') has been investigated using isotope ratio mass spectrometry (IRMS) for its utility in authenticating agricultural products, including wine, beer and cheeses [12], but also for pharmaceuticals [13] as well as in forensic studies on drugs [14]. These isotopic 'fingerprints' are intrinsic characteristics and built-in in all organic compounds and therefore largely insensitive to adulteration. Other successful applications have been described where stable isotopes serve as internal standards for quantitative Liquid Chromatography-MS [15], as tracers in source-sink studies [16–18] and in identifying functional micro-organisms [18,19].

It is well recognized that direct rapid headspace techniques measured by mass spectrometry without chromatographic separation can effectively represent a 'fingerprint' of the sample being analyzed and can provide distinct chemical information in relation to product odor, flavor, shelf-life, geographic or genetic origin, processing, and presence of micro-organisms. In the last decade, several non-chromatographic instrumental approaches, such as electronic noses with different types of chemical sensors, headspace mass spectrometry, or real time monitoring using techniques, such as Atmospheric pressure chemical ionization mass spectrometry (APCI-MS), proton transfer reaction mass spectrometry (PTR-MS) or selected ion flow tube mass spectrometry (SIFT-MS) have been applied for the characterization of the volatile compounds of food [20,21].

The aim of this study is a first exploration of two types of analytical fingerprints for their capabilities of differentiating rhizome samples of 'Lily of the Valley', *Convallaria majalis*, of different provenance for which isotope ratio analysis by IRMS and volatile organic compound (VOC) analysis by PTR-MS were selected. Isotope ratio analysis was selected because of the known impact of geology and climatology and thus geography on isotope ratio fingerprints of plant material. PTR-MS was selected because of the known impact of environmental conditions on volatile metabolites in plants and the technique's sensitivity, rapidity and non-destructive nature. In order to distinguish *C. majalis* cultivated in The Netherlands, other European countries and in China stable isotope ratios $^{13}C/^{12}C$, $^{15}N/^{14}N$ and $^{18}O/^{16}O$ determined by IRMS were determined as well as the integral VOC profiles, which were examined in conjunction with advanced statistical methods.

2. Materials and Methods

2.1. Sample Material and Study Design

A total of 34 *C. majalis* rhizomes samples were collected. The set of samples consisted of samples from The Netherlands (18), three neighboring EU countries (2 from Belgium, 6 from France, 4 from Germany) and from China (4). Representative samples of dry rhizome materials were ground to a powder. These rhizome samples were subjected to stable isotope analysis by IRMS and VOC analysis by PTR-MS. For comparison flowers of a *C. majalis* plant grown in The Netherlands was analyzed for their volatile composition by PTR-MS as well.

2.2. Stable Isotope Analysis by IRMS

The principles of natural abundance stable IRMS have been described in detail [22], as well as the application of IRMS for geographical origin authentication of foods [23]. This method employs the natural variation in isotope ratios of the chemical elements and is suited for accurate (<0.0002 atom%) analysis of the ratios of isotopes (IR) occurring -in this case- in organic matter, like carbon (C, $^{13}C/^{12}C$; 1.1% ^{13}C), nitrogen (N, $^{15}N/^{14}N$; 0.3% ^{15}N) and oxygen (O, $^{18}O/^{16}O$; 0.2% ^{18}O). In other studies, hydrogen $^{2}H/^{1}H$ and sulfur $^{34}S/^{33}S/^{32}S$ have been used [12,23].

Oven-dry samples (1–2 mg) were packed in tin (^{13}C, ^{15}N) or silver foil (^{18}O) for oxidation and analysis using a PDZ Europa ANCA-GSL elemental analyzer interfaced to a PDZ Europa 20-20 isotope ratio mass spectrometer (Sercon Ltd., Cheshire, UK; C and N). Freeze-drying increased the $^{18}O/^{16}O$ ratio by ~5 per mil. Oxygen isotope analysis was performed using an elementar PyroCube (Elementar Analysensysteme GmbH, Hanau, Germany) interfaced to a PDZ Europa 20-20 isotope ratio mass spectrometer (Sercon Ltd., Cheshire, UK). The long-term standard deviation is 0.2 per mil for ^{13}C (0.0002 atom%) and 0.3 per mil for ^{15}N (0.0003 atom%).

In accordance with international agreements delta-values (δ, in per mil; ‰) provide the deviation relative to the ratio of isotopes of international standards, V-PDB (Vienna Pee Dee Belemnite), V-SMOW (Vienna Standard Mean Ocean Water) and Air for carbon, oxygen and nitrogen, respectively. For information on delta notation and the international references, see e.g., Sharp 2006 [22]. Single analyses were carried out on the 34 samples.

2.3. Volatile Organic Compound Analysis by PTR-MS

The VOC fingerprints were measured by PTR-MS. For this type of analysis the sample headspace is continuously introduced into a drift tube, where it is mixed with H_3O^+ ions formed in a hollow cathode ion source. VOCs that have proton affinities higher than water (>166.58 kcal/mol) are ionized by proton transfer from H_3O^+, mass analyzed in a quadrupole mass spectrometer and eventually detected as ion counts/s (cps) by a secondary electron multiplier. By using H_3O^+ as the proton source, the ionization of most of the common inorganic constituents of air (N_2, O_2 or CO_2) is avoided since they have proton affinities lower than H_2O. Furthermore, this soft ionization avoids excessive fragmentation of ions, which makes multicomponent analyte mass spectra simpler and easier to interpret [24].

A total of 2 g of sample (dry powder of rhizomes, or flowering stems) were placed in a 250 mL screw cap glass vial (Figure 1). Samples were equilibrated at 35 °C for at least 30 min in a water bath, in order to assure equilibrium of the VOC between sample and headspace. No further sample preparation was required. Measurements were performed using a commercial High Sensitivity PTR-MS system (Ionicon GmbH, Innsbruck, Austria). The headspace of the samples was delivered directly to the inlet of the PTR-MS system with a flow rate of 50 mL/min. The temperature of the inlet and drift chamber were both maintained at 60 °C to prevent loss of volatiles along the sampling inlet line for on-line analysis. A blank was measured before each sample. Measurements were carried out in the mass full-scan mode and the mass spectra were collected in the range of 20–160 atomic mass units (amu). A dwell time of 0.2 s/mass unit was used, resulting in a cycle time just under 30 s. Sample analyses were carried out in triplicate. For each replicate, a full mass scan was recorded. The data were background and transmission corrected, yielding one corrected mass spectrum per replicate. Then, the three mass spectra of the three replicates of each sample were averaged to obtain a mean mass spectrum per sample. In this manner, a dataset containing mean mass spectra per sample analyzed was compiled for the 34 samples.

Figure 1. Photograph of the sample flask and example mass spectrum of the flowers of *C. majalis*.

2.4. Statistical Analysis

Multi-factor analysis of variance (MANOVA, factors provenance and samples) and Fisher's least significant difference (LSD) tests were carried out to determine significant differences among groups for isotope ratio measurements using XLSTAT 2 March 2014 (Addinsoft, Paris, France). MANOVA can assess two or more independent variables for significance of effects on two or more metric dependents. It allows a joint analysis of each dependent rather than performing several univariate tests, thus avoiding multiple testing risks.

For further multivariate modeling and classification, Pirouette 4.0 (Infometrix, Seattle, WA, USA) was used. Principal Component Analysis (PCA) was performed on the PTR-MS data of the 34 samples to screen the multivariate data for outliers and to explore the presence of any natural clustering in the data. PCA performs a reduction in the data dimensionality in order to facilitate the visualization of the multivariate data retaining as much as possible the information present in the original data. Then, Partial Least Squares-Discriminant Analysis (PLS-DA) was used to develop a classification model for samples from The Netherlands *versus* other origins. PLS-DA is a supervised classification technique that is often used for high dimensional data, especially when the amount of variables greatly exceeds the number of samples. It performs a variable reduction on the data set by calculating new variables (called latent components or factors) combining the variables in the data set, in order to find the maximum correlation between them and the class variable, and thus, the maximum separation among two classes (The Netherlands *vs.* other provenance). Then, linear discriminant analysis is applied on the reduced variable set (the latent components) to provide the final classification model. Since data pre-processing can have a profound effect on the model results, several ways of data pre-processing were evaluated: none (raw data), auto-scaling (scaling to unit variance), mean centering, and log transformation. The optimal PLS-DA model was then selected and its performance examined by leave-one-out cross validation because of the limited size of the exploratory sample set.

3. Results and Discussion

3.1. Stable Isotope Analysis by IRMS

The nitrogen, carbon, and oxygen isotope ratios (delta values) of the 34 samples were determined by IRMS. The isotope ratios were compared for the various provenances (Table 1). Overall, δ-^{15}N varied by 11 per mil, δ-^{13}C varied by 7 per mil, and δ-^{18}O by 12 per mil (sample extremes). MANOVA indicated significant differences in δ-^{15}N values between the *C. majalis* from China and their European counterparts (France, Germany, The Netherlands) whereas the Belgian products revealed intermediate values. For the δ-^{13}C values mostly overlapping ranges were observed. The δ-^{18}O values showed relatively high values for the rhizomes originating from The Netherlands and low values for those from China, with the other European countries showing intermediate values.

Naturally occurring stable isotope ratios of organic elements like N, C and O measured in this study have characteristic values in relation to their geographical origin. This is due to systematic effects of climate factors like temperature, humidity, precipitation, and geographical factors such as distance to the sea, plant physiological processes, plant genotype, and soil or substrate factors such as fertilizer type. The three isotope ratios that are dealt with here differ in their relation to the main environmental factors related to plant growth. ^{13}C/^{12}C is related plant water use efficiency, via water conditions, including water supply and relative humidity interacting with plant physiological characteristics [25]. ^{15}N/^{14}N varies with the chemical type and origin of the mineral N fertilizer [26], and increases with trophic level in N of organisms [27]. The ^{18}O/^{16}O isotope ratio of atmospheric water is directly related to temperature and inversely related to the dominant down-wind distance from the sea through the mechanism of preferential precipitation of the heavier water molecules (*i.e.*, H_2^{18}O; [28]), while this oxygen isotope ratio in leaf water (and thus, plant biomass) is affected by local climatic conditions affecting plant transpiration, *i.e.*, mainly temperature and humidity [29].

Table 1. Nitrogen, carbon and oxygen isotope ratio's (δ values, per mil) for *C. majalis* of various provenance [a]; means ± SE.

Provenance	δ-^{15}N	δ-^{13}C	δ-^{18}O
The Netherlands	1.5 ± 2.4 [x]	−26.6 ± 2.6 [y]	36.0 ± 3.2 [x]
Belgium	−0.9 ± 0.1 [xy]	−25.5 ± 3.4 [xy]	35.5 ± 2.7 [xy]
France	0.4 ± 1.2 [x]	−24.2 ± 0.7 [x]	34.7 ± 3.1 [xy]
Germany	1.8 ± 0.2 [x]	−24.5 ± 1.3 [xy]	33.0 ± 1.4 [xy]
China	−2.8 ± 2.2 [y]	−26.4 ± 2.0 [xy]	30.4 ± 4.0 [y]

[a] Different superscripts (x, y, xy) in a column indicate significant differences (Multi-factor analysis of variance (MANOVA) and Fisher's least significant difference (LSD) test, $p < 0.05$).

Since the individual isotopes would not allow full discrimination between *C. majalis* from The Netherlands, other European countries, and those of Chinese origin, multiple isotopes were compared for their discriminatory properties. The data of the two most discriminating isotopes, ^{15}N and ^{18}O were combined in a 2D plot (Figure 2). The two isotopes display a clear distinction between the rhizomes from Europe and China (red-colored symbols *versus* others). In the study sample numbers are low and do not represent all variation in real-life, but the results are promising though. Different nitrogen sources for fertilization and water sources are likely to have contributed to the consistencies observed.

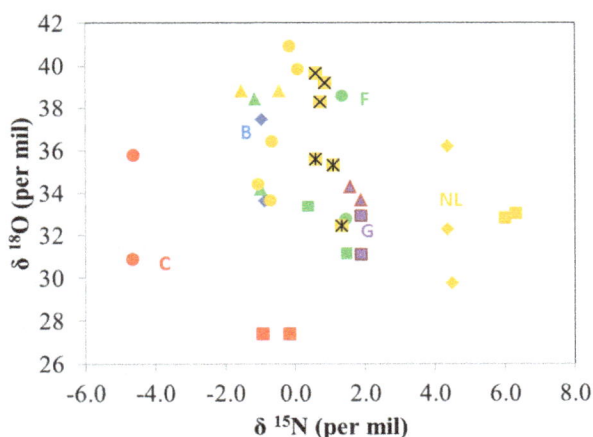

Figure 2. 2D-plot of N and O isotopic data of *C. majalis* rhizomes originating from five countries (Belgium (B, blue), China (C, red), France (F, green), Germany (G, purple) and The Netherlands (NL, yellow)), *i.e.*, stable isotope ratios ^{18}O/^{16}O (δ ^{18}O) and ^{15}N/^{14}N (δ ^{15}N). Different colors indicate different countries, symbol shapes indicate different origins within countries; identical symbols indicate replicate provenance samples; least significant difference (LSD) ($p < 0.05$) values: 0.7 (δ ^{15}N), 4.6 (δ ^{18}O).

3.2. Volatile Organic Compound Analysis by PTR-MS

VOCs in the headspace of the *C. majalis* rhizomes and flower samples were analyzed by PTR-MS without prior chromatographic separation and resulted in a spectrum of the VOCs (their mass-to-charge ratios and their intensities). The mean mass spectra of the samples from The Netherlands are presented in Figure 3 for both rhizomes and flowers. Some similarities in groupings of more predominant ions for both types of samples are observed in the lower molecular weight range. On the other hand, the flowers present clearly more higher molecular weight volatile compounds than the rhizomes. In decreasing order, the most abundant mass-to-charge ratios found in the flowers were m/z 37 (water cluster), m/z 45, m/z 33, m/z 81, m/z 137, m/z 91 and m/z 83.

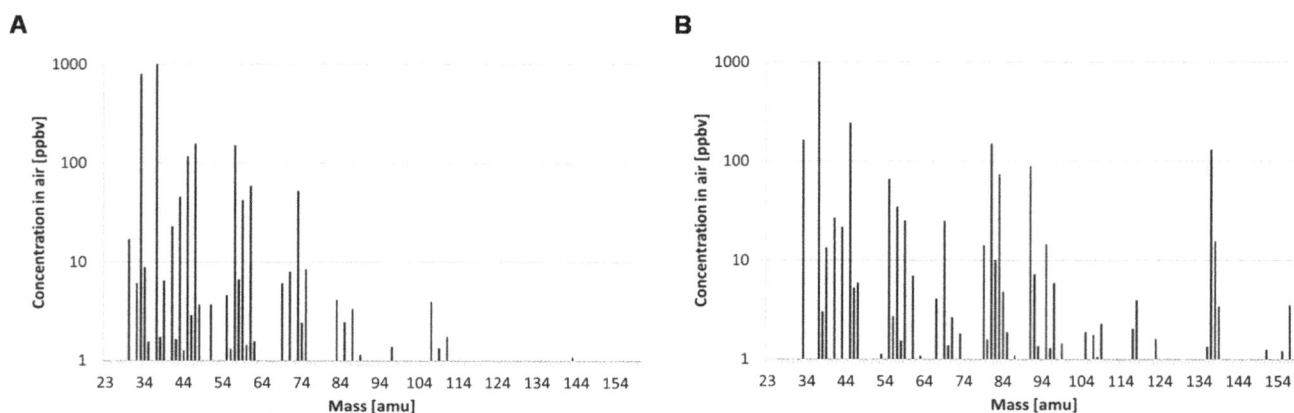

Figure 3. Mean PTR-MS (proton transfer reaction mass spectrometry) spectrum of *C. majalis* cultivated in The Netherlands: rhizomes (**A**) and flowers (**B**).

Flowers contain a large variety of volatile organic compounds, just as there are a multitude of colors and forms of flowers. Those compounds with a sufficiently high vapor pressure may have an odor, which is approximately for compounds with a low polarity and up to a molecular weight of around 300 [30]. The odor of Lily of the Valley plays an important role in perfumery. However, very few studies have identified the responsible volatile compounds of the plant. Brunke and co-workers [31] reported the following predominant volatile compounds in the extract of *C. majalis* analyzed: benzyl alcohol (35%), *cis*-3-hexen-1-ol (11%), citronellol (10%), geraniol (8%) and *cis*-3-hexenyl acetate (8%). Benzyl alcohol, citronellol, geraniol would carry floral-rosy-citrus notes, whereas the *cis*-3-hexen-1-ol and *cis*-3-hexenyl acetate would be responsible for green-grassy odor notes. They are all compounds with molecular weights of 100 and over, and therefore expected to appear (if present) in the VOC mass spectrum in the mass range of 100–160 amu. The response for mass peak 157 in the flower analysis is likely to be citronellol (mw = 156, ionized response expected at 157 amu). Takahiro *et al.* [32] identified 148 compounds. They investigated also the enantiomeric ratio of the chiral compounds in *C. majalis* using multidimensional GC-MS and determined predominantly the (S)-form of citronel, citronellyl acetate, citronellal, and dihydrofarnesol. Due to its pleasant odour various attempts have been made to synthesize Lily of the Valley-like fragrances. Lilial® (3-(4-*t*-butylphenyl)-2-methylpropanal), Lyrial®, and hydroxycitronellal are the commercially most important compounds of this class of odorants and they show apparent molecular similarities. Dupical® and Mugetanol® are two more recently developed Lily of the Valley odorants. Lillial® is a powerful, fresh, floral note reminiscent of lily of the valley, linden blossom, and cyclamen. Besides its use as a fragrance and fragrance intermediate it is used as an intermediate for the production of fenpropimorph, a biodegradable fungicide. It is produced on a kiloton scale by a multistep synthesis [28,33].

PTR-MS is not primarily used to identify the volatile compounds as isobaric compounds might yield the same m/z signal. Fragmentation in PTR-MS is mostly reduced due to the soft chemical ionization of the compounds; thus, the molecular structure of most volatile compounds is preserved. This allows the spectra to be used as fingerprints. Geographical provenancing using a univariate approach is not always a robust strategy. For instance, even if we found significant differences in the content of a particular volatile among the origins of the rhizomes, basing the verification of them on only this volatile could on the one hand easily lead to incorrect assignment to a particular origin and on the other hand may be the recipe to fraud. Therefore, a multivariate approach was selected for the VOC data.

The VOC fingerprints were examined using chemometric analysis, which is particularly suitable for handling large data sets. PCA was conducted on the VOC data of the 34 samples. The data matrix consisted of 34 rows (samples) and 138 variables (ions). The first three dimensions of the PCA on the normalized and auto-scaled data are presented in Figure 4.

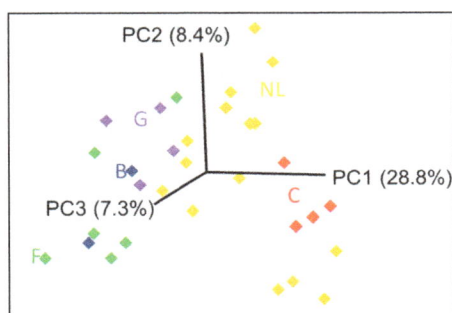

Figure 4. PCA scores plot of the PTR-MS (proton transfer reaction mass spectrometry) data (normalized and auto-scaled) of *C. majalis* from the five countries (Belgium (B, blue), China (C, red), France (F, green), Germany (G, purple) and The Netherlands (NL, yellow)).

Some natural clustering is observed: with groupings of the samples from The Netherlands, the other European samples, and the samples of Chinese origin. Subsequently the data for these three provenance groups were subjected to PLS-DA in order to optimize the classification of the samples by their VOC profiles. The data matrix was the same as for the PCA. A model was built to classify the samples from The Netherlands, from other European countries and from China using the normalized, auto-scaled data. The model's performance was evaluated by a leave-one-out cross validation and indicated 100% accuracy for the prediction of the origin for the samples from The Netherlands, other European countries and China (Table 2). In order to evaluate the masses contributing to the classification, a correlation spectrum for the masses and the model classifying samples from The Netherlands and the other origins was generated (Figure 5). It shows that many masses are positively or negatively correlated with the origin of the *C. majalis* samples origin.

Table 2. PLS-DA (Partial Least Squares-Discriminant Analysis) leave-one-out cross validation results: classification of *C. majalis* from The Netherlands, other EU countries and China by their normalized and auto-scaled PTR-MS (proton transfer reaction mass spectrometry) spectral data.

Provenance	Prediction		
	The Netherlands	Other EU Countries [a]	China
The Netherlands	100%	0%	0%
Other EU countries	0%	100%	0%
China	0%	0%	100%

[a] *C. majalis* from Belgium, France, and Germany.

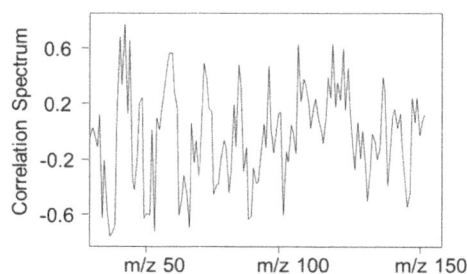

Figure 5. Correlation spectrum associated with the PLS-DA (Partial Least Squares-Discriminant Analysis) classification of *C. majalis* from The Netherlands and other countries presenting the relevance of particular masses analyzed by PTR-MS for the classification.

4. Conclusions

Consistent differences in composition between Lily of the Valley rhizomes of different provenances by both their nitrogen and oxygen isotope ratios and their VOC fingerprints were observed in the current set of samples. The isotope ratios allowed discrimination between samples cultivated in Europe and China. The VOC fingerprints differentiated between the samples from The Netherlands, from the other European countries and the samples from China. PTR-MS is a rapid, non-destructive technique that may be suitable as a first on site, screening technique, while the more time-consuming isotope ratio analysis could be used for further confirmation. Since origin may add value to agricultural produce, produce from particular origins may be susceptible to mix-up and substitution. In order to assure genuine trade, purchaser confidence, and therefore future production this type of analytical techniques may have a positive effect in addition to administrative controls. This study is a first step in that direction but the sample set would need to be further extended to build a reliable database for practical use.

Acknowledgments

The authors are grateful to Naktuinbouw for providing the Lily-of-the-Valley sample material, to Alex Koot for technical help with the PTR-MS analyses, and to Sara N. de Visser for graphics support. Furthermore, the authors would like to acknowledge strategic Wageningen University and Research

Centre funding from the Dutch Ministry of Economic Affairs in the KB-project Authenticity-traceability, and the EU Operationeel Programma Oost-Nederland EFRO 2007-2013 (GO).

Author Contributions

Ries de Visser contributed primarily to the generation and interpretation of the IRMS data and Saskia van Ruth was mostly involved in the PTR-MS part of the study as well as the statistical calculations. The current manuscript was drafted together.

Conflicts of Interest

The authors declare no conflict of interest.

References

1. Olsen, P.; Borit, M. How to define traceability. *Trends Food Sci. Technol.* **2012**, *29*, 142–150.
2. Monahana, F.J.; Moloneyb, A.P.; Osorioa, M.T.; Röhrlea, F.T.; Schmidta, O.; Brennanc, L. Authentication of grass-fed beef using bovine muscle, hair or urine. *Trends Food Sci. Technol.* **2012**, *28*, 69–76.
3. Mattarucchi, E.; Stocchero, M.; Moreno-Rojas, J.M.; Giordano, G.; Reniero, F.; Guillou, C. Authentication of Trappist beers by LC-MS fingerprints and multivariate data analysis. *J. Agric. Food Chem.* **2010**, *58*, 12089–12095.
4. Berkowitz, S.A.; Engen, J.R.; Mazzeo, J.R.; Jones, G.B. Analytical tools for characterizing biopharmaceuticals and the implications for biosimilars. *Nat. Rev. Drug Discov.* **2012**, *11*, 527–540.
5. Schama, S. *The Embarrassment of Riches: An Interpretation of Dutch Culture in the Golden Age*; Alfred A. Knopf: New York, NY, USA, 1987; p. 698.
6. Burrows, G.E.; Tyrl, R.I. The *Liliaceae*. In *Toxic Plants of North America*; Burrows, G.E., Tyrl, R.I., Eds.; Iowa State Press: Ames, IA, USA, 2001; pp. 271–295.
7. Fitzgerald, K.T. Lily toxicity in the cat. *Top. Companion Anim. Med.* **2010**, *25*, 213–217.
8. Ogorodnikova, A.V.; Latypova, L.R.; Mukhitova, F.K.; Mukhtarova, L.S.; Grechkin, A.N. Detection of divinyl ether synthase in Lily-of-the-Valley (*Convallaria majalis*) roots. *Phytochemistry* **2008**, *69*, 2793–2798.
9. Delamont, S. *Appetites and Identities: An Introduction to the Social Anthropology of Western Europe*; Routledge: London, UK, 1994; p. 272.
10. Dimara, E.; Skuras, D. Consumer evaluations of product certification, geographic association and traceability in Greece. *Eur. J. Market.* **2003**, *37*, 690–705.
11. Luykx, D.M.A.M.; van Ruth, S.M. A review of analytical methods for determining the geographical origin of food products. *Food Chem.* **2008**, *107*, 897–911.
12. Kelly, S.; Heaton, K.; Hoogewerff, J. Review: Tracing the geographical origin of food: The application of multi-element and multi-isotope analysis. *Trends Food Sci. Technol.* **2005**, *16*, 555–567.
13. Jasper, J.P.; Westenberger, B.J.; Spencer, J.A.; Buhse, L.F.; Nasr, M. Stable isotopic characterization of active pharmaceutical ingredients. *J. Pharm. Biomed. Anal.* **2004**, *35*, 21–30.

14. Aguilera, R.; Hatton, C.K.; Catlin, D.H. Detection of epitestosterone doping by isotope ratio mass spectrometry. *Clin. Chem.* **2002**, *48*, 629–636.

15. Nakabayashi, R.; Sawada, Y.; Yamada, Y.; Suzuki, M.; Hirai, M.Y.; Sakurai, T.; Saito, K. Combination of liquid chromatography-Fourier transform ion cyclotron resonance-mass spectrometry with ^{13}C-labeling for chemical assignment of sulfur-containing metabolites in onion bulbs. *Anal. Chem.* **2013**, *85*, 1310–1315.

16. De Visser, R.; Vianden, H.; Schnyder, H. Kinetics and relative significance of remobilized and current C and N incorporation in leaf and root growth zones of *Lolium perenne* after defoliation: Assessment by ^{13}C and ^{15}N steady-state labelling. *Plant Cell Environ.* **1997**, *20*, 37–46.

17. Frost, G.; Sleeth, M.L.; Sahuri-Arisoylu, M.; Lizarbe, B.; Cerdan, S.; Brody, L.; Anastasovska, J.; Ghourab, S.; Hankir, M.; Zhang, S.; *et al.* The short-chain fatty acid acetate reduces appetite via a central homeostatic mechanism. *Nat. Commun.* **2014**, *29*, doi:10.1038/ncomms4611.

18. Tannock, G.W.; Lawley, B.; Munro, K.; Sims, I.M.; Lee, J.; Butts, C.A.; Roy, N. RNA-stable isotope probing (RNA-SIP) shows carbon utilization from inulin by specific bacterial populations in the large bowel of rats. *Appl. Environ. Microbiol.* **2014**, *80*, doi:10.1128/AEM.03799-13.

19. Schmidt, O.; Horn, M.A.; Kolb, S.; Drake, H.L. Temperature impacts differentially on the methanogenic food web of cellulose-supplemented peatland soil. *Environ. Microbiol.* **2014**, *16*, doi:10.1111/1462-2920.12507.

20. Reineccius, G.; Peterson, D. Principles of food flavor analysis. In *Instrumental Assessment of Food Sensory Quality*; Kilkast, D., Ed.; Woodhead Publishing Ltd.: Cambridge, UK, 2013; pp. 53–102.

21. Heenan, S.; van Ruth, S.M. Emerging flavour analysis methods in food authentication. In *Instrumental Assessment of Food Sensory Quality*; Kilkast, D., Ed.; Woodhead Publishing Ltd.: Cambridge, UK, 2013; pp. 284–312.

22. Sharp, Z.D. *Principles of Stable Isotope Geochemistry*; Prentice Hall: Upper Saddle River, NJ, USA, 2006; p. 344.

23. Drivelos, S.A.; Georgiou, C.A. Multi-element and multi-isotope-ratio analysis to determine the geographical origin of foods in the European Union. *Trends Anal. Chem.* **2012**, *40*, 38–51.

24. Hansel, A.; Jordan, A.; Holzinger, R.; Prazeller, P.; Vogel, W.; Lindinger, W. Proton transfer reaction mass spectrometry: On-line trace gas analysis at the PPB level. *Int. J. Mass Spectrom. Ion Process.* **1995**, *149*, 609–619.

25. Seibt, U.; Rajabi, A.; Griffiths, H.; Berry, J.A. Carbon isotopes and water use efficiency: Sense and sensitivity. *Oecologia* **2008**, *155*, 441–454.

26. Flores, P.; Fenoll, J.; Hellín, P. The feasibility of using δ ^{15}N and δ ^{13}C values for discriminating between conventionally and organically fertilized pepper (*Capsicum annuum* L.). *J. Agric. Food Chem.* **2007**, *55*, 5740–5745.

27. De Visser, S.N.; Freymann, B.P.; Schnyder, H. Trophic interactions among invertebrates in Termitaria in the African savanna: A stable isotope approach. *Ecol. Entomol.* **2008**, *33*, 758–764.

28. Miller, D.L.; Mora, C.I.; Grissino-Mayer, H.D.; Mock, C.J.; Uhle, M.E.; Sharp, Z. Tree-ring isotope records of tropical cyclone activity. *Proc. Natl. Acad. Sci. USA* **2006**, *103*, 14294–14297.

29. Helliker, B.R. On the controls of leaf-water oxygen isotope ratios in the atmospheric crassulacean acid metabolism epiphyte *Tillandsia usneoides*. *Plant Physiol.* **2011**, *155*, 2096–2107.

30. Fráter, G.; Bajgrowicz, J.A.; Kraft, P. Fragrance chemistry. *Tetrahedron* **1998**, *54*, 7633–7703.

31. Takahiro, I.; Tsuyoshi, K.; Kazuhiko, T.; Fumio, M. Volatile components of Lily of the Valley (*Convallaria majalis*)-CHIRAROMA analysis. *Koryo, Terupen oyobi Seiyu Kagaku ni kansuru Toronkai Koen Yoshishu* **2006**, *50*, 79–81.

32. Brunke, E.J.; Ritter, F.; Schmaus, G. New data on trace components with sensory relevance in flower scents. *Dragoco Rep.* **1996**, *1*, 5–21.

33. Forsyth, S.A.; Gunaratne, H.Q.N.; Hardacre, C.; McKeown, A.; Rooney, D.W.; Seddon, K.R. Utilisation of ionic liquid solvents for the synthesis of Lily-of-the-Valley fragrance {β-Lilial®; 3-(4-*t*-butylphenyl)-2-methylpropanal}. *J. Mol. Cat. A Chem.* **2005**, *231*, 61–66.

6

A Contemporary Introduction to Essential Oils: Chemistry, Bioactivity and Prospects for Australian Agriculture

Nicholas Sadgrove * and Graham Jones

Pharmaceuticals and Nutraceuticals Group, Centre for Bioactive Discovery in Health and Ageing, University of New England, S & T McClymont Building UNE, Armidale NSW 2351, Australia; E-Mail: nsadgrov@une.edu.au

* Author to whom correspondence should be addressed; E-Mail: nsadgrov@une.edu.au

Academic Editor: Muraleedharan G. Nair

Abstract: This review is a comprehensive introduction to pertinent aspects of the extraction methodology, chemistry, analysis and pharmacology of essential oils, whilst providing a background of general organic chemistry concepts to readers from non-chemistry oriented backgrounds. Furthermore, it describes the historical aspects of essential oil research whilst exploring contentious issues of terminology. This follows with an examination of essential oil producing plants in the Australian context with particular attention to Aboriginal custom use, historical successes and contemporary commercial prospects. Due to the harsh dry environment of the Australian landmass, particularly to the cyclical climatic variation attendant upon repeated glaciation/post-glaciation cycles, the arid regions have evolved a rich assortment of unique endemic essential oil yielding plants. Though some of these aromatic plants (particularly myrtaceous species) have given birth to commercially valuable industries, much remains to be discovered. Given the market potential, it is likely that recent discoveries in our laboratory and elsewhere will lead to new product development. This review concludes with an emphasis on the use of chemotaxonomy in selection of commercially viable cultivar chemotypes from the Australian continent. Finally, drawing largely from our own results we propose a list of Australian endemic species with novel commercial potential.

Keywords: essential oil; organic chemistry; pharmacology; Australian; cultivation; chemotype; cultivar; history; Aboriginal

1. Introduction

1.1. Terminology of Essential Oils and Methodologies of Production

Essential oils are a mixture of volatile lipophilic (fat loving, *i.e.*, soluble in fat) constituents, most commonly sourced from leaf, twig, wood pulp or bark tissue of higher plants, but also widely found in bryophytes, such as the liverworts [1]. Although essential oils are only slightly soluble in water, the aqueous solubility of individual essential oil components varies with respect to polarity (magnetic activity). Generally, components with more polar functional groups are expected to be more soluble in water relative to other components.

Essential oils are most commonly produced using hydrodistillation; however prior to this, individual components of the whole essential oil are present within the source tissue, either in the same molecular form or as a heat labile precursor. The process of hydrodistillation involves heating in the presence of water to temperatures higher than boiling point, to produce mixed gases that expand and travel into a condenser. A variation of this is steam distillation, which places the source tissue (leaves, stem or bark) in the path of steam and not in the boiling water itself, as in hydrodistillation.

During hydrodistillation, mixed gases (steam and oil vapour) are produced and expand into a condenser where they are cooled to below 30 °C and condensed into two separated (non-mixing) liquid phases; one phase being a hydrosol and the other an essential oil. The two condensed liquids are gravity fed into a separation funnel, where they are separated. Problems occur when hydrodistillation is performed at higher temperatures, because the subsequent temperature of the hydrosol is not sufficiently lowered before entering the separation funnel. The consequence is fractionation of the essential oil, with a greater representation of components with higher boiling points. In addition, there may also be a failure to condense any essential oil at all; or if condensed oils are observed they may be subject to re-evaporation if the hydrosol temperature is too high. Thus, it is generally a priority to regulate the boiling temperature in order to optimise the hydrodistillation to maximise essential oil yield.

Most authorities contend that if a process other than hydro- or steam distillation is used, such as solvent extraction or mechanical pressing, to collect liquids containing volatile compounds, the product should not be regarded as an essential oil and may instead be called an "absolute" or an "expressed oil", respectively. In the preparation of an "absolute" a hexane extract is first taken of the raw material and evaporated to produce a "concrete". The concrete is dissolved into ethanol and cooled to −20 °C. At this lower temperature waxes, sterols and other lipids solidify and are removed, which concentrates the volatile compounds in the product; the absolute.

A host of other names can be used to describe aromatic preparations, such as "tincture", "spice oleoresin", *et cetera*. The term "volatile oil" is commonly used if reference is made to the volatile fraction of any of these extracts, but this expression also encompasses essential oils. Having said this, some authorities still refer to the mechanically expressed oil from citrus peel as an essential oil and this is upheld by the International Organisation for Standardisation [2,3].

The specific definition of an essential oil may be subject to disputation among interested individuals including scientists, aromatherapists or lay people. Although a consensus has generally been agreed upon, extraction methods are still evolving, and this has the capacity to introduce further confusion in terminology. Essential oils can now be extracted using modern microwave-assisted hydrodistillation [4,5] or microwave-assisted distillation techniques that require no additional water, other than cytosolic and vascular fluids already present in the source tissue [6]. These methods result in differences, both qualitative and quantitative, in the composition and yield of the subsequent essential oil [4,6]. Strictly speaking the latter technique, requiring no additional water, is not hydrodistilled but merely distilled.

In this regard, the International Organisation for Standardization (ISO) defines an essential oil as a:

"product obtained from natural raw material, either by distillation with water and steam, or from the epicarp of citrus fruits by mechanical processing, or by dry distillation" [2,3].

With regard to the classification of expressed oil from citrus peel, such as orange or bergamot oil, these are commonly referred to as essential oils [7,8]. However, using this terminology they may be confused with the essential oils produced using hydrodistillation. In the former case of expressed oils, the source tissue is not hydrodistilled and the subsequent oil contains dissolved lipids (waxes and sterols) and other larger compounds that are not volatile, such as the coumarin bergaptene (**31**) (see all chemical structure diagrams in Figure A1) in the case of bergamot oil (*Citrus bergamia* Risso.) [8,9].

Further implications of definition of essential oils appear when considering the traditional medicinal applications of aromatic plants. This is particularly relevant when plant material is heated to produce an acrid steamy smoke and then either condensed onto the skin [10] or in the lungs through inhalation [11]. In this context, medicinal effects may sometimes be produced via a molecular interaction between multiple compounds that could be of both lipophilic and hydrophilic character. Such potentially synergistic interactions will not occur using only the pure essential oil as produced in hydrodistillation [10]. However, more often than not there is a single active compound involved that produces the greater part of the medicinal effect [12], which can be of either lipophilic or hydrophilic in character.

In this regard, slightly larger intact or modified compounds are evaporated, in significant quantities, when higher temperatures are involved, such as in the Australian Aboriginal smoke fumigation practices [12], or indeed in microwave assisted distillation [4,6]. These slightly larger compounds can be found in simulated smoking extracts or dissolved in either the essential oil or hydrosol, when microwave assisted distillation is used.

Some may propose that volatile oils produced using microwave-assisted distillation or hydrodistillation technology should correctly be called essential oils because of the chemical alteration of heat labile constituents that become part of an essential oil with both natural ingredients and these derived "artefacts". This is clearly an area of contention and the essential oil industry may need to embark upon the development of a new system for communicating information related to the distillation method used to produce essential oil products, to establish consumer awareness of potential qualitative differences. A similar approach may also need to include expressed oils from citrus peel, to avoid confusion with hydrodistilled oils, also produced from citrus peel. Furthermore, in the case of heating plant material to produce therapeutic gases in ethnomedicinal contexts, this may be recognised as a mixture of essential oils and other larger compounds, such as diterpenes, together with the more hydrophilic components that are not usually detected in significant quantities in the essential oil *per se*.

To avoid further confusion researchers and academics often use the word "hydrodistilled" or more recently "distilled" instead of "extracted" if they are referring to an essential oil. This is to clarify that essential oils are being described and not the same volatile components (plus non-volatiles) that have been separated using other techniques. Such oils are either produced using solvent extraction or are mechanically expressed from the source material to produced volatiles dissolved in lipid oils or *vice versa*. Systematic avoidance of the term "extracted" can become contentious, cumbersome and impractical, so it should be avoided only when there is a possibility of confusion. As far as we can tell, using the term "produced" is acceptable with reference to essential oils.

Disputes regarding the nomenclature of essential oils also impact on received history of essential oil usage, because volatile oils in earlier use may not fall into the modern definition of an "essential oil", since they were not hydrodistilled in the conventional sense. For instance, there is no evidence that modern hydrodistillation technology was available in biblical times or in ancient Egypt, meaning that the medicinal applications described in these earlier references most likely used expressed or absolute oils with a mixture of volatile and fixed components and were therefore not essential oils *per se* [13], as is commonly accepted [7,14].

The earliest authentic description of an essential oil, produced by a method resembling conventional hydrodistillation, was compiled by Arnald de Villanova sometime during the late 12th or early 13th century (1235–1311 AD). Prior to this, details of a primitive form of distillation, used to produce turpentine and camphor (17), were described by the ancient Romans and Greeks in the first century [13]. However, because no other essential oil was produced in this manner it is unclear if this can be taken as evidence of essential oil production at that earlier time. Although there is clear evidence that a primitive form of distillation technology was in use from 400 BC (Terracotta distillation apparatus dated to approx. 400 BC, now stored in the Taxila Museum, Pakistan) until the ninth century, this method was primarily used to produce distilled waters where fractionated essential oils, such as camphor (17), were often produced as a by-product [13]. Using such primitive hydrodistillation, distilled waters or "hydrosols" could be achieved without any difficulty, but intact essential oils could not be captured without modern methods enabling the cooling of steam to the required lower temperatures. Thus, only essential oil components with higher boiling points, such as camphor (17), could be retained.

Distillation technology was improved in the ninth century by earlier Arabic scientists [15,16], but again it is not clear if they used this technology to deliberately produce essential oils or if the primary focus was for floral waters. Therefore, historians currently agree that the essential oil technology that was adopted into therapeutic use in Europe in the Middle Ages was from the 13th century work of Villanova, who provided the earliest record that can be reliably authenticated [13].

Essential oil components are usually no larger than 300 Daltons (amu) in size [17], except in unusual cases involving larger diterpenoids such as incensole acetate from *Boswellia* spp. [18–20] and these require longer periods of hydrodistillation (perhaps higher temperatures) with cohobation before they are measured in the whole essential oil. This general observation may change with the advent of new distillation technology that produces slightly heavier molecules (approx. 350 Daltons), such as the microwave-assisted distillation method aforementioned.

With regard to the production of floral waters or hydrosols, the hydrophobic character of essential oil causes phase separation of oil and water, but trace quantities of the essential oils dissolve as mentioned before. Usually, due to a relatively low saturation point, the hydrosol dissolves only small amounts of the essential oils, but occasionally volatile components can be dissolved in hydrosols at relatively high concentrations [12]. In such cases these constituents have greater polarity than other essential oil components, making them more soluble in water. Minimisation of distillation waters and recycling of the hydrosol can significantly enhance the yield of oil obtained from a hydrodistillation. Cohobation is one method used to reduce the loss of essential oil via solubility into the hydrosol, where the hydrosol is manually returned to the still throughout the duration of the hydrodistillation.

Another method used to reduce the loss of essential oil to dissolution in the hydrosol, employs the Clevenger-type apparatus (Figure 1), which returns the hydrosol to the still in real-time, during the course of the distillation. This also reduces the overall volume of water, initially required for the distillation. Unfortunately one of the challenges in using the Clevenger-type apparatus is in maintaining the hydrosol at a lower temperature, as higher temperatures can lead to a reduction in oil yield or an emulsion of the oil in the hydrosol. A disadvantage of the Clevenger apparatus is that essential oils must float (be less dense than water), otherwise they will be lost back into the still with the hydrosol.

As depicted in Figure 1, the positioning of the condenser (D) is directly above the Clevenger-type apparatus (C). Here we have tilted the condenser so that the condensates run along the sides of the glass and meet the liquid at a reduced velocity. In the traditional Clevenger-type spatial configuration, the condenser is positioned vertical to the liquid surface, but in our experience the fall of liquid from the condenser disrupts the phase separation of the essential oil and the hydrosol. In addition, the positioning of the condenser so that condensed liquids return via the passage of steam, is why there is difficulty in reducing the temperature of the hydrosol. To combat this we have adjusted the heating mantle (B) to a lower temperature, using a power regulator (A). However the best measure would involve reinventing the Clevenger-apparatus to include a water jacket around the essential oil and hydrosol phases to maintain a lower temperature and therefore prevent re-evaporation.

The essential oil phase typically floats over the hydrosol, but in fewer cases the essential oil is denser than water so it settles below the hydrosol [21]. For example, some phenylpropanoids, such as safrole (14) and methyl eugenol (15), are denser than water and will settle below the hydrosol, but only if they occupy sufficiently high relative abundance in the whole essential oil. An example of this is the essential oil produced from one of the chemotypes of *Eremophila longifolia* (Scrophulariaceae) in Western Australia [22], which is a mixture of safrole (14) and methyl eugenol (15), comprising approximately 97% of the entire essential oil. In this particular case, a hydrodistillation of this species using the Clevenger-type apparatus would fail to capture the essential oil because it would return to the still with the hydrosol.

Essential oils are biologically regarded as metabolites of secondary importance to the organism because, in contrast to primary metabolites, they are not universal across the plant kingdom, nor do they constitute any of the basic building blocks of life [17]. Although such secondary metabolites are generally regarded as metabolic by-products, it is widely acknowledged that they provide an evolutionary advantage to the plant (or liverwort), which may involve protection against grazers such as fungi, insects or herbivores. Alternatively, the essential oils may play a less obvious ecological role, such as in fire tolerance, attracting pollinators and/or herbivores for seed dispersal, drought tolerance or plant-to-plant biosemiosis (pheromones).

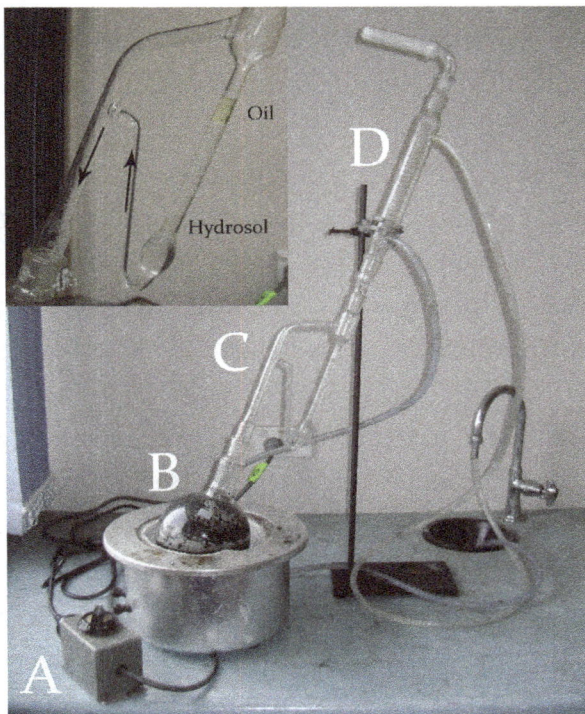

Figure 1. Hydrodistillation using the Clevenger-type apparatus. (**A**) Power regulator; (**B**) Heating mantle with round bottom flask containing water and aromatic leaves; (**C**) Clevenger-type apparatus which returns the hydrosol to the still and maintains the essential oil phase, but only for essential oils that are less dense than water and therefore float; (**D**) The condenser.

Although essential oils may contribute significantly toward the evolutionary survival of the respective organism, the term "essential oil" did not derive from this function. A common misconception is that essential oils are called "essential" oils to highlight their importance in the biological survival of the organism. However, the term "essential oil" actually has its origin from the word "quintessence", the English rendering of *Quinta essentia*. This term means the fifth element in the earlier alchemical constellation, used for essential oils in the early 16th century by the Swiss medical pioneer, Bombastus Paracelsus von Hohenheim [13]. At the time von Hohenheim believed that the essential oil was the most pure and concentrated form of the medicinal principle of any plant, produced by hydrodistillation of the plant tissue.

Use of the term "quintessence" by von Hohenheim is a reflection of the Aristolean paradigm, which described matter as being composed of the five elements: earth, fire, water, air and spirit. Quintessence (literally the fifth essence) was regarded as the latter of these; the spirit or life force of the plant, which could be removed and contained by the distillation process. Use of the modern term "spirits" to describe various liquors, specifically those produced by distillation, is again a reflection of this ancient concept [17].

A variety of other names are given to the essential oil. These include essence, fragrant oil, volatile oil, etheric oil, aetheroleum or aromatic oil [21]. The latter term "aromatic" is another term that generates a lot of confusion and contention. Although the term "aromatic" in modern usage describes the quality of giving off an aroma that is either pleasant or odious to the nose, an aromatic compound or moiety, in the language of chemistry, has a chemical arrangement that results in delocalisation of electrons, producing greater molecular stability. Thus, essential oils may be a mixture of aromatic and aliphatic (non-aromatic) compounds, all of which contribute to the perceived aroma. This is obvious to professional chemists but leads to confusion with other non-scientific users of essential oils.

In strictly chemical terms, aromatic compounds, also often called arenes, contain an aromatic group. An aromatic group is planar, cyclic with overlapping p electron orbitals and an odd number of electron pairs within the π bond formation $((4n + 2)/2)$. Although the benzene moiety is the most commonly cited example [21]; other aromatic groups include the heterocycles pyrrole, pyrans, furans and thiophenes.

The term aromatic (or arene) first entered the language of chemistry when Augustus W. Hofmann (1855) used it in reference to a series of volatile mono- and "bibasic [sic] acids", including the provisionally named insolinic acid. Because all of the compounds in Hofmann's series contained a benzene moiety, the term aromatic came to be associated with arene compounds [23]. Because all of the compounds in Hofmann's aromatic series contained a benzene moiety and have odour, the term aromatic came to be associated with essential oils and other odour causing molecules. When the advancement of chemistry eventually demonstrated that odour causing compounds were mostly terpenes and other non-benzenoid chemical groups, use of the term aromatic to describe these respective compounds persisted. Thus, although the term "aromatic plants" is now widely used to describe essential oil yielding varieties, most essential oil compounds are aliphatic in the strict chemical sense.

1.2. Chemistry, Chirality and Stereochemistry of Essential Oils

Essential oil conferences attract attendees from a multitude of professions with a diversity of expertise. Some of the participants will be expert chemists whilst others will have no previous exposure to chemistry at all. Generally it is expected that our readers will have at least a basic level of chemistry, but for readers with no previous exposure, we have included an image and explanation in the Appendix to help in the understanding of molecular diagrams, in Figure A2; lessons A–E. Furthermore, information related to understanding chirality and stereochemistry has also been included in the Appendix.

In the chemistry of essential oils, chirality and stereochemistry of components are of considerable importance. This is because the spatial orientation of connective parts of a molecule can significantly influence the chemical behaviour and pharmacological activity of the compound. In this regard, molecules with the same molecular formula and the same bonds between atoms, but different spatial arrangements of these atoms, are called stereoisomers. The two main types of stereoisomers that are relevant to the discussion of essential oils are diastereomers and enantiomers.

Generally a pair of stereoisomers are called diastereomers, which are distinguished as separate entities in routine chemical analysis, such as in gas chromatography (GC) or nuclear magnetic resonance spectroscopy (NMR). However, some stereoisomers are exact mirror images of each other that cannot be superimposed, like a left and right hand (Figure 2). Each of these are called enantiomers of a chiral molecule. An example of a chiral compound is carvone (1) (Figure 2). Because carvone (1) is a chiral molecule, differences between enantiomers cannot be observed using routine GC or NMR [24].

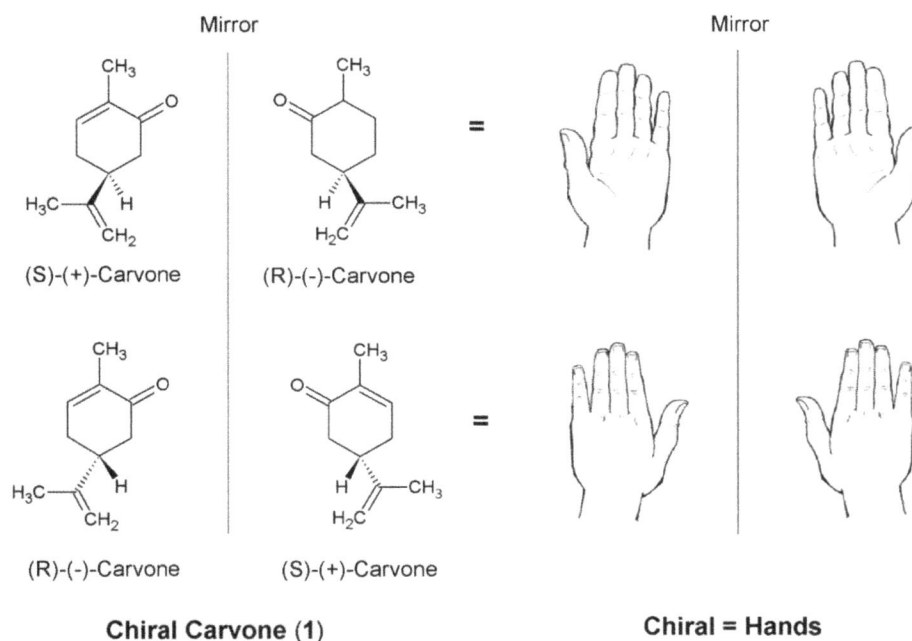

Figure 2. The two enantiomers of a carvone (**1**).

The term "chiral" first joined the language of chemistry after it was coined by Lord Kelvin in 1893 [25]. It derives from the Greek word for hand, the most familiar chiral object in nature. Before the technology was available to elucidate the absolute stereochemistry of chiral compounds they were identified on the basis of being able to rotate plane polarised light. Compounds that are able to rotate plane polarised light are known as "optically active" and are assigned a "specific rotation" measurement that is generally unique for each chiral compound. However, each enantiomer in a chiral compound will rotate equally in opposite directions, one to the right and the other to the left. Thus, one will have a negative specific rotation, which is to the left, and the other will have an equal positive value, which is to the right. Using terminology deriving from Latin, rotation to the right is *dextrorotatory* (*dextro-* derived from *dexter* for "right") and rotation to the left is *laevorotatory* (*laevo-* derived from *laevus* for left). Thus, in the older language of chemistry the prefix to specify enantiomers was either D- or L-. Examples of where this language has survived include D-alpha-pinene (**2**) and D-limonene (**3**) (Figure A2), which are sometimes shown as D-(+)-alpha-pinene and D-(+)-limonene, respectively. However, the use of D- and L- has more recently been dropped and replaced by the symbols for positive (+) and negative (−) [21,24].

An example of an achiral (not chiral) molecule is ρ-cymene (**4**). This molecule is achiral because it is superimposable on its mirror image but it does not have a chiral centre (bear in mind that an aromatic ring has delocalised electrons, so the placing of double bonds is arbitrary). A clearer picture of what a chiral centre looks like is elucidated in Figure 3.

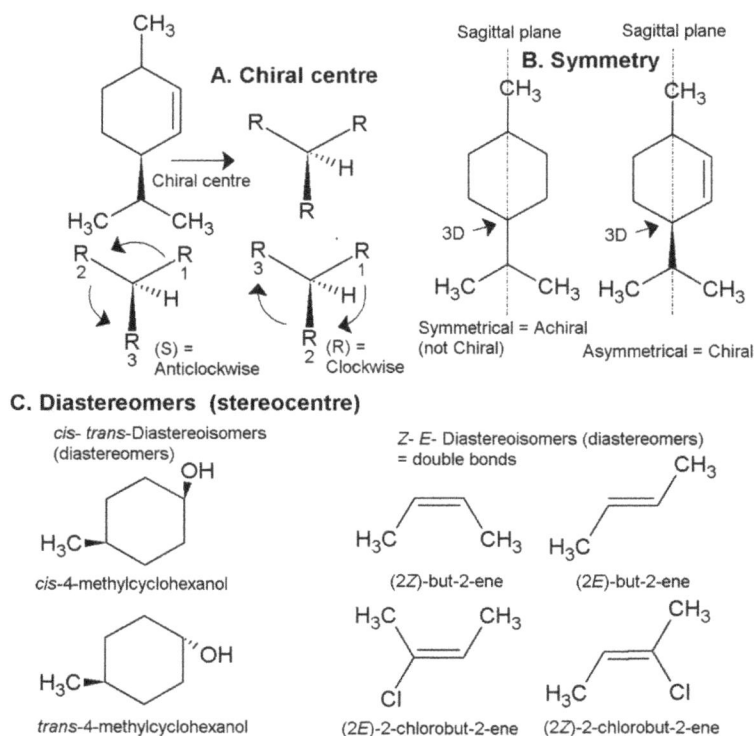

Figure 3. The stereochemistry of a molecule where the rotation of priority bonds around a chiral centre are defined as either *S* or *R* (**A**), where symmetry can influence whether something is chiral or achiral (**B**) and where two chiral centres (or stereocenter for double bonds) can result in either *cis-* or *trans-* isomerism (**C**), which can be more accurately denoted with *Z-* and *E-* if isomerism is over a double bond.

The specific rotation of a chiral compound is measured using a polarimeter. Although this technology is quite old, its use in chemistry has continued to this day. However, because chemists are now aware of the absolute stereochemistry of each enantiomer (the exact 3D configuration of bonds), the convention for describing an enantiomer is complemented with either *S* or *R* descriptors (Figure 3A). This helps chemists to communicate the 3D spatial arrangement of atoms or groups around the bond without resorting to drawing a diagram. Such descriptors are generally only used in the common name of an essential oil component where only one chiral centre is present, otherwise each chiral centre should be assigned an *S* or *R* configuration, *i.e.*, (1*R*,5*R*)-(+)-alpha-Pinene (**2**). However, where the configurations are identical, it is conventional to use the descriptor just once, *i.e.*, (*R*)-(+)-alpha-Pinene (**2**) [24].

A common misconception is to believe that *S* or *R* indicate the direction of rotation of plane polarised light, but this is wrong. A chiral centre that is in the *S* configuration can be either the positive (+) or the negative (−) enantiomer, but will always be opposite to the *R* configuration. Thus, when *S* = (+) then *R* = (−) and *vice versa* [24].

A chiral centre is defined as either *S* or *R* by the rotation of priority groups around the main carbon in the bond (Figure 3A); it is purely a constitutional concept. Priority groups are defined by the Cahn-Ingold-Prelog priority rules (CIP), which prioritises groups based on higher atomic number (higher numbers = higher priority). The CIP rules are explained in detail in any modern organic chemistry textbook [24].

In a chiral centre, the rotation of priority bonds in an anticlockwise direction (to the left from 12 O'clock) is called *S*, which derives from the Latin word *sinister* (Figure 3A) meaning wrong or to the left. In a clockwise direction it is called *R* for *rectus*, which means straight or correct, but in this context it could mean to the right or correct. Because influences from the left side were regarded as evil in ancient societies, the latin word *sinister* for left, survives today in the English language to mean evil or threatening, no longer associated with the left. However, it is clear that the modern chemical use of these words is reflective of the ancient concept that left and right had moral connotations [24].

In essential oil chemistry it is also common for differences at a chiral centre to result in an entirely different molecule, not just a different enantiomer. Such compounds are asymmetrical but not chiral (no mirror image), so they are diastereomers as mentioned previously. Diastereomers can result from a variation in one of two or more chiral centres, or alternatively from the stereochemistry of substituent groups about a double bond. For example, a compound with two chiral centres can lead to epimers where the two compounds differ at just one chiral centre (globulol (**52**) and ledol (**53**) are epimers). This has the capacity to substantially alter the chemical and pharmacological properties of the compound (Figure 3C). Such changes will be more pronounced between diastereomers than between enantiomers. However, if groups at both chiral centres are changed, the resulting compound is its enantiomer [24].

A compound with a double bond can occur as either a *cis*- and *trans*- isomer, which makes it one of two diastereomers. This convention can also be used on single bonds but only in an alicyclic molecule (non-aromatic ring structure) where rotation about the single bond cannot occur. These isomers are defined by substituent groups, which are the non-hydrogen attachment at the chiral centre or stereocenter (sterecenter is a more general expression that includes stereochemistry on a double bond as well as chiral centres). Where the substituent group at one stereocenter is on the same side of the molecule as the group on the other stereocenter, it is called *cis*- and on opposite sides called *trans*- (Figure 3C). Specifically where the stereocenters are on opposite sides of a double bond, *Z*- and *E*- notation can be used to replace *cis*- and *trans*-, respectively (*Z*- is from the German word *zusammen*, meaning together; *E*- is from the German word *entgegen*, meaning opposite) [24].

However, *E*- and *Z*- notation uses the CIP priority rules mentioned earlier. *Cis*- and *trans*- isomerism does not, but alternatively prioritises for non-hydrogen groups that are in a different position compared to its known diastereomer. Furthermore, *cis*- and *trans*- substituent groups are not always obvious, making this convention rather arbitrary. Such ambiguity is more common in complex molecules or where more than two different types of elements are present at the stereocenter. That is why *cis*- and *trans*- isomerism is regarded as relative stereochemistry (relative to its known diastereomer), whereas *Z*- and *E*- notation is regarded as absolute. Thus, in a similar way to the *R* and *S* descriptors, *E*- and *Z*- notation in alkenes (molecules with double bonds) does not always translate to the *cis*- and *trans*- isomerism of the same alkene. However, this inconsistency should only occur if more than two different types of elements are present at the stereocenter, such as in the case of 2-chlorobut-2-ene depicted in Figure 3C. Therefore such ambiguity is avoided by using only *E*- and *Z*- notation on alkenes, where the convention of using *cis*- and *trans*- has not already been established [24].

As previously mentioned, generally diastereomers can be detected using basic gas chromatography or NMR to the level of its relative (*cis*- or *trans*-) or absolute (*Z*- or *E*-) stereochemistry if it has previously been reported, but the absolute stereochemistry of known chiral molecules (*S* or *R* configuration of chiral centres) cannot be realised without further more comprehensive investigation.

Therefore, in using such non-specialist instrumentation, it is not possible to differentiate between enantiomers in an essential oil [24].

The importance of chirality arises in essential oil chemistry when samples are corrupted by profit oriented manufacturers who adulterate natural products with cheaper synthetics. Consequently measures need to be routinely undertaken to prove or disprove claims of authenticity made by sellers and manufacturers in the marketplace. It is also common practice for producers of essential oils to optimise profit by adulterating natural essential oils with more common natural but cheaper essential oils. Among other things, the consequence of any kind of adulteration is that the composition of enantiomers is not reflective of the naturally occurring essential oil. Cheap manufacturers are dependent on the fact that no specialist methods will be employed to investigate the composition of their essential oil products, but times are changing [21,26].

Essential oils that are biosynthesised by plants are composed of a variety of components that may include chiral compounds. Generally if these chiral compounds are synthesised in a laboratory a 50:50 mixture of the enantiomers will be produced, but only if the chiral centre is a part of the reaction or if the precursor is racemic, meaning an equal proportion (50:50) of enantiomers. Such a mixture is very difficult to separate using common inexpensive chromatography, in contrast to the separation of diastereomers, which are also produced in synthetic reactions and are more easily separated using basic chromatography. Because it is expensive and difficult to separate enantiomers, the occurrence of racemic mixtures in natural products is therefore a reliable indication of cost cutting and adulteration [21].

The specific rotation of a racemate is zero because the rotation from each of the enantiomers sums to zero. Although such a mixture is called a racemate, this should not to be mistaken for a mixture with unequal proportions of enantiomers, which can occur in natural products, often at a ratio of 40:60 [27]. Such a mixture is referred to as enantioenriched or can be defined as having an enantiomeric excess (ee) which is a figure that shows the amount of unpaired enantiomer as a mass percentage of the whole (g/g), or how much one enantiomer outweighs the other. Thus, enantiopure samples (just one enantiomer) have an ee of 100%, whereas a racemate has an ee of 0%. Furthermore, a mixture with 75% of the R or S enantiomer will have an ee of 50% ($75 - (100 - 75) = 50$). Another way to look at this is to imagine that enantiomeric excess describes the part of the mixture that is not racemic (not a one to one mixture of enantiomers) [21,24].

As high stereospecificity is achieved in enzyme-catalysed reactions (biological reactions), such chiral compounds from nature are generally enantiopure, but not always. Regardless, studies focused on elucidating the enantiomeric excess of essential oils provide guidelines as to the expected enantiotypes from the selected species under investigation. This rising field of enantiotaxonomy has already identified single or multiple enantiotypes within the one species, which may serve as fingerprints where authentication is an issue [21].

The consequence of adulterated essential oils is that they won't display the same pharmacological or aesthetic qualities of the natural essential oils. This was made implicit in earlier studies where double blind olfactometry studies demonstrated that enantiomers were differentiated by the character of the odour as perceived by a specialist odour panel. In one such study, the two enantiomers of carvone (**1**) were synthesised from enantiopure samples of Limonene (**3**), where the chiral centres were not altered in the reaction, so high enantiomeric purity of the carvones was achieved. Using a panel of 21 reliable odor specialists, the odor of (+)-carvone was characterised as caraway-like and (−)-carvone as spearmint-like [28]. This is in agreement with the predominating enantiomers found in caraway (*Carum carvi* L.: Apiaceae) and spearmint (*Mentha spicata* L.: Lamiaceae), respectively [21].

1.3. Chemical Analysis and Standardisation/Legislation of Essential Oils

The most common technique employed in the chemical characterisation of essential oils is gas chromatography coupled with mass spectrometry (GC-MS). The relatively small size of essential oil components means that they are all volatile and can therefore be separated according to boiling points. This process takes place in a long thin column (*i.e.*, 30 m) that has the appearance of coiled wire. This column is prepacked with a porous stationary phase that is either polar (slightly charged), such as a wax column (polyethylene glycol; *i.e.*, DB-wax, Carbowax 20M, PEG-20M) or apolar, such as polymethylsiloxane (*i.e.*, HP-1MS or 5% diphenyl- *i.e.*, HP-5MS). The apolar HP-5MS column is the most commonly used [29].

The essential oil is diluted into a solvent and injected into a heated injection chamber (*i.e.*, 300 °C) so all components of the essential oil are evaporated and delivered by a benign (non-reactive) gas (*i.e.*, nitrogen or helium) to the start of the column, which is usually at a lower temperature (*i.e.*, 60 °C), so essential oil components precipitate onto the column [29].

In gas chromatography, separation is performed by heating the column in an oven, most commonly employing a programmed temperature ramp, from a lower to a higher temperature; however, occasionally isothermal (constant temperature) programs are used. When a temperature program is used, the separation of components occurs when the temperature is raised to each of the component's individual boiling points. At this point the component vaporises and is carried by the non-reactive gas to the detector [29]. The most common detector used in gas chromatography is a mass spectrometer, but flame ionisation detection (FID) is often used where the accuracy of quantification is a concern.

In using FID the identity of the compound cannot be known. However, if the identity of the compound and its retention time (how long it takes to come out of the column) is already known from a previous experiment, then GC-FID can be used to calculate an accurate relative abundance of each component in the essential oil. Generally prior to GC-FID, GC-MS is used to identify components in the essential oil. Although GC-FID gives more accurate quantification data, it is more common to use the less accurate quantification method, which is calculated from the GC-MS chromatogram (Figure 4).

In a GC-MS chromatogram the retention time of components generally reflects their sizes and the presence of functional groups (Figure 4). In Figure 4 the elution of components starts with monoterpenes (10 carbon compounds), which is then followed by oxygenated monoterpenes and finishes with sesquiterpenes (15 carbon compounds). Although Figure 4 shows the mass spectrum of a few selected components, such information is always depicted in a separate window.

In mass spectrometry the separated component (say, Limonene, **3**) is fragmented by electron impact ionisation, which produces a spectrum of ions that are separated according to mass, with heavier components exerting a greater inertial resistance to a magnetic field than lighter components. In the process of magnetic deflection of their paths, ions are diverted onto a detector. The result is a spectrum showing ions of different sizes with different relative abundances. This spectrum can be generally considered a fingerprint of each component in the essential oil. Because of the reproducibility of this experiment, each mass spectrum can be compared across a spectral library. Using other pieces of information, such as retention time, a fairly reliable match can be made, with minor exceptions [29].

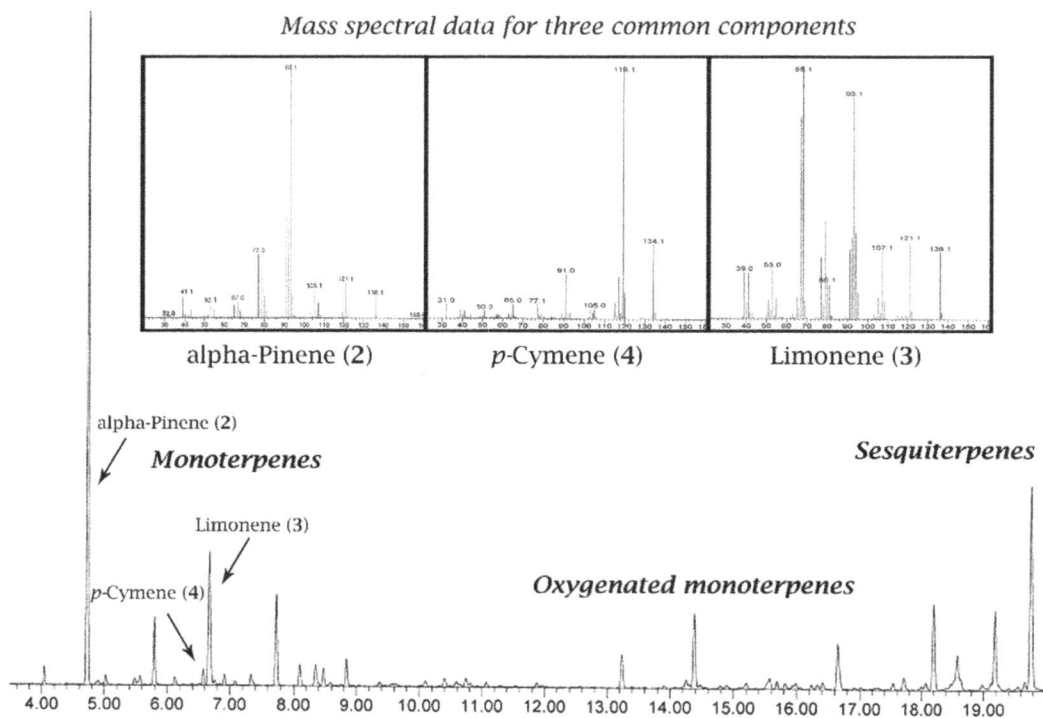

Figure 4. A gas chromatogram with mass spectral data superimposed for three common essential oil components. On the chromatogram *x-axis* is retention time (RT) in minutes and on the *y-axis* is abundance in arbitrary units.

Another method worthy of mention here is solid phase microextraction (SPME) or to put it more simply, head-space analysis or absorption [30]. This method delivers volatiles to the GC-MS but by-passes the solvent extraction or distillation process. This technique employs an adsorptive solid phase that is positioned in the headspace of the aromatic preparation. Following this either the solid phase can be injected directly into the GC-MS or the volatiles can be desorbed into a solvent before injection. A number of adsorbents are used in SPME and these are listed by Reineccius [30].

After GC-MS analysis is performed, where discrepancies in identification are involved, essential oil components can be purified using flash chromatography before analysis using nuclear magnetic resonance spectroscope (NMR). Flash chromatography is achieved by packing silica gel (polar phase) into a vertical column with a solvent mixture. The essential oil is added to the top of the column. Because each component has a different binding affinity to the silica gel, the components can be subsequently washed out individually by the same solvent under pressure. The solvent mixture is customised relative to the polarity of the target compound to optimise separation [29].

The NMR of the purified component will provide a unique spectrum of ^{13}C or ^{1}H shifts that can be matched to a published value if it is a known compound. If not, more comprehensive structural elucidation can be performed using 2D-NMR experiments. In addition, by using 2D-NMR experiments new structures can be discovered. The theory behind NMR is explained in detail in any modern organic chemistry textbook [24].

The spectral data generated by NMR, as well as mass spectral data from GC-MS, will not help in differentiating between enantiomers. In addition, enantiomers cannot be separated by either a polar or apolar stationary phase in chromatography, unless a different binding affinity is employed. To achieve this, an enantiopure additive is included in the stationary phase, with the most common being enantiopure aryl or alkyl derivatives of a cyclodextrin (α-, β- or γ-cyclodextrins). Such a stationary phase is called a "chiral column", which can be fitted to a GC-MS for chromatographic separation of enantiomers. This methodology, called enantioselective chiral gas chromatography (enantio-cGC) [31] is the preferred choice in authenticating essential oils because it requires no prior separations. The whole essential oil can be injected and the chiral compound will separate into two peaks. Without access to a chiral column a more primitive method involving flash chromatography and a polarimeter can be used, but this proves to be very time consuming and resource wasteful.

Knowledge of the chemical composition of an essential oil, as well as the enantiomeric excess of chiral components, comprises the most significant part of the standardisation process. If both of these chemical aspects conform to the published standard, other physical parameters of the essential oil should fall into line. Thus, with the correct number of components at relative abundances within the defined range and the correct enantiomeric composition, the optical rotation and refractive index of the whole essential oil, its colour, density and appearance should conform to the defined standard [3].

However, these other parameters may be elevated to become more significant authenticating tools when essential oils are adulterated with carrier oils or diluted into alcohol. Additionally, this can also be of relevance if the essential oil is produced using a method other than hydro- or steam distillation. This is pertinent in the production of absolutes or expressed oils, such as the previous mentioned example of Bergamot (*Citrus bergamia* Risso and Poit). In this case components other than those identified in GC-MS will be present; components that are not volatile or gaseous. Thus, all of the authentication parameters are necessary for standardisation purposes, not just the chemical character as deduced from gas chromatography.

Another factor worthy of consideration when preparing or examining published standards of essential oils is the occurrence of chemotypes. It is extremely common for a single essential oil yielding species to produce varieties of essential oils, called chemotypes. Chemotypes often occur where a geographical or geological difference influences diversification of biosynthetic pathways. Chemotypes may result from diverging evolutionary pathways, or from environmental cues, such as soil type or altitude. Where chemotypes occur in a species, published standards are generally specific, *i.e.*, the standard of Tea Tree Oil (*Melaleuca alternifolia* Cheel) as specified by the International Standards Organisation (ISO: Geneva Switzerland) clarifies the chemotype being described: "Oil of Melaleuca, terpinene-4-ol type (Tea Tree Oil)' (ISO 4730:2004) [3]. The standard described by ISO is identical to that described by Standards Australia (AS 2782-2009) [32] (Figure 5).

Requirements:		
Appearance	Clear, mobile liquid	
Colour	Colourless to pale yellow	
Odour	Characteristic	
Relative density (20° C)	Min: 0.885	Max: 0.906
Refractive index (20° C)	Min: 1.475	Max: 1.482
Optical rotation (20° C)	between + 5° and + 15°	
Miscibility in ethanol (20° C)	not necessary to use more than 2 volumes of ethanol, 85% (volume fraction) to obtain a clear solution with 1 volume of essential oil	
Flashpoint (closed cup)	mean value	mean value
Min volume of test sample	50 ml	

Chromatographic profile:		
Component	Min %	Max %
α-Pinene	1	6
Sabinene	trace	3.5
α-Terpinene	5	13
Limonene	0.5	1.5
p-Cymene	0.5	8
1,8-Cineole	trace	15
γ-Terpinene	10	28
Terpinolene	1.5	5
Terpinen-4-ol	30	48
α-Terpineol	1.5	8
Aromadendrene	trace	3
Ledene (syn. viridiflorene)	trace	3
δ-Cadinene	trace	3
Globulol	trace	1
Viridiflorol	trace	1

Figure 5. The Australian standard of chemical components in Tea Tree Oil (AS 2782: 2009), the terpinen-4-ol chemotype, Standards Australia [32].

The ISO reference defines Tea Tree Oil as:

"Essential oil obtained by steam distillation of the foliage and terminal branchlets of *Melaleuca alternifolia* (Maiden et Betche) Cheel, *Melaleuca linariifolia* Smith, and *Melaleuca dissitiflora* F. Mueller, as well as other species of *Melaleuca* provided that the oil obtained conforms to the requirements given in this International Standard".

With regard to the occurrence of chemotypes also requiring published standards, as well as information related to enantiotaxonomy, the standardisation bodies world-wide have much catching up to do. However, those standardisation bodies regarded as most reliable in terms of the description of chemical components of essential oils, are the previous mentioned ISO and the "Association Française de Normalisation" (AFNOR: France) [33]. Other standardization bodies include the British [34], International, European and United States Pharmacopoeia. Although Australia does not have a pharmacopoeia *per se*, Australian standards are kept by Standards Australia [32].

Briefly, all standards are kept under a reference number, which includes a file number and the year it was last revised. Published standards must also conform to standards of measurement and presentation, which are also defined by ISO, however Australia's equivalent references are listed alongside the ISO references in Table 1. To view any of these standards they must first be purchased from the organisation. Although standards are not enforced by law, manufacturers of essential oils can provide guarantee of reproducibility and reliability if their product conforms to a widely accepted standard.

Table 1. References equivalent between the International Standards Organisation (ISO) and Australian Standards (AS) in relation to essential oils and preparation of standards for publication.

Reference to International Standard (ISO)		Australian Standard (AS)	
212	Essential oils—Sampling	4550	Essential oils—Sampling
11024	Essential oils—General guidance on chromatographic profiles	5025	Essential oils—General guidance on chromatographic profiles
11024-1	Part 1: Preparation of chromatographic profiles for presentation in standards	5025.1	Part 1: Preparation of chromatographic profiles for presentation in standards
11024-2	Part 2: Utilization of chromatographic profiles of samples of essential oils	5025.2	Part 2: Utilization of chromatographic profiles of samples of essential oils

Recent challenges in the authentication of Tea Tree Oil were met when it was revealed that enantiomeric excess of most of the chiral components of Tea Tree Oil were similar to Eucalyptus oil. Thus, manufacturers of Tea Tree Oil could fraudulently add components from Eucalyptus oil to bring levels into accordance with the ISO standard and enantio-cGC had limited capability to detect this. In this case enantio-cGC could be complemented with isotope ratio mass spectrometry. Because of varying kinetic and thermodynamic factors across species, during primary carbon dioxide fixation in photosynthesis, isotopic ratios of carbon vary between Melaleuca and Eucalyptus species [31]. Therefore, where isotopic ratios of carbon fail to meet known ratios for Melaleuca, the essential oil could be regarded as adulterated.

It is not only standardisation that governs essential oil manufacture and production. Standardisation merely ensures reproducibility and conformity, but where such standards are used in the formulation of products, legislation provides the framework. The primary objective of legislation is to govern the amount of biologically active material added to a product, at what concentration and the amount of information provided to the consumer (as ingredients), so that health and safety controls are in place. Although legislation may be regarded with distaste by many, with conflicting views and indecision across the board, such legislation does impact upon essential oil usage and therefore should be given due attention.

Generally the safety of an essential oil is difficult to predict from merely examining its chemical composition. This is because naturally occurring combinations rarely demonstrate the same biological activity as the individual separated components. However, the first indicator of an essential oil's biological activity utilises a constituent based approach [35]. Using the constituent based approach, the most common cited examples of potentially dangerous compounds are related to hepatotoxicity, phototoxicity and skin sensitisation. By far the vast majority of these contain aromatic moieties, such as a phenyl, phenol or methoxy phenol group. Phototoxicity is quite common in essential oils derived from *Citrus* species, which is attributed to the UV sensitising components belonging to the coumarin or furanocoumarin groups (*i.e.*, bergaptene (**31**)) [36]. Within the context of hepatotoxicity, phenylpropanoid derived compounds such as the methoxy phenols safrole (**14**) and methyl eugenol (**15**) are considered dangerous and carcinogenic [37]. Indeed, within the Australian context the occurrence of a rare chemotype of *Eremophila longifolia* F. Muell which yields an essential oil made up entirely of these two components, cast doubt over the safe use of this species for a long time, before other safe chemotypes were described [38].

Phenols such as carvacrol (**13**) and thymol are considered dangerous if consumed in higher quantities or at lower dosages but over long time periods, leading to cases of hepatotoxicity. Oils high in ketones, such as the abortifacient pulegone (**40**), or aldehydes should also be used cautiously, but of course there are many exceptions to the rule, such as the essential oil from *Mentha piperita* L, which contains high quantities of menthone (**39**) [37].

The main message from texts describing essential oil safety is that although most of them may be considered safe if used in an informed way, by merely being "natural" it does not mean that they can be utilised without consideration of dose, "… no substance is safe independent of consideration of dose" [35]. The first comprehensive text to examine essential oil safety was compiled in 1995 by Tisserand and Balac [36] and in it are described several cases of hospitalisation of adults or children who have consumed "safe" essential oils at unsafe dosages. This book also includes a multitude of anecdotes, such as the case of an 11 year old child carrying a bottle of *Cinnamomum zeylanicum* Blume essential oil in his pocket. When the bottle broke, the boy wore the same pants for another 48 h and sustained a severe burn around the pocket area. In another case, camphor (**17**) applied to a child's nostrils resulted in instant collapse, thereby demonstrating the effects of location of application [36].

A concept of considerable importance in essential oil usage is the occurrence of sensitisation, where allergic reactions occur in particular locations. A list of these allergens has been compiled in a recent study, where the frequency of sensitisation earns it a grouping classification. In order of descending priority, Group 1 allergens are more serious than Group 2 or 3. According to the first amendment of the "Detergent Regulation and Allergen Labelling", if any of these allergens are present in a detergent or cosmetic product it is mandatory for them to be declared, irrespective of the way they are added (such as part of an essential oil). Under the flavouring directive, various maximum and minimum levels are specified for biologically active substances [35].

Whilst the European Union controls legislation of therapeutic goods in Europe, in the Australian context it is the Therapeutic Goods Administration (TGA) that governs such approval [39]. When the therapeutic goods act of 1989 was enacted, a number of household natural products were "grandfathered" onto the "Australian Register of Therapeutic Goods" (ARTG), such as Eucalyptus or Tea Tree Oil. However, Australian natural products that had not become a household name, such as other Aboriginal medicines, could not be included on the ARTG without 75 years of documented use, with clear guidelines as to the exact method of preparation and safe use. Because the Aboriginal *materia medica* of Australia has primarily been transmitted orally, or documented in books with minimal details, the process of getting ARTG approval for such natural products is a herculean task. Thus, comprehensive documentation of such use should start in an altruistic sense, for the benefit of future generations.

1.4. Biosynthesis and Subjective Classification of Essential Oils

Essential oils are classed as secondary metabolites derived by various biosynthetic processes, starting with phosphoenolpyruvate, a glycolysis product of the glucose produced in photosynthesis. Secondary metabolites can be defined as components that are present in some species but not others, which is in stark contrast to the four primary metabolites that constitute the basic building blocks of life and are therefore found in all life forms, consisting of proteins, carbohydrates, nucleic acids and lipids. Of the secondary metabolites the most significant with regard to essential oils are terpenoids and shikimates, although polyketides also occur in essential oils and also rarely alkaloids [17].

Phosphoenolpyruvate, is the precursor to both shikimic acid and acetyl coenzyme-A. Therefore, phosphoenolpyruvate marks a major crossroad in the synthesis of secondary metabolites and some primary metabolites, such as lipids. At this crossroad, those metabolites along the acetyl coenzyme-A path include lipids, polyketides and terpenes, whilst those along the shikimic acid path include coumarins, flavonoids (colour compounds) and lignin. Of particular interest, lignin is a complex monomer of aromatic alcohols called monolignols, an integral part of the secondary cell wall of plants. The lipids also include free fatty acids that appear in fixed oils and sometimes in essential oils. Furthermore, some essential oils are decomposition products of lipids [17].

Where the path also splits at acetyl coenzyme-A, on one side mevalonic acid is produced, made from three molecules of acetyl coenzyme-A. This serves as the precursor to the isoprenes, which are the building block of terpenes. On the other side there is the carboxylation of acetyl coenzyme-A to give malonyl coenzyme-A. This combines with acetic acid then decarboxylates to give a β-ketoester. If this process is repeated a molecule will form with a carbonyl group on every alternating carbon, hence the name polyketide. Alternatively the ketone function can be reduced to an alcohol, which is then eliminated with the corresponding carbon hydrogenated and therefore giving a higher homologue, giving the start of a fatty acid (lipid). Since this lipid pathway corresponds to polyketide synthesis, there lies an explanation for why fatty acids are mostly even numbered [17].

The biosynthetic origin of shikimates (or the phenylpropanoids) is from the *shikimic acid* pathway. The biosynthesis of shikimic acid itself starts with the previously mentioned phosphoenolpyruvate and erythrose-4 phosphate, which is a precursor to carbohydrates. This means that the biosynthesis of shikimates diverges from the carbohydrate pathway [17].

As previously mentioned, terpenoid essential oils are biosynthesised via the *mevalonate pathway* involving the derivatisation and polymerisation of 5-membered isoprene alkenes from isoprenyldiphosphate (IPP) and dimethylallyldiphosphate (DMAPP). Isoprene units therefore combine to build terpenes, involving repeated carbon chains in multiples of five. There are currently more than 30,000 known terpenoids, isolated from plants, microorganisms and animals, many of which occur in essential oils. Within this array of known terpenoids are multiple chemical classes divided into groups of size and elemental/structural composition. The monoterpenes are known to comprise 25 different classes of terpenoids, 147 classes exist for sesquiterpenes, and diterpenes occur in 118 classes [21].

The term "terpene" was coined by Kekulé in 1880, because terpenes were first discovered in turpentine oil, as the main constituents [19]. A single terpene, called a monoterpene, is formed from two isoprene units, typically connected head to tail. Hemi- (1 isoprene), mono- (2 isoprenes), sesqui- (3 isoprenes) and di- (4 isoprenes) terpenes are the most common essential oil components, followed by the non-terpenoid group, phenylpropanoids. Although in the earlier literature the term "terpene" was often used to describe terpenoid compounds (including oxygenated terpenes), in modern terminology "terpene" only describes monoterpene hydrocarbons (two isoprenes with only carbon and hydrogen) [17].

With regard to the conventions for qualitatively or subjectively describing the character of whole essential oils, they can be described as terpenoid if they are predominantly composed of components of terpenic character. An essential oil is of monoterpene character if it is dominated by monoterpene hydrocarbons and of monoterpenol character if components are predominantly monoterpene alcohols. The same convention is used for sesquiterpene or sesquiterpenol rich essential oils [40]. This convention is not commonly used to denote other chemical classes, such as ketones or ethers, as the suffix may be confused with coumarins/lactones or ether/oxides, respectively.

Essential oils can also be described and to an extent classified according to their aroma. Oils dominated by monoterpene components could be described as "top note" because the aroma is sharp and perceived immediately upon application. Well-known top notes are citrus (*Citrus bergamia* Risso) and ginger (*Zingiber officinale* Roscoe). Such top note (or head note) oils contain small components that evaporate quickly. However, although not apparent at first, the middle notes (heart notes) and base notes strongly affect the aesthetics of the odour and mask the sharpness of the head notes. In addition, they are perceived immediately after the top notes dissipate. Well-known heart note oils are lavender (*Lavandula angustifolia* Mill.) and rose (*Rosa damascena* Mill.), which are described as having a more mellow or rounded smell. These oils are composed of components that are slightly larger than simple monoterpenes, such as monoterpenols and esters. Oils that are dominated by sesquiterpenes are generally regarded as producing base note odour. Base notes are typically regarded as rich, earthy and deep, with the most common example being musk (*Angelica archangelica* L.). Generally perfume designers seek to combine notes to achieve an optimisation and orchestration of ingredients that please the senses [41].

In returning to the concept of biosynthesis of essential oils, microbial endophytes can also play a significant role in their synthesis. Although not a great deal is yet known about exactly how involved endophytes can be in this process, it is clear that at the very least such endophytes can give the final chemical alteration before the essential oil becomes the end product. Such microbial reactions are referred to as biotransformation, which is a process that can be utilised *in vitro* to create less common components from precursor compounds that are available in abundance. It is already known that endophytes are responsible for the biosynthesis of an array of natural products, some of which provide defence against herbivores or other competing plants, such as the cytotoxic quinoline alkaloid camptothecin, biosynthesised *de novo* (from the start) by the endophytic fungus *Fusarium solani* and accumulated in the tissues of *Camptotheca acuminata* Decaisne [42].

In the case of bacterial endophytes it is becoming increasingly evident that they influence plant physiology, nutrient uptake and plant growth vigour via the biosynthesis of phytohormones, such as ethylene, indol-3-acetic acid and acetoin, 2,3-butanediol [43]. Such bacterial endophytes from *Lavandula angustifolia* Mill were isolated and demonstrated to secrete metabolites that were inhibitory to human pathogens. Due to the similarity of inhibition demonstrated between the endophytes and the essential oil it was hypothesised that the endophytes may have been involved in the synthesis of the essential oil. However, the metabolites from the endophytes were not chemically characterised in that study [43]. In a ground breaking study of the fungal endophytes of *Mentha piperita* L. it was demonstrated that the fungal organism itself was biosynthesising all of the essential oil components *de novo* in the rhizosphere of the plant, although the composition of components was regulated by a plant interaction [44].

In the Lamiaceae the essential oil secretory structures, called glandular trichomes, are well studied and regarded as the site of biosynthesis, accumulation and secretion of the essential oils [45]. Very little research has been dedicated to identifying possible endophytic bacterial communities that may be involved in this biosynthetic path and in what capacity.

Currently there are approximately 100 or so molecules in the flavour industry derived from enzymatic or microbial processes [46]. This is because ingredients derived from transformation of natural raw materials by microbial or enzymatic processes can be labelled as "natural" under European and United States legislation. Of course such flavouring compounds could be sourced more expensively from plants or crops. In the classical flavour industry, the predominant source of ingredients was from plants, but as synthetic chemistry evolved botanical sources became out-dated, particularly in larger industries where cost cutting takes precedence. This phase eventually faded when "natural" products started to attract significantly higher market prices of up to two orders of magnitude. At that time research on microbial biotransformations started to expand until the industry was dominated by flavours of microbial origin. The rise of biotransformation and *de novo* biosynthetic products of microbial origin in the industrial flavours realm owes its success to the biological or "natural" origin of products, which yields such flavours at a substantially lower cost than classical methods of production [46]. A comprehensive coverage of microbial transformations are provided by Noma and Asakawa [47] for biotransformation of monoterpenes and Asakawa and Noma [48] for sesquiterpenes.

Although such microbial processes are dominating the market where individual isolated flavour compounds are in demand, whole essential oils are still predominantly sourced from classical agriculture or wild harvest. However, in an attempt to lower the cost and challenges of traditional agriculture and reduce dependence on wild harvesting, researchers are undertaking to develop essential oil industries where large-scale plant tissue cultures are involved in an attempt to create suitable alternatives to essential oil production [49]. At such a large scale this technology is still in its infancy. Unknown variables such as the role of endophytes and their survival in these culture environments, needs to be taken into consideration. Additionally, if the essential oils biosynthesised by the respective "plant" are shown to be synthesised *de novo* from the bacterial endophytes, such plant culturing would be purely a waste of resources, unless synthetic precursors are made by the plant itself.

1.5. Essential Oils in Agriculture

When it comes to the discussion of essential oils in agriculture, the major focus of course is of growing essential oil yielding species to maximise yield whilst still producing an essential oil that is in accordance with published standards. There is minimal discussion on how such oils could be employed more generally in agriculture itself. This is perhaps due to the perceived costs involved in applications involving essential oils as pesticides or herbicides. However, one might argue that the decline in popular demand for synthetic flavours in human foods could mean that such products will find a use in agriculture as a substitute for much less popular pesticides and herbicides. Where agriculture aligns with organic methods of production, such compounds could be sourced from bacterial or fungal cultures.

The vast majority of essential oil research today is confined to the pharmacological or drug discovery laboratory. Perhaps this is inspired by the financial stimulus from large pharmaceutical firms that seek to buy out patents to biologically active substances, with the promise of huge financial gain to the patent holder. Alternatively it may be a consequence of a gravitation toward popular science, which seeks to celebrate highly specialised research and purely academic discoveries. However, areas of more practical significance have subsequently been neglected, as explained by Murray;

"Academic researchers tend to be more concerned about maintaining the rigour of science, judged by their peers in journals and conference proceedings, rather than research that contributes directly to the exploitation of essential oils and development of the industry" [50].

The advantage of using volatile "natural" herbicides in the field is that they don't persist and become part of the post-harvest product and in some cases the product may be marketed as "organic" and therefore attract a premium price. Although this may be an advantage, efforts need to be taken to prolong the residence time during the growth phase, by using surfactants or encapsulation technology. The possibility of using essential oils as herbicides has already been demonstrated, particularly where seed germination has been inhibited, but again there is minimal movement toward employing such technology in broad acre farms due to costs. Another advantage of the use of volatiles as a complement to herbicides is that they may facilitate pollination by acting as specific desirable insect attractants, while repelling others [51].

Essential oils may also serve agriculture by acting as antimicrobial compounds. Recently EU legislation has prohibited the use of antibiotics in the rearing of animals for slaughter. This is partly because of the growing resistance of pathogens [51] but primarily concerns the quality of end-of-use meat products. In this regard, natural products provide the best alternative, particularly essential oils as they serve other advantages. In conjunction with advantages gained in using essential oils as antimicrobials, they also act as appetite stimulants and as stimulants of saliva production, gastric and pancreatic juices [52].

The antimicrobial activity of essential oils can also be exploited to combat fungal or bacterial spoilage of shelf foods or as a treatment of fruits and vegetables at the harvest stages of production. A challenge in using this methodology is that the concentrations required to achieve such inhibition may add other flavours and fragrances unfamiliar to the product [51]. However, recent innovative techniques have utilised low pressures and warm air flow as a means to significantly enhance antimicrobial activity of essential oils applied to fruit and vegetable produce [53].

In returning to the optimization of agricultural production of essential oils, much attention has shifted toward genetic modification to enhance yield. Traditionally optimisation of yield focused on plant ontogeny, which means selection at a particular growth stage or for particular plant organ, cell or tissue structures. A suggested biotechnological innovation involves modification of plants to optimise for phenotypic characters that support essential oil production, such as trichomes or epidermal hairs *et cetera* [54]. Generally where genetic engineering of the plant aims to upregulate the biosynthetic pathway the limiting factor is a precursor compound, making things difficult. Whilst some have attempted to recreate the entire terpenoid biosynthetic pathway in recombinant microbes, thereby overcoming precursor supply by adding them manually after biosynthesis in another microbe [55], the next challenge lies in using a microbe that has resistance to the target compound. In practice, a microbe sourced from the plant itself is the answer [56].

In light of the greater interest and easier manipulation of microbes as biosynthetic factories, in particular the ease by which genes can be cloned from plants into microbes, there is not a great deal of interest in modifying plant organisms themselves to enhance essential oil production. Whilst this might seem threatening at first to the agricultural industry, the feasibility of employing a single microbe to biosynthesise a complex mixture of components identical to an essential oil is still out of reach.

Contrarily, using a set of microbes to individually biosynthesise the ingredients and therefore create the mixture, will probably not be met with approval by the vast majority who enjoy having essential oils of plant origin.

2. Pharmacology of Well-Known Essential Oil Components of World-Wide Origin

2.1. Bioactivity Testing

Essential oils may be characterized by either high or low biological activities, but this subjective description is of relative importance, exclusively within the context of essential oils. An inhibitory concentration of a "highly" antimicrobial essential oil may not necessarily be as low as an over the counter antibiotic. Therefore, the use of such terminology to describe the biological activities of essential oils must always be complemented with data values to show context.

The standardization of methods for biological testing is still evolving, which means that outcomes described in the literature may not necessarily be easily reproducible. In this regard, the more commonly described bioactivities in the literature are related to antimicrobial, antiviral, antinociceptive (analgaesic), anticancer, anti-inflammatory (antiphlogistic), digestive, semiochemical and free radical scavenging activity. The methodology behind these tests will be briefly described here [57,58].

Although there is a range of biological activities demonstrated in the literature, by far the most commonly cited activity for an essential oil is its antimicrobial activity. This is perhaps because antimicrobial susceptibility testing is a simple, inexpensive and straightforward technique. Alternatively, the growing popularity of essential oils as antimicrobial substances could be related to concerns about the growing resistance of pathogenic organisms against mainstream antibiotics.

In a similar way to the generalizations about essential oil toxicity to humans, a constituent-based approach can provide a simple guide in predicting the antimicrobial activity of an essential oil, but the actual activity cannot be known until a sample is tested and even then the results can be surprising. This simple generalization gives highest priority (highest activity) to essential oil components with high lipophilic character on the hydrocarbon skeleton, but high hydrophilic character on its functional group, with a ranking as follows: phenols > aldehydes > ketones > alcohols > esters > hydrocarbons [58].

To test antimicrobial activity two methods are primarily used, which are the disc diffusion and broth dilution assays. A variation of the disc diffusion assay is the well diffusion assay, but they both convey the same type of result, which is a more qualitative inhibition assay. A disc diffusion is performed on an agar plate, whilst a broth dilution is performed in a nutrient broth. Briefly, in microbiology agar plates are utilized in general as a medium for the growth of bacterial or fungal organisms (Figure 6A). The agar itself is a gelatinous substance discovered as early as 1650–60 AD in Japan, derived from the cell walls of algae. Its primary purpose in an agar plate is to act as a solidifying agent in a nutrient-enriched medium, to create a flat moist nutrient-rich surface for easy manipulation of growing microbial or fungal colonies. In a disc diffusion assay this surface is completely covered in a thin layer of microbes and paper discs are placed onto the surface. Each paper disc is inoculated with the treatment, in this case being a volume of the essential oil. If the essential oil is inhibitory to the chosen microbe, there will be a zone of clearance around the disc, representing the diffusion of the antimicrobial substance across and into the agar (Figure 6A) [59].

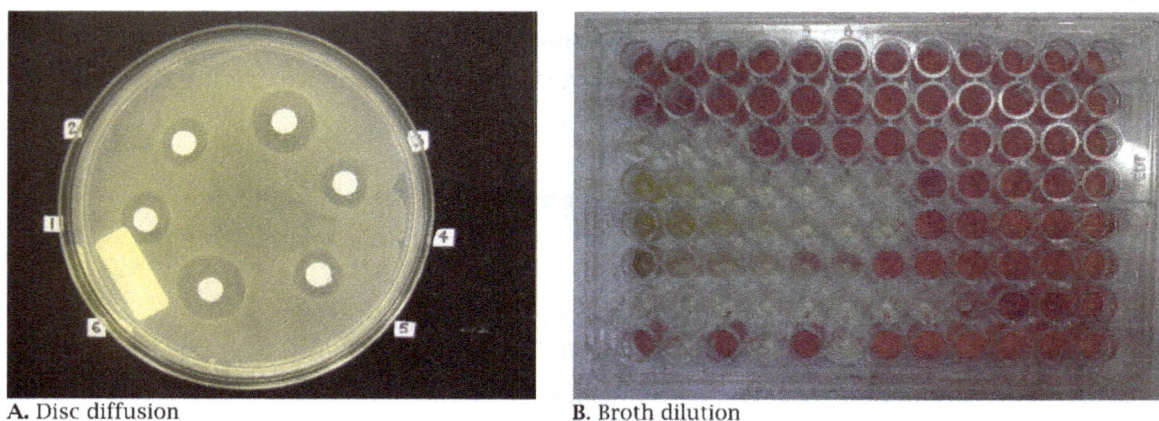

A. Disc diffusion   B. Broth dilution

Figure 6. Examples of common methods employed for antimicrobial testing. Subjective antimicrobial activity is demonstrated by a disc diffusion assay (**A**) and mean inhibitory concentrations (MIC) are calculated from a broth dilution assay performed in a 96-well microtitre plate, with colorimetric detection of organism growth, with red indicating organism growth (**B**).

The disc diffusion method is not regarded as an accurate method for the determination of a representative inhibitory concentration. This technique is merely employed as a primary screening tool before a more rigorous method is employed, which will be the broth dilution assay. The broth dilution assay is almost always performed in a 96-well microtitre plate (Figure 6B), where the antimicrobial substance is mixed with a nutrient rich broth and serially diluted to progressively lower concentrations (*i.e.*, 5%, 2.5%, 1.25% *et cetera*), usually moving from left to right. In essential oil assays it is common practice to add something to help with the formation of an emulsion, due to the usual phase separation of oil and water. In some cases a small amount of agar is used to make the broth "sloppy" [60], but more often a detergent is used, such as Tween 20 [61].

Although the bacterial growth can be visualized by turbidity in the broth, it is common practice to add a metabolisable salt to the medium (*i.e.*, p-iodonitrotetrazolium) before recording the results. The salt is converted to a coloured compound by live bacteria and the results can be read according to the appearance of colour (Figure 6B). The concentration at which no colour is observed is reported as the mean inhibitory concentration (MIC). At this concentration it is not clear if the organism was inhibited or destroyed. However, if the broth from this well is spread onto an agar plate and no organisms grow on the agar, then this concentration can also be reported as the mean bactericidal concentration (MBC). The lower the MIC or MBC, the higher the activity of the oil [61].

A problem often faced in broth dilution experiments using essential oils is that they are continuously evaporating. This has the effect of lowering the concentrations being measured and leads to fractionation of the oils due to some components having a higher vapor pressure than others. Perhaps more problematic is the diffusion of components from one well to another, which has the capacity to create the appearance of synergistic or antagonistic interactions, which again reduces the reproducibility of the results. Whilst using parafilm to lock all volatiles into the 96-well plate has the advantage of slowing the evaporation of essential oils during the experiment [62], the disadvantage is that synergistic or antagonistic interactions can be enhanced. Thus, a more reliable method uses a sterile plastic sticky sheet to cover all wells individually [63].

In both cases, often authors try to report their data with an average and standard error but due to variability and lack of reproducibility, this is misleading. This is because the starting concentration will influence the atmospheric concentration of essential oil vapours, which has the capacity to influence the observed MIC. Furthermore, MIC data is ordinal, which requires a different type of statistical analysis. Generally if a particular MIC value can be repeated at least 3 out of 4 times it is regarded as significant and can be reported.

Whilst the use of a sticky sterile plastic sheet to cover all wells on the 96-well microtitre plate could be considered the most reliable in terms of the reproducibility of results, there still is the challenge of translating results achieved in experimentation (viz, *in vitro*) to actual effects in an application on an animal or human being (viz, *in vivo*). This problem of *in vitro versus in vivo* is encountered by all biological assays, not just antimicrobial tests.

Aside from antimicrobial assays, other biological tests generally use animal models or cell or tissue culture assays where known pathways toward pathogenesis are inhibited. For example, in the inflammatory model a number of known inflammatory pathways can be induced in an assay (or kit) and the test substance (essential oil) can be introduced to attempt to inhibit transcription or activity of the relevant factors. Some well-known inflammatory pathways include the NO (nitric oxide), TNF-α or PGE2 just to name a few [57]. An interesting development for testing the TNF-α pathway is by transfecting the RAW264.7 murine monocytic macrophage cell line with pDNA that encodes for reporter proteins (proteins that can indicate the outcome of the experiment) [64]. In one particular experiment, activation of NF-κB (that leads to TNF-α production) can be visualized via the simultaneous transcription of the same phosphorescent luciferase protein that is better known from a species of firefly (*Photinus pyralis*). The cell "glows" when triggered by the inflammatory signal compound lipopolysaccharide. If the cell does not glow then inhibition of the TNF-α pathway may have occurred [65].

In some tests for antiviral activity mammalian cells are transfected with the virus and cell survival detected by any of a number of colourimetric or flow-cytometric methods [66]. In the most basic test for anticancer activity immortalized cancer cells are grown in a culture and examined for survival or suppression under conditions of the respective treatment [67]. For antioxidant or free radical scavenging activity, a coloured free radical can be reduced by the treatment and made invisible to the naked eye, with the results quantified by a spectrophotometer on the visible spectrum [68]. A variation of this is that the free radical becomes coloured when reduced, also measured with a spectrophotometer [69]. For digestive activity related to blockage of Ca^{2+} gated ionic channels, the caecum segment of an animal intestine can be placed into a buffer and attached to an assembly that measures spasmolytic activity related to contraction of the caecum. An excess of a particular ion in the buffer demonstrates a blockage of the channel [70].

In semiochemical activity a number of insect species are shown to be repelled by a particular treatment or essential oil. For example, the Australian essential oil from *Eremophila mitchellii* Benth proved to be a strong repellent against termites [56]. A common animal model used to demonstrate the topical analgaesic properties of an essential oil is the "hot plate test" where the response of a mouse or rat after placing its paw on a hot plate can be delayed by treatment with a particular essential oil [71].

Analgaesia is also commonly related to activity on the central nervous system but in this regard it is called antinociceptive activity. Antinociceptive activity can be more broadly associated with psychological effects where other diseased states include depression, anxiety or hysteria. Antinociceptive activity is not regarded as a biological property *per se* [57] but is nevertheless of importance in biological activity of essential oils. Methods commonly employed to measure such effects extensively use animal models where pain or intestinal writhing from intraperitoneal injections is induced. The treatment aims to reduce these observed effects [57].

A comprehensive coverage of the results of such biological activities in the context of essential oils is provided by Buchbauer [57] and Koroch *et al.* [58]. In this regard, due to the overwhelmingly high amount of data that is now available related to biological activities, there is currently a paradigm shift taking place in the biological research of essential oils. Increasing concern for the translation of *in vitro* results to *in vivo* use has prompted a number of experiments aimed at modifying the application of essential oils to reproduce *in vitro* results in the human or animal model.

Of particular consequence to essential oils are the problems of evaporation, solubility and absorption. There are now several studies that seek to enhance encapsulation of essential oils, which may include encapsulation of essential oils into various substrates, such as chitosan-coated liposomes, to slow evaporation and increase the antimicrobial activity [72]. Another approach involves the entrapment of essential oils into dissolved cyclodextrins, which can be used as a feed additive to disguise taste, as well as to slow evaporation of essential oils and therefore increase the shelf life of topical ointments and creams [73]. Penetration experiments where essential oils act as the carrier to another antimicrobial drug have also been undertaken with demonstrated success, which may be related to the interaction of the oils with liquid crystals of skin lipids [57]. It could be argued that such innovative techniques will comprise a large part of the future of biological research on essential oils.

2.2. Pharmacological Character of Internationally Recognized Essential Oils

During the late 1800's when essential oil chemistry was in its infancy, the chemical character of essential oils was communicated in broad generic categories, such as terpenoid or phenylpropanoid, or otherwise specialist chemical nomenclature was used. Terminology to express essential oil character was later expanded upon and improved following the proposition by Belaiche to assign chemical classes that could be used to predict the biological activity of the oils themselves [2]. This took place when the well-known French authors Pierre Franchomme and Daniel Penoel published *L'Aromathérapie éxactement* [2], which provided a framework for essential oil classification that continues to be used to this day. Franchomme and Penoel provided a list of these types according to structural functional, which are listed in Table 2 [2]. Some of these pharmacological groups do not occur as common components in essential oils, and these have been italicised to aid the reader. A pictorial representation of common groups is provided in Figure 7.

Table 2. Essential oil pharmacological types defined by functional group [2].

Essential Oil Types Described by Franchomme and Penoel		
Alcohols and Phenols (hydroxyl group)	Coumarins	*Ether-Oxides*
Methoxycoumarins	*Acetophenones*	*Hydroquinones*
Non-Terpenoid Hydrocarbons	Acids	Oxides
Terpenoid and Non-Terpenoid esters	Ketones;	Lactones
Phenol and Methyl-Ether	*Phthalides*	Aldehydes
Bi- or Multifunctional Compositions	Acids and Esters	Terpenes (hydrocarbons)
Nitrogen Compositions	*Sulfur Compounds*	-

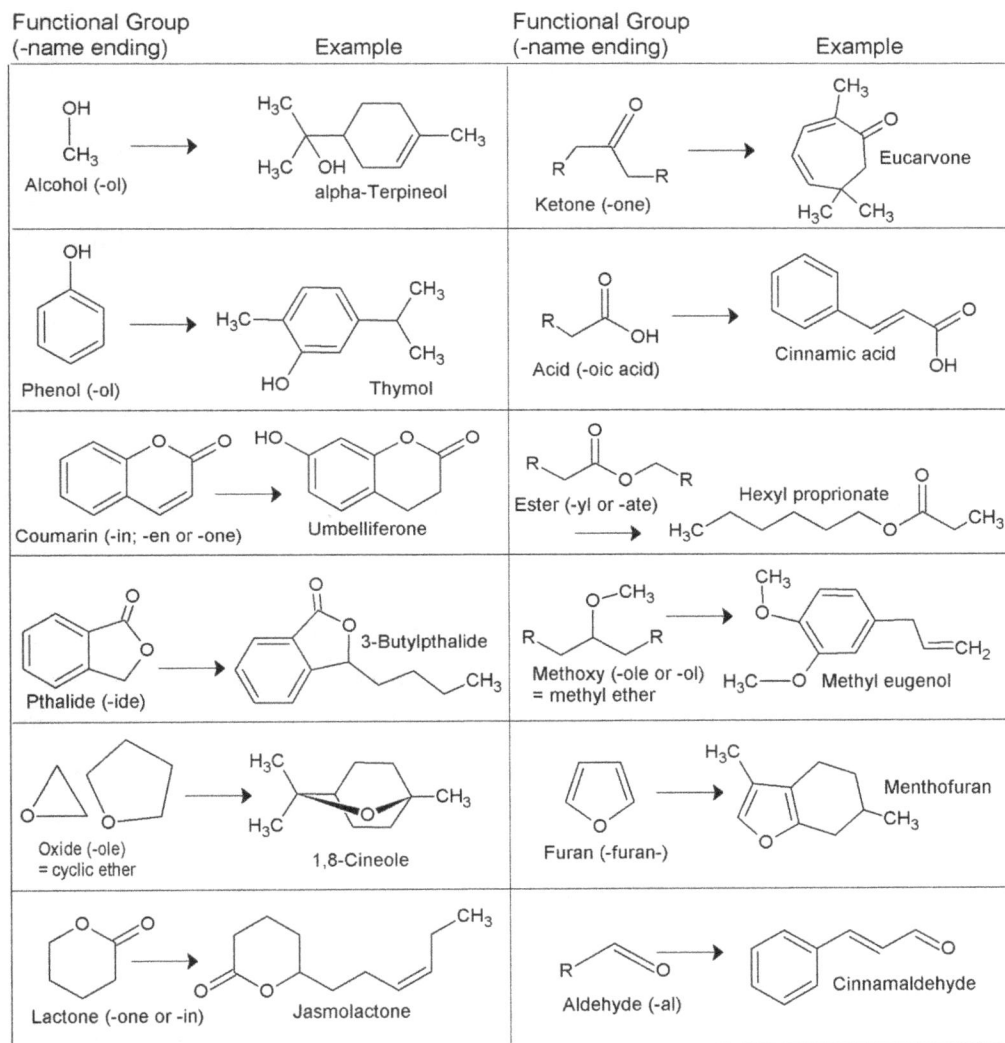

R = H or R = Remainder of molecule

Figure 7. Examples of common functional groups.

Pharmacologically significant essential oil functional groups are of both terpenoid and non-terpenoid origin unless specified. With regard to commercially significant essential oils, monoterpene hydrocarbons, such as α-pinene (**2**), limonene (**3**) or ρ-cymene (**4**) are major components in grapefruit (*Citrus paradise* Macfad*: Rutaceae*), pine (*Pinus pinaster* Aiton: *Pinaceae*), juniper berry (*Juniperus communis* L.: *Cupressaceae*) and frankincense (*Boswellia carteri* Birdw: *Burseraceae*), respectively.

The *S*-enantiomer of limonene (**3**) is best known from Citrus, whereas the *R*-enantiomer is known from Turpentine [26].

Such monoterpene dominated essential oils have pronounced antiviral effects and produce a drying effect on the skin [2]. The phenyl hydrocarbon ρ-cymene (**4**) has been demonstrated to have skin sensitising effects, so essential oils rich in ρ-cymene are therefore avoided in topical applications [40]. Phenols can also demonstrate such effects, together with hepatotoxicity if ingested in high concentrations or moderate concentrations over a long period of time [2].

With regard to sesquiterpene hydrocarbons, β-caryophyllene (**5**) is known from Black Pepper and chamazulene (**6**) from German chamomile (*Chamomilla recutita* (L.) Rauschert: Asteraceae) [40]. Azulene sesquiterpenes, such as chamazulene (**6**) or guaiazulene from *Callitris intratropica* R.T.Baker and H.G.Sm (*Cupressaceae*), are responsible for the blue colour of their respective essential oils, if present at sufficient concentrations [40,74].

Monoterpenols such as linalool (**7**) from Lavender (*Lavandula angustifolia* Mill: *Lamiaceae*), menthol (**8**) from peppermint (*Mentha piperita* L.: *Lamiaceae*) or α-terpineol (**9**) from Tea Tree (*Melaleuca alternifolia* Cheel: *Myrtaceae*) are known for slight analgaesic effects if applied topically [40]. Furthermore, linalool (**7**) has been associated with possible sedative effects as well [75]. Strictly within the context of essential oils, monoterpene alcohols are generally of high inhibitory character against bacterial pathogens [2]. In the context of adulteration of essential oils, the *R*-enantiomer of linalool (**7**) predominates in bergamot oil, so its adulteration is signalled by the presence of the *S*-enantiomer. The opposite is true for coriander oil, which has an excess of the *S*-enantiomer [26].

Well known sesquiterpenols are α-bisabolol (**10**), again from German chamomile, α-eudesmol (**11**) from cedarwood (*Juniperus virginiana* L.: *Cupressaceae*) or β-santalol (**12**) from Indian sandalwood (*Santalum album* L.: *Santalaceae*). The sesquiterpenol α-bisabolol and the sesquiterpene chamazulene (**6**), have been associated with anti-inflammatory activity, particularly α-bisabolol [40]. The (−)-enantiomer of α-bisabolol is a signature for chamomile oil (*Chamomilla recutita* (L.) Rauschert: *Asteraceae*) [26].

In vitro blockage of neuronal Ca^{2+} channels by α-eudesmol (**11**) has been linked to potential psychoactive effects [76]. This may be significant with regard to anecdotal accounts of Cedarwood essential oil associated with enhanced memory and creativity [77]. Psychoactive and physiological effects consistent with sedation were observed when Indian sandalwood (*S. album* L.) was transdermally absorbed, with activity attributed to α-santalol [78]. The (−)-enantiomer of β-santalol (**12**) is a signature compound for Sandalwood oil (*S. album* L.). Sandalwood oil has also been associated with potential inhibition of the *Herpes simplex* virus [40].

Other well-known examples from the chemical groups described by Franchomme and Penoel (Table 2) include the phenol carvacrol (**13**) from oregano (*Origanum vulgare* L.: *Lamiaceae*), which has been potentially implicated in liver damage, along with a host of other phenols and more specifically, phenylpropanoids, such as the aforementioned carvacrol (**13**) and the potentially hepatotoxic safrole (**14**) and methyl eugenol (**15**), known to be available in high yields from various Australian *Zieria* (*Rutaceae*) species [79] and an unusual and rare chemotype of *Eremophila longifolia* F.Muell (*Scrophulariaceae*) [22,62], as mentioned previously.

An example of a well-known component from the aldehyde class is citronellal (**16**) from *Eucalyptus citriodora* Hook (Myrtaceae), which is used as an insect repellent with mosquitocidal activities [80].

The (−)-enantiomer of citronellal is sourced in enantiopure form from balm oil (*Melissa officinalis* L.), making it a useful for establishing authenticity [26].

Camphor (**17**) is the best-known example of a ketone, which is the major component in essential oils from the Spanish chemotype (CT1) of rosemary. Although the use of camphor (**17**) is treated with suspicion, after studies demonstrated potential convulsant activity and liver/central nervous system damage, the camphor (**17**) and α-pinene (**2**) rich chemotype of rosemary continues to be used as a liniment for muscle aches and pains [40]. Camphor (**17**) is the ketone of the alcohol borneol (**18**), which occurs abundantly in a specific essential oil chemotype of the Australian species *E. longifolia* that demonstrated moderate antimicrobial activity [62].

Because acids are more soluble in water, they do not often become part of an essential oil. An example of this is the boswellic acids group from various Frankincense species (*Boswellia* spp.). Small amounts of boswellic acids do appear in the essential oils but the majority are dissolved in the hydrosol. Thus, Frankincense oils produced using supercritical CO_2 extraction have much higher concentrations [40].

Acids and alcohols are usually precursors to esters and when esters form into closed rings they become lactones [17]. Typically when alcohols are esterified by acetic acid, or another larger mass molecule, the resulting esters are named according to the parent alcohol, thus, linalool (**7**) becomes linalyl acetate (**19**), borneol (**18**) becomes bornyl acetate (**20**) and fenchol (**21**) becomes fenchyl acetate (**22**). The ketone of fenchol is fenchone (**23**).

Linalyl acetate (**19**) is another of the major components of Lavender oil and is additionally a significant component in the essential oil from Clary Sage (*Salvia sclarea* L.: *Lamiaceae*). Linalyl acetate was associated with the previous mentioned analgaesia, along with linalool (**7**) in Lavender oil [40]. An oil rich in fenchyl- (**22**) and bornyl acetate (**20**) is that from the Australian species *Eremophila bignoniiflora* F.Muell (*Scrophulariaceae*), and these components are probably responsible for the demonstrated moderate to high activity against the yeast *Candida albicans* and the bacteria *Staphylococcus epidermidis* [81]. Additionally, *E. bignoniiflora* was used in traditional medicinal applications by Australian Aboriginal people to treat headaches using volatile gases, and extracts from leaves as a laxative. Schnaubelt [37] lists ester rich essential oils as having antispasmodic activity and are also effective in the treatment of central nervous system and stress related ailments. Thus, ester-rich essential oils from *E. bignoniiflora* may have been significantly involved in traditional medicinal uses.

Another well-known ester, making up approximately 98%–99% of the whole essential oil of wintergreen (*Gaultheria procumbens* L: *Ericaceae*), is methyl salicylate (**24**), which is thought to have analgaesic, anti-inflammatory and counter-irritant effects comparable to aspirin. Methyl salicylate is often used as a positive control in various pharmacological assays for analgaesia and anti-inflammatory activity [82,83].

In essential oils it is extremely rare for an ether to occur in any other form than as a methoxy group or closed into a ring structure (cyclic ether). With regard to methoxy groups (methyl ethers) in essential oils, they typically occur as phenyl ethers, such as the phenylpropanoids eugenol (**25**), known from clove bud oil (*Syzygium aromaticum* L: *Myrtaceae*) in concentrations as high as 75%, and methyl chavicol (estragole) (**26**) from Comoro Island basil oil (*Ocimum basilicum* L: *Lamiaceae*) at approximately 85% of the whole. In general such ethers are commonly associated with psychotropic effects, which can lead to death if taken in high dosages. The best known examples of these are the phenylpropanoids, myristicin

(27) and elemicin (28), highly concentrated in essential oil produced from nutmeg seed (*Myristica fragrans* Houtt: *Myristaceae*) [40,84,85].

When ethers occur in closed cyclic structures, they are called oxides. Perhaps the best known of these is 1,8-cineole (29), also known as eucalyptol. In the Australian flora, Eucalyptus species are not the only ones exhibiting high yields of this compound, as 1,8-cineole (29) also occurs in high concentrations in the essential oil of many other endemic genera, including *Prostanthera* spp. (*Lamiaceae*) along with a host of other sesquiterpenols. Species such as *P. ovalifolia* R.Br., *P. rotundifolia* R.Br., *P. caerulea* R.Br., *P. lasianthos* Labill., *P. cineolifera* R.T.Baker and H.G.Sm and *P. incisa* R.Br. have high concentrations of 1,8-cineole (29) in their essential oils [45,86–89]. As 1,8-cineole (29) produces expectorant effects, it is not surprising that a large number of plants, rich in this compound, were used ethnomedicinally for decongestion by sufferers of coughs and colds.

Lactones are constituents of many essential oils [21]. Lactones are produced by an intramolecular esterification reaction, where an aliphatic alcohol joins with an acid and closes into the respective cyclic ester [40]. The lactones are named after, and derived from, lactic acid ($C_3H_6O_3$). They usually occur in five- or six-membered heterocyclic rings in saturated or unsaturated forms, bonded to a carbonyl group. Lactones occurring in five-member rings are referred to as γ-lactones; those occurring in six-member rings they are referred to as δ-lactones [21], and those occurring in four-member ring as β-lactones [21].

Constituent γ-lactones, some with a peach-like flavour, are found in fenugreek, coffee and sake; representatives of δ-lactones are found in cheese, fruits and dairy products, typically with a creamy-coconut or peach-like odour. Lactones with larger carbon rings are found in essential oils from ambrette seed or angelica root. Angelica also contains phthalides, which are a lactone of 2-hydroxymethyl benzoic acid. Phthalides are restricted to the Apiaceae family, typically in celery, lovage and angelica [21].

Lactones also have demonstrated possible expectorant effects, but it is not yet clear if topical applications should be contraindicated, as some studies have highlighted the potential for skin-sensitisation to occur. Despite this, lactones have also demonstrated high *in vitro* activity consistent with anti-inflammatory effects, meaning lactone rich essential oils may be suitably used for topical applications to treat inflammation [40]. In this context then, given the widespread chronic nature of gastric inflammatory disease, it may be worth investigating the potential for lactones to treat inflammation of the bowel or alimentary canal.

When an aromatic lactone is adjacent to a benzenoid moiety it becomes a coumarin. In its simplest structural form it is simply called coumarin (30), which is the principal odour compound responsible for the aroma of freshly cut hay [90]. Perhaps the best known coumarin is the furanocoumarin bergaptene (31), found in bergamot oil (*Citrus bergamia* Risso: *Rutaceae*) and also in Australian species, such as *Philotheca trachyphylla* (F.Muell) Paul G. Wilson (*Rutaceae*) (previously *Eriostemon*) [91]. Bergaptene (31) has a UV-sensitising effect, linking to melanin in the skin if applied topically in the sun [40]. This has the effect of intensifying the effects of the sun's rays. Interestingly, this bergaptene (31) rich oil is most likely a consequence of the method of extraction, being mechanical processing. Bergamot essential oils produced by hydrodistillation, as opposed to those expressed oils, are not likely to have significant quantities of bergaptene (31). This is because these types of coumarins are not easy to evaporate at the relatively lower temperatures employed in hydrodistillation, so they are generally present in trace quantities only, unless the oil is expressed or produced by a solvent extraction technique leading to an absolute.

Coumarins are also potentially associated with anticoagulant activity, but this has not yet been fully investigated. It is well known that the double coumarin "dicoumarol" (32) is related to the occurrence of internal bleeding when herbivores consume large amounts of Yellow Sweet Clover (*Melilotus officinalis* Lam: *Fabaceae*). If other coumarins could be associated with anticoagulant activity, this effect may be employed in the treatment of cardiovascular disease [40]. To the best of our knowledge dicoumarol (32) has not been observed in an essential oil. However, in one such study the biologically active coumarins isopsoralen (33), xanthyletine (34) and osthole (35) were discovered in trace quantities in the hydrosol and essential oil following hydrodistillation of leaves from *Geijera parviflora* Lindl. (*Rutaceae*) [38]. Such coumarins can be implicated in the traditional therapeutic uses of these plants where smoking modalities were used by the Australian Aboriginal peoples.

A wide selection of both furano- and pyranocoumarins are known to produce bioactive effects *in vitro* and should form the basis for further pharmacological investigations in Australian plants used medicinally by Aboriginal people. The furanocoumarin geiparvarin and the methoxycoumarin dehydrogeijerin (36) are potentially responsible for differences in sheep palatability of leaves from *G. parviflora* [92]. The chemically similar osthole (35) has already been demonstrated in essential oils from *G. parviflora* [38], which makes it plausible that dehydrogeijerin (36) may occur in essential oils as well. However, although it is not possible for geiparvarin to appear in a hydrodistilled essential oil, this coumarin would no doubt be abundantly present in a hexane derived extract or concrete.

Novel, as well as known, coumarins were identified in *P. trachyphylla* (as *Eriostemon* in that study) [91]. A host of others are known from Australian plants, but it is not yet known if any of these have been observed in a hydrodistilled essential oil, however they would certainly be abundantly present in aromatic preparations as absolutes.

Because *G. parviflora* was used in various medicinal, ceremonial and recreational activities by Aboriginal Australian people, the involvement of coumarins in these types of activities should be investigated. For example, the desmethyl congener of geiparvarin has already been demonstrated to have *in vitro* effects consistent with psychoactive sedation [93]. Due to the relatively large size of this molecule it is not clear if this effect could be related to psychoactivity achieved in traditional smoking activities [90].

With respect to using coumarins medicinally, as with other compounds, the relative and absolute stereochemistry strongly influences subjective and pharmacological effects [94], which obviously make synthesis more expensive. A corollary of this is the toxic *cis*-anethole (37) diastereomer, which is not produced in nature, but is rather a consequence of synthetically producing the medicinal compound, *trans*-anethole (37), which is sourced in stereo-pure forms from aniseed oil (*Pimpinella anisum* L.: *Apiaceae*) or fennel seed oil (*Foeniculum vulgare* Mill: *Apiaceae*) [90]. The effect of stereochemistry and chirality on the pharmacokinetics and pharmacodynamics of drugs is now fully appreciated [95] and researchers are beginning to seriously investigate the effects in natural product and synthetic medicine.

3. More on Essential Oils in the Australian Context

3.1. Historical Uses of the Australian Essential Oils

The first eucalyptus oil to enter British pharmacopoeia, under the name *Oleum Eucalypti*, was the cineole rich form known widely today [96]. The most common eucalyptus (*Myrtaceae*) species used to produce the well-known 1,8-cineole (**29**) rich essential oil are, among others, the "blue mallee" (*E. polybractea* R.T.Baker), the "broad leaf peppermint" (*E. dives* Schauer var C) and *E. leucoxylon* F.Muell, *E. sideroxylon* A.Cunn, *E. oleosa* F.Muell, *E. radiata* Sieber var australiana *et cetera* [90]. Currently much of the global production of Eucalyptus oil from Eucalypts is carried out in Portugal and Spain, which have established *E. globulus* Labill. as the favoured cultivar; however, in terms of gross production China leads the way with Chinese Eucalyptus oil, a by-product of camphor (**17**) production from *Cinnamomum camphora* L. (*Lauraceae*).

Another form of eucalyptus oil, rich in the ketone piperitone (**38**), is produced in Australia from commercial plantations of another chemotype of *E. dives*. In terms of volume, the major supplier of this oil is based in Swaziland in South Africa. This piperitone (**38**) rich oil can also be produced from *E. piperita* Sm (*Myrtaceae*), which was first distilled by First fleet Surgeon Denis Considen in 1788. The basis of Considen's attraction to this species was its odorous resemblance to *Mentha piperita* L. (*Lamiaceae*), hence the botanical name [97,98]. Although the subjective comparison is correct, *M. piperita* essential oils are dominated by menthol (**8**), menthone (**39**) and pulegone (**40**) [99], but contain no piperitone (**38**).

This account reflects the natural tendency of early Australian colonialists to focus on species that resembled in some manner those already described in British or European pharmacopoeia. This could be seen as an impediment in terms of accessing the rich existing tradition of Aboriginal Australian medicines. Having said this, there are many examples of colonial medicines, taken from the Australian environment, that were not in fact used by Aboriginal people. In many such cases the Aboriginal people were aware of medicines that more effectively targeted the respective ailments than those chosen by early colonial settlers.

The piperitone (**38**) rich oil produced by Considen from *E. piperita* apparently constitutes the first recorded distillation of an essential oil from an Australian *Eucalyptus* species. The resultant product was one of the first useful exports from the colony to Britain. Although, for over 100 years it was wrongly believed that the credit was owed to "Surgeon-General to the Colony", John White, it was later clarified by Maiden that this was wrong, when he examined a letter addressed to Sir Joseph Banks from Considen, who had posted him a sample of the oil for use in medicinal applications [90].

With regard to the cineole-rich oils from *Eucalyptus* species, apart from medicinal applications consistent with decongestion in coughs and colds, there are traditional reports of using the Tasmanian blue gum (*E. globulus*) in applications consistent with mosquitoe repellence. Accordingly, it was given the colloquial name "fever tree" or "fever prevention tree" as the leaves were hung in and around homes to prevent the occurrence of malaria and other mosquitoe borne diseases. Interestingly, the absence of malaria in New Caledonia at the time was attributed to the high occurrence of the cineole rich chemotype of *Melaleuca quinquenervia* (Myrtaceae) [90]. Furthermore, *Prostanthera cineolifera* (Lamiaceae), named for its higher yield of 1,8-cineole (**29**), was also used as an insect repellent by early colonialists [86].

As an aside, Maiden [96] reported that eucalyptus oil may be useful for treating malarial symptoms, albeit less effective than quinine but nevertheless, capable of providing relief. However, it was the insect repellent activity of 1,8-cineole (**29**) that formed the basis for its use in and around homes.

Some of Australia's best0known essential oils were listed by Maiden as early as 1889. These included species such as *Eucalyptus globulus, E. citriodora, Backhousia citriodora* (*Myrtaceae*) and *M. alternifolia* (as *M. linarifolia: Myrtaceae*). It is apparent from the text that Maiden [96] favoured a number of species, which for a range of reasons have not achieved significant commercial value. For example, a potentially hepatotoxic essential oil from *Zieria smithii* (*Rutaceae*) is listed for its flavour enhancing activity, although it has more recently been demonstrated to have potentially carcinogenic phenylpropanoids, safrole (**14**) and methyl eugenol (**15**), in its essential oil [79]. *Eucalyptus* species provided the greater part of Maiden's focus, but *Melaleuca* species were also given due attention.

Although the well-known *Melaleuca alternifolia* (as *M. linarifolia*) is only given brief mention by Maiden [96], *M. leucadendra* (as *M. leucadendron*) is probably the best described by him. Because of the similarity of this essential oil to that of the Malay cajeput (*M. cajuputi: Myrtaceae*), the tree has been given the vernacular name "cajeput tree". Maiden's description of the preferred method for preparing leaves for hydrodistillation somewhat resembles the modern post-harvest leaf wilting technique in contemporary use for commercial production of TTO from *M. alternifolia*. Prior to hydrodistillation this method involved collection of the leaves, storing in a sack and wilting for approximately a day, before maceration of the leaves and soaking in water for fermentation, taking yet another day [96].

Maiden recommended this method for essential oil extraction from any of the *Melaleuca* species. The latter part of this method, involving the fermentation of macerated leaves in water, is not commonly in use today, but it may be worth investigating the possibility that it can improve essential oil yield by facilitating the hydrolysis of glycosidically bonded essential oils. Leaf fermentation preparation may still be in common use for production of essential oils from *M. cajuputi* in Asia and India.

By the 1980s, only two Australian essential oils had achieved significant international market success. These were the cineole-rich *Eucalyptus* and *M. alternifolia* essential oils. With respect to addressing the international market place, Cribb and Cribb [96] hypothesise that the limiting factors include the availability of commercial scale plantations and the lack of anecdotal reference describing traditional use modalities of the oil. The adoption of essential oils, addressing specific uses, into the international market place, would be fuelled tremendously from such anecdotal accounts. Contemporary pharmacological investigations, informed by traditional medicinal uses by Australian Aboriginal people, also facilitate the emergence of this market niche. To a large extent, the research that follows, in this thesis, is an attempt to do exactly that. Having said this, it is primarily the availability of viable plantations that is the limiting factor. At the moment moves to involve Aboriginal communities in plantation and harvest of suitable cultivars will address both these factors as well as providing a source of much needed employment.

An object lesson supporting the above hypothesis is provided by the history of the Australian Sandalwood (*Santalum spicatum* R.Br: *Santalaceae*) industry. The sesquiterpenol dominated essential oil is known for medicinal activity, as demonstrated firstly by Aboriginal Australian people, who made use of concoctions for coughs and colds, or as a liniment (from the nuts) for muscle stiffness [90]. Smoke from the Eastern Australian species of Sandalwood (*Santalum lanceolatum* R.Br: *Santalaceae*) was used to drive away mosquitoes in New South Wales [98] or in aromatherapy applications for babies in the Northern Territory [100]. Subsequent pharmacological testing has revealed good antimicrobial activities against such microbial species as *Candida albicans* or *Staphylococcus aureus* [101].

Because the distillation of *S. spicatum* required destruction of the heartwood of the tree, the procurement of essential oils had a negative impact on wild populations. Due to the tree's growth habit as a parasite on other trees, regrowth was very slow, so sustainability of the industry was threatened by the disappearance of wild populations [90]. Thus, the limiting factor was primarily the lack of viable plantations. Recently the industry recovered with the formation of initiatives such as the Australian Sandalwood Network or WA Sandalwood Plantations, so the product is once again available for consumers.

Another scenario which demonstrates how the availability of plantations is a limiting factor in establishing a commercial niche is the recent emergence of the *Callitris intratropica* Benth. (*Cupressaceae*) essential oil industry. At one time, *C. intratropica* was botanically classified as *C. columellaris* F.Muell, together with *C. glaucophylla* Joy Thomps. & L.A.S.Johnson [102]. Because these species were known under the one name, many ethnobotanic records describing traditional Australian Aboriginal medicinal uses of *C. columellaris* included those for *C. intratropica*. Medicinal uses included topical applications using hydrophilic or animal fat extracts, as well as smoke fumigation treatments for various ailments [90]. Barr [100] provides clearer details of traditional medicinal use, specifically of *C. intratropica*, which apart from topical applications for effects consistent with antimicrobial activity, also involved the use of a concoction of the inner bark, and applied topically for relief from abdominal pains and cramps, perhaps achieved via transdermal absorption of the relevant medicinal principles.

Early 19th century colonial settlers were also aware of medicinal uses of *Callitris* species and the needles were steamed and inhaled for chills and pains. Maiden declared that;

> "there is nothing more delightful in the approach, on a winter evening, to a township where Cypress pine is used as a fuel. Its delicious perfume is borne on the air for miles, and is often the first intimation that the weary traveller experiences that he is approaching a human habitation, and that his long journey is drawing to a close" [103].

After it was observed that houses built using *Callitris* timbers, had resisted termite infestation over several decades, an attempt was made to develop a timber industry for international export [96]. The formation of a plantation of *C. endlicheri* F.M.Bailey (*Cupressaceae*) was the plan. Much to the disappointment of Maiden, this proved not to be economically viable, due to the high costs involved in transporting timber from the proposed plantations in New South Wales, to the Northern Territory, where ships would transport further into south-east Asia and beyond.

It wasn't until the 1960's, long after Maiden had passed on, that a timber plantation of *C. intratropica* had been established in the Northern Territory. After the disastrous occurrence of Cyclone Tracey in Darwin during 1974, it was observed that structures built with *Callitris* timbers were not as resilient. The timbers were therefore not considered strong enough to be used in infrastructure and the plantations were abandoned. In 1995 the blue essential oil, from the timber of *C. intratropica*, was first discovered and an essential oil industry was quickly established, supported by pharmacological studies demonstrating antibacterial and possible anti-inflammatory activities and supplied by the existing plantations [104].

With regard to the pioneering efforts of earlier Australians to examine essential oil yielding flora of Australia, another of the names frequently highlighted in the literature is Arthur de Ramon Penfold (1890–1980), a phytochemist with a special interest in the Australian essential oils [105,106]. Penfold achieved an international reputation for his work in chemistry when he started to characterise unusual essential oil components, unique to the Australian flora. Penfold was also the one to elucidate the structure of piperitone (**38**) and demonstrated how menthol (**8**) and thymol could be synthesised from it [105,106]. Penfold substantially contributed to Ernest Guenther's six-volume work, *The Essential Oils* [107].

In 1915 Penfold became a research chemist and assistant work manager to the eucalyptus oil distillers, Gillard Gordon Ltd. [105,106]. *Eucalyptus* species have been a part of European pharmacopoeia for well over 100 years. In relative terms, *Melaleuca alternifolia* Cheel. essential oil has only recently acquired an international niche. It was Penfold who demonstrated significant antibacterial activities of *M. alternifolia* essential oil in a series of papers published in the 1920s and 1930s [108].

Prior to this, antimicrobial activities of *M. alternifolia* were familiar to the Bundjalung people from north-eastern New South Wales, who didn't use essential oils *per se*, in medicinal applications, but rather inhaled the vapours from crushed leaves for coughs and colds [108]. Additionally, a topical compress was used for skin infections and so forth, or a concoction to achieve a similar effect, or as a gargle for sore throats. Interestingly, according to the oral history of the Aboriginal people, lakes that received large amounts of fallen leaf matter from riparian *M. alternifolia*, developed medicinal properties [108].

Today the essential oil from *M. alternifolia* is officially known as Tea Tree Oil (TTO); however, a large number of *Melaleucas* and *Leptospermum* species are also called Tea Trees, which can confuse the nomenclature. The description Tea Tree in fact arises from the tannins which can cause a brownish colour in lakes and water courses; hence the name Tea Tree Lake on the north coast of New South Wales [56].

After the antimicrobial properties of TTO were promulgated by Penfold, its first significant documented use was in the mid-1920s when it was applied as an antiseptic in surgery and dentistry. Following this, during World War II, it was used as a surface disinfectant in munitions factories, to curb infections to the workers following skin injuries. Additionally, the WWII soldiers were also issued TTO in their first aid kids. Following the advent of antibiotics, TTO was eventually forgotten and by the 1960s the oil became a rare commodity. In 1976 Eric White, convinced of a resurgence of interest in TTO in modern society, established a plantation near Coraki in northern New South Wales, after a crown lease was granted on a Thursday. The company therefore became known as the Thursday Plantation. Today TTO is used in a selection of soaps, shampoos and disinfectant products. The oil is sourced from commercial scale plantations in New South Wales, Queensland and Western Australia [108].

Another essential oil worth mentioning here, because of its long history in Australia, is from the Western Australian species, *Boronia megastigma* Bartl. (Rutaceae) [98]. Although it is better known commercially for its fragrant flowers, an essential oil industry was trialled in the early 1900s and declined, as it was wild harvested at that time and faced similar problems to the industry centred on *S. spicatum*. In the recent 20 years plantations have been established in Tasmania, which have had varying success, but essential oils from *B. megastigma*, rich in β-ionone and dodecyl acetate, as well as their absolutes produced from the flowers for food flavouring, are now available under the name "Brown Boronia" [109].

3.2. Today's Essential Oil Industry

Due to a recent surge in interest in healthy living, complementary therapies and the non-synthetic health product sector, coupled with concerns raised about the growing resistance of pathogens to conventional antibiotics, the market for essential oils and suitably formulated creams and lotions has initiated surprising new developments. Although a significant number of antibiotic compounds have been isolated from Australian plants, the greater focus has been on essential oils [56].

Essential oils today are either sourced from plantations or wild harvested from populations that have grown to apparent unnatural densities because of a change in fire regime. A good example of this would be *Eremophila mitchellii* Benth. (Scrophulariaceae). In the early 20th century, when the *S. spicatum* (Santalaceae) populations started to decline from over harvesting, the fragrant eremophilane rich essential oils from heartwood of *E. mitchellii* were temporarily used as an alternative, but the subjective and chemical differences between the two essential oils prevented this change from taking effect [56].

Although some vernacular names include "Buddah Wood", "False Sandalwood" and "Native Sandalwood", the other name "Bastard Sandalwood" is perhaps the most cognisant of previous attempts to use it as a *S. spicatum* alternative. The *E. mitchellii* essential oil industry today owes its viability to the overgrowth of populations in the South Australian Flinders Ranges. Although the timber and essential oils are known for anti-termite activity [110], the essential oil is marketed as an aromatherapy complement to meditation [56].

Another plant known for termite resistant timbers is the Tasmanian native *Kunzea ambigua* Sm. (Myrtaceae). In a similar way to the discovery of termite resistance in *E. mitchellii*, Tasmanian farmers observed that fence posts produced from *K. ambigua* remained intact when others did not. Most famously, in 1993 John Hood produced an essential oil from the species when he noted that his north boundary fence, constructed from *K. ambigua* wood, remained intact after 35 years. Interestingly, the vernacular name "Tick Bush" derives from observations by early colonialists of the preference that wild animals had for sleeping under the bush, eventually demonstrated to reflect protective benefits from tick infestation [56].

The essential oils produced from *K. ambigua* leaves show a high degree of variation from predominantly monoterpenoid to predominantly sesquiterpenoid compositions, characterised by components such as α-pinene (2), 1,8-cineole (29), spathulenol, bicyclogermacrene, globulol (52), ledol (53) and viridiflorol [111]. Some of these oils are unusual because of the higher abundance of sesquiterpenes. In terms of the biological activity, the oil is best known anecdotally for its anti-inflammatory activity, which has led to its involvement in topical applications for the treatment of insect bites, itching and irritation [56].

Like *E. mitchellii*, commercial quantities of *Kunzea* oils, known as "Ducane Kunzea", are also produced from wild harvest. However, unlike the previous mentioned *S. spicatum* and *E. mitchellii*, essential oils are produced from the leaves, not the heartwood. In ecological terms, wild populations have been grazed by wild animals for millennia, so leaf harvesting is not a new occurrence and is therefore sustainable. Thus, commercial growth of the Kunzea industry is not expected to be restricted by a decline in species density or a threat to the density of wild populations, but rather to the rejuvenation rate of the leaves [56].

Apart from the previously mentioned *Eucalyptus* spp., as well as *E. mitchellii, S. spicatum* and *M. alternifolia*, other examples of commercial scale essential oil plantations in full production today include; anise myrtle (*Syzygium anisatum* Craven and Biffin: Myrtaceae), fragonia (*Agonis fragrans* J.R.Wheeler and N.G.Marchant: *Myrtaceae*), lemon myrtle (*Backhousia citriodora* F.Muell: *Myrtaceae*), lemon tea tree (*Leptospermum petersonii* F.M.Bailey: *Myrtaceae*), bracelet honey myrtle (*Melaleuca armillaris* Sm. *Myrtaceae*) [112,113], nerolina (*Melaleuca quinquenervia* S.T.Blake CT Nerolina: *Myrtaceae*), niaoulina (*M. quinquenervia* S.T.Blake CT Niaouli: *Myrtaceae*) and rosalina or lavender tea tree (*Melaleuca ericifolia* Sm: *Myrtaceae*) [111]. This latter essential oil, rosalina, is produced from both wild harvest and commercial plantations [56].

3.3. Recent Innovation in Australian Essential Oils

As previously mentioned, possibly the most important factor, with regard to the establishment of viable industries focused on select essential oils and natural products, is the establishment of commercial plantations. As a necessary prelude, it is important to perform chemical character studies and pharmacological activities, to complement ethnobotanical records of traditional use by Australian Aboriginal people. Chemogeographic studies demonstrate the variation of naturally occurring chemotypes, and in concert with respective pharmacological activities, they aid in the identification and promotion of significant cultivar chemotypes. Botanical and chemotaxonomic investigations are also significant with regard to identifying these cultivar chemotypes.

With regard to recent research focused on ethnopharmacological investigations of Australian plants, a significant number of novel chemical structures have been elucidated since the 1960s. A significant proportion of these novel structures were extracted from *Eremophila* and *Myoporum* (*Scrophylariaceae*) species. With regard to essential oils, for several decades all wild specimens of *Eremophila longifolia* were wrongly considered within the context of an essential oil hydrodistilled from a rare chemotype occurring in north-west Western Australia [22], which yielded 5.5% w/w wet leaves, of an essential oil comprised almost entirely of the potentially hepatotoxic phenylpropanoids safrole (**14**) and methyl eugenol (**15**). This served to put a dampener on medicinal research of the other essential oils from *E. longifolia*.

In conjunction with reports of another chemotype in the Northern Territory, identified by Barr [100], with a monoterpenoid character predominantly made up of α-pinene (**2**) and limonene (**3**), it is surprising that the initiatives to implement a commercial crop of *E. longifolia* for essential oil production, are to an extent still compromised by claims that the species in general yields the potentially harmful safrole (**14**)/methyl eugenol (**15**) essential oil. Of course as previously mentioned plants yielding this oil have a relatively restricted geographic range (Murchison area, Western Australia). Clearly, misconceptions regarding the oil of *E. longifolia* should be brought up to date.

Several years after Barr [100] characterised the limonene (**3**) chemotype of *E. longifolia*, Smith *et al.* [114] identified three other essential oil chemotypes, occurring in New South Wales. One of these chemotypes produces a particularly high yield of a monoterpene ketone dominated essential oil (isomenthone (**41**)/menthone (**39**); CT.A) that shows considerable promise on a commercial level, given the high oil yield and localised abundance (Table 3). This isomenthone (**41**)/menthone (**39**) rich oil (CT.A) is hydrodistilled to produce a yield ranging from 3% to 8% w/w of fresh leaves. The other two chemotypes firstly included CT.B, made up predominantly of karahanaenone (**42**), and secondly CT.C, made up predominantly of monoterpenes, such as α-pinene (**2**), limonene (**3**), α-terpinolene and significant amounts of borneol (**18**) [114].

With regard to the identification and delineation of essential oil chemotypes of *Eremophila longifolia*, it is now clear that the first such chemotype identified in 1971 by Della and Jefferies, with an essential oil made up predominantly of the potentially hepatotoxic carcinogenic phenyopropanoids safrole (**14**) and methyl eugenol (**15**), is confined to a small geographic region in Australia's far west, in central-west Western Australia. This is important since although *E. longifolia* has a widespread distribution throughout the Australian landmass, perceptions still prevail that this single chemotype reflects the constituents of all individuals of the species. This is simply not true. Further clarification reveals that this chemotype is an unusual biotype with diploid cytology [115].

In all, a total of three diploid populations of *E. longifolia* were identified in Australia, the other two being geographically clustered in western New South Wales and producing terpenoid based essential oils via the *mevalonate pathway*. These ketone rich chemotypes, as is the case for the phenylpropanoid type, produce significantly high yields of essential oils, making them potentially suitable for commercial development. The first of these types is the isomenthone (**41**)/menthone (**39**) type (CT.A), described above.

The second is a recently discovered high yielding karahanaenone (**42**) type (CT.B), yielding at a range of 1%–5% for diploid specimens [115]. The previously known tetraploid karyotype yields at 0.3%–0.7% [113]. Both these high yielding diploid types are good candidates as cultivars for commercial plantations. Should such plantations be established and developed this would make a significant contribution to Australia's essential oil industry. Essential oils and or/extracts from the high yielding CT.A isomenthone (**41**)/menthone (**39**) type could be used to make ointments and lotions suitable for topical, antifungal, aromatherapeutic and cosmeceutical/aesthetic applications (Table 3). At present it is unclear how CT.B could be utilised, but karahanaenone (**42**) is already in demand as a feedstock in the flavour and fragrance industry and may also be useful as a chemical scaffold for further drug development.

In addition to the five essential oil chemotypes of *E. longifolia* described above, another four have also been discovered [115]. One of these new essential oils, with dominant components of bornyl- (**20**) and fenchyl-acetate (**22**), is similar in composition to the antimicrobial essential oil produced from *Eremophila bignoniiflora* [81]. Traditional ethnomedicinal use of *E. bignoniiflora* by Australian Aboriginal people involved applications consistent with antispasmodic activity and headache therapy. Because essential oils rich in esters are often associated with antispasmodic and nervous calming activity, the essential oils from *E. bignoniiflora* may have contributed to this effect. The same essential oil produced from the new chemotype of *E. longifolia*, in significantly higher yields, could be marketed for treatment of headaches, nervous tension or gastrointestinal disorders.

Interestingly, another of the newly characterised chemotypes of *E. longifolia* produces an essential oil comprised predominantly of fenchone (**23**) and camphor (2-bornanone) (**17**), which are analogues of the previous mentioned fenchyl- (**22**) and bornyl acetate (**20**), respectively, after removal of the acetate groups [115]. In the case of fenchone (**23**) and camphor (**17**), a ketone is in the place of the ester; however, in the case of the other known chemotype, dominated by fenchol (**21**) and borneol (**18**), an alcohol functional group is in the place of the ester. Clearly, the oils produced by these three chemotypes are of very similar biosynthetic provenance. The structural resemblances are depicted in the following image.

The essential oils dominated by the alcohols, fenchol (**21**) and borneol (**18**), demonstrated high antimicrobial activity against the yeast *C. albicans*, bacterial species, such as *Staphylococcus aureus*, *S. epidermidis*, and the human pathogenic fungal species *Trichophyton rubrum*, *T. mentagrophytes* and *T. interdigitalis* [62]. Similar activity was demonstrated by the fenchyl- (**22**) and bornyl acetate (**20**) oils against *C. albicans* and *S. epidermidis* [81]. The fenchone (**23**) rich essential oil is yet to be tested for antimicrobial activity.

Another of the new essential oil chemotypes of *E. longifolia* is rich in α-pinene (**2**), sabinene, limonene (**3**) and α-terpinolene [115]. At first this essential oil appeared to be consistent with an earlier type reported from an individual *E. longifolia* collected from Alice Springs, in the Northern Territory. However, the unusually high concentration of α-terpinolene in the former, makes this new essential oil unique. To date, the last of the new chemotypes identified by Sadgrove and Jones [115] is dominated by ρ-cymen-8-ol, along with a host of other unidentified compounds.

Currently then, at least nine chemotypes of *E. longifolia* have been characterised but preliminary results suggest that others wait to be confirmed. All essential oil chemotypes occurring outside the small region of the safrole (**14**)/methyl eugenol (**15**) diploid type, the isomenthone (**41**)/menthone (**39**) diploid type and the karahanaenone (**42**) diploid type show tetraploid cytology. The karahanaenone (**42**) and isomenthone (**41**)/menthone (**39**) types also exist as tetraploid forms but produce relatively low essential oil yields by comparison with the diploid varieties. Such tetraploid types appear as randomly emerging individuals in isolated patches throughout the range of *E. longifolia*, probably emerging as a result of sexual reproduction and assortment of recessive allelic traits related to biosynthesis [115].

Considered within the context of proposals to cultivate commercial scale crops of *E. longifolia* species, quality control of plantations of tetraploid chemotypes may involve the elimination of karahanaenone (**42**) and isomenthone (**41**)/menthone (**39**) chemotypes emerging in plantations from sexual reproduction. However, in any case, this is not expected to occur with any great frequency since this species has a preference for reproduction by root suckers.

With regard to the emergence of unintended chemotypes in populations of known chemotypes, one may consider the emergence of the safrole (**14**)/methyl eugenol (**15**) type a potential risk in a commercial scale plantation, particularly since safrole (**14**) and methyl eugenol (**15**) have been red flagged as potential hepatotoxic carcinogens. Our research indicates that the risk of this occurring is vanishingly small. Thus far the safrole (**14**)/methyl eugenol (**15**) type has not been demonstrated to occur in the tetraploid form. However, even if this did occur, the parent chemotype would produce essential oils via the *shikimic acid pathway*, because emergent chemotypes may not contradict the biosynthetic origins of the parent chemotype. However, if a tetraploid chemotype is discovered with both phenylpropanoid components, such as safrole (**14**) or methyl eugenol (**15**), and terpenoid components in the essential oil, then this genealogy would be diligently avoided during the development of a cultivar chemotype.

With regard to the role of volatiles in the medicinal efficacy of smoke or steam fumigation rituals, using *E. longifolia*, both partially pyrolysed essential oils and the more hydrophilic component "(−)-genifuranal" (**43**) may be involved [12]. Most of the essential oil components are present in the leaf tissue before heating, but are accompanied by other derived artefacts in the steamy smoke, produced when the leaves are placed on hot embers for use in medicinal applications consistent with antibacterial or antifungal applications, as well as lactagogue activity. The smoking procedure was also used to prepare surgical tools, no doubt for sterilization but conceptualised as a type of exorcism ritual. The essential oils and artefacts were also accompanied by pyrolysed derivatives including radical essential oil fragments and lignin decomposition products such as phenolic or benzoid constituents; together producing significantly enhanced antimicrobial activity [10].

The heat derivative "genifuranal" (**43**) itself exhibited significant antimicrobial activity, with a mean inhibitory concentration as low as 100 µg/mL against some species [12]. In traditional Aboriginal medicinal fumigation rituals using *E. longifolia*, "genifuranal" (**43**) and partially pyrolysed essential oils are delivered in warm air to the patient. Although the transdermal absorption of components such as "genifuranal" (**43**) is expected to produce significant biological activity, the first application with warm air is itself expected to have enhanced activity, relative to cooler applications [10].

It is proposed that genifuranal (**43**) derives from the cleavage of geniposidic acid (**44**) [12] at the glycosidic bond, to produce glucose and a hemiacetal. The hemiacetal transforms into the product genifuranal, (**43**) a stable furan aldehyde. Geniposidic acid is one of two non-volatile cardioactive glycosides that occur in *E. longifolia*, the other being verbascoside. The occurrence of volatile heat derivatives from verbascoside has not yet been demonstrated. However, in light of the occurrence of genifuranal (**43**), with demonstrated biological activity, there is potential for the development of a therapeutic lotion or use as a chemical scaffold for further drug development.

3.4. Ethnopharmacology of Aromatic Medicinal Plants Used Traditionally by Aboriginal Australians

The medicinal potential of the essential oil of *E. bignoniiflora* has already been summarised above. In other studies a dichloromethane extract of the leaves of this plant demonstrated calcium channel blockage that may be consistent with a number of traditional medicinal uses. Because the calcium channel subtype was not clarified in this earlier study, the results have implications for both therapeutic activity related to headaches and spasmolysis of the intestine [81].

According to ethnobotanical accounts, therapeutic activity from the use of *E. bignoniiflora* is expected to vary from use of the leaves to the fruits. The leaves were reportedly used as a laxative and the fruits as a purgative. In the modern context a laxative often means something that restores elimination activity to the colon, however in the historic language used by colonial ethnobotanists it may have referred to merely correcting digestive complaints or treating/reversing diarrhoea. Activity as a spasmolytic in this context could therefore be related to the abundance of the two fragrant esters bornyl- (**20**) and fenchyl-acetate (**22**). Interestingly this reflects advice given to us by an elder of the Kamillaroi tribe, who reported seeing his grandfather forage for the most aromatic specimen when employed in therapeutic use [116]. However, the biological activity of the fruit as a purgative requires a more comprehensive investigation.

Although essential oils from the fruit of *Pittosporum undulatum* Vent. (Pittosporaceae) have already been partly characterised in an earlier study completed in 1905, a recent characterisations enhanced and extended this earlier study [117]. In this study Sadgrove and Jones [117] tentatively identified the optically inactive compound referred to in the earlier study from 1905, conducted over a hundred years ago. It was believed to be bicyclogermacrene.

Unlike *P. undulatum*, *Pittosporum angustifolium* Lodd. was involved in a significant number of traditional medicinal applications [117]. The most common of these to be recorded in the literature is related to the treatment of coughs and colds, for lactagogue activity or in the treatment of eczema.

More recently, a number of anecdotal reports have surfaced related to *ia* cancer inhibition, autoimmune conditions in the intestines and antimicrobial activity. Previous studies have supported potential anticancer activity [118–120], as well as possible antiviral activity (using the older name *P. phylliraeoides*), particularly the Ross River Fever virus [121].

Jones and Sadgrove [117] examined the chemical character of volatiles from *P. angustifolium*, demonstrating a degree of variation. Compounds with structural similarities to previously described chemosemiotic compounds identified in mother-infant communications, were also noted, including acetic acid decyl ester and 1-dodecanol. These compounds may be involved in the traditional application as a lactagogue, particularly because the modality of usage involved heating a compress of leaves to produce such volatiles, which were then used to fumigate the breasts of the nursing woman.

In another study examining the essential oils from *Geijera parviflora* and *G. salicifolia*, the chemical character was consistent with previous identified chemotypes; however some variation was noted and new potential chemotypes were identified [38]. One of these, from a specimen of *G. parviflora*, yielded oil comprised of a larger abundance of bicyclogermacrene and *trans*-caryophyllene (and unknown B), which may be the first known sesquiterpene dominated essential oil from *Geijera* species. As previously mentioned, following hydrodistillation performed on this specimen a dichloromethane partition of the hydrosol produced a residue that was rich in pyranocoumarin xanthyletine (**34**), furanocoumarin isopsoralen (**33**) and the methoxy coumarin osthole (**35**). This hydrosol partition was attempted using other chemotypes but they did not yield these same coumarins [38].

This comprehensive study of the essential oils from species of *Geijera* also presented antimicrobial and free radical scavenging activity of these essential oils. The most active of these oils was the green oil from *G. parviflora*, made up predominantly of green compounds pregeijerene (not to be confused with the methoxy coumarin) (**45**)/geijerene (**46**) and linalool (**7**). Previous studies on these components indicate that this green essential oil may have applications as an insect repellent (particularly mosquitoes) as well as a topical analgaesic agent. Another interesting chemotype from this species is that dominated by the acetophenone xanthoxylin (**47**). Although this compound is known to possess cytotoxic and fungicidal activity, it is not clear if this was ever utilised in the *materia medica* of Aboriginal Australians [38].

3.5. Phytochemical and Chemotaxonomic Investigations

Phytochemical investigations of *Zieria* species corroborated previously published data on representatives of this genus [79]. Sadgrove and Jones [79] expanded the available information on this genus by, for the first time, detailing the chemical character of essential oils from the two species, *Z. odorifera* J.A.Armstr. subsp. *williamsii* and *Z. floydii* J.A.Armstr. Considered within the context of the chemotaxonomic approach undertaken in earlier studies, the remarks of the discoverer of the species, A. G. Floyd, now seem somewhat prescient.

"This is a quite oddity! This specimen does not match any known *Zieria* taxon. It appears to be allied to 3 closely related species; *Z. furfuracea*, *Z. granulata* and *Z. smithii*".

The former two species mentioned above, being *Z. furfuracea* R.Br. and *Z. granulate* C.Moore, produce an essential oil rich in car-3-en-2-one (**48**). The essential oil from *Z. floydii* was also dominated by this component.

Although essential oils from species of *Zieria* have been previously examined for antimicrobial activity, extracts and essential oils were tested against a broader range of organisms [79]. In that particular study the activity of essential oils with solvent extracts from the same species. High antimicrobial activity in both solvent extracts and essential oils was found. Therefore, a putative essential oil industry based on species of *Zieria* would provide a novel range of essential oils, attractive to the aromatherapy community, as well as providing purified compounds useful as scaffolds in pharmaceutical development.

In a further chemotaxanomic study addressing existing taxanomic concerns regarding the *Phebalium squamulosum* Vent. heterogenous species aggregate, some headway was made using the chemical character of essential oils to demonstrate specific differences between so-called subspecies [122]. The first species examined was *P. squamulosum* subsp. *verrucosum* Paul G.Wilson, which was regarded as having greater morphological alliance with the *Phebalium glandulosum* Hook. complex. Essential oils of *P. squamulosum* subsp. *verrucosum* were dominated by dihydrotagetone (**49**) at concentrations ranging from 95% to 98% [27]. An identical essential oil, with the same yield g/g wet weight of leaves, was produced in an earlier study from *P. glandulosum* subsp. *macrocalyx* R.L.Giles. In another study this was also demonstrated to be the case with *P. glandulosum* subsp. *glandulosum*. An almost identical essential oil was characterised from *P. glandulosum* subsp. *nitudum* Paul G.Wilson and *P. squamulsoum* subsp. *eglandulosum* Paul G.Wilson. Therefore, dihydrotagetone (**49**) dominated essential oils is a general characteristic of the *P. glandulosum* subspecies complex.

In a subsequent study other members of the *P. squamulosum* heterogenous species aggregate were phytochemically investigated [122]. It was demonstrated that all apparent subspecies currently assigned to this assemblage are characterised by separate individual essential oil chemotypes. Interestingly, several separate chemotypes were demonstrated from specimens currently assigned to *P. squamulosum* subsp. *squamulosum* Paul G.Wilson. In this regard, a notable chemical characteristic of oils from southern specimens (collected near Sydney and in the Hunter Valley) was the almost total predominance of a tricyclic sesquiterpene ketone; squamulosone (**50**). By contrast, northern specimens were characterised by essential oils rich in the heat derivative elemol (**51**); derived from the hedycaryol precursor [122].

A study as yet unpublished demonstrated significant potential for the use of essential oils from specimens in the genus *Prostanthera*. Essential oils from species of *Prostanthera* (in particular series racemosae) are almost always characterised by a major representation of 1,8-cineole (**29**). However the differentiating factor is the existence and relative abundances of tricyclic sesquiterpene alcohols, such as globulol (**52**), its epimer ledol (**53**), prostantherol (**54**) and maaliol (**55**), which are characterised by a cyclopropane moiety, attached to either a decahydro-napthalene or -azulene structure. Again, further significant differentiating components of some essential oils were the tricyclic sesquiterpenes, but with heterocycle substituents in place of the cyclopropane moiety. Examples include cis-dihydroagarofuran (**56**) or kessane (**57**), also on a decahydro-napthalene or -azulene structure, respectively [87].

As with other genera, *Prostanthera* essential oils were considered within the context of possible pharmacological activities. It was demonstrate that oils dominated by the sesquiterpene alcohols provided the greatest antimicrobial activity against a range of organisms, most pronounced against some Gram-positive species (results unpublished). Individual components found in significant amounts in the essential oils were related to this enhanced antimicrobial activity, particularly prostantherol (**54**). In a separate study of *P. centralis* a prostantherol-rich essential oil demonstrated significantly low antimicrobial activity against Gram-positive bacterial organisms and the yeast *Candida albicans* [123]. In one specimen currently assigned to *P. prunelloides* maaliol (**55**) was found in significant amounts. This is of considerable potential pharmacological interest, given the importance of maaliol (**55**) in the antinociceptive activity of a widely used Indian medicinal plant species (*Valeriana wallichii*) [124]. This antinociceptive activity is therefore expected to also be produced by oils from maaliol (**55**) rich species of *Prostanthera*.

In a study conducted by Lassak [125], the occurrence of maaliol (**55**) was demonstrated in a specimen of "*P. ovalifolia*" R.Br., at approx. 2% of the whole essential oil. This tricyclic sesquiterpene alcohol was formerly only known to occur in one other *Prostanthera* species, *P. prunellioides* R.Br. [126] at approximately 60%. However, it may also be known from *P. ringens* Benth. (as *P. lepidota*) [127]. Maaliol (**55**) was first characterized from the oleoresin "maali", from *Canarium samonensa* (*C. vitiense* A.Gray: Burseraceae), a Samoan native plant [128]. Members of the *Canarium* genus have been used extensively in traditional medicinal applications by Polynesian people to the north and north-east of the Australian landmass. The Philippine oleoresin "elemi" from *Canarium luzonicum* Miq. is the best known of these traditional medicines, used in applications to treat bronchitis, catarrh, extreme coughing, aged, damaged or injured skin and generalised stress [56].

The other maaliol (**55**) rich traditional medicine mentioned previously, from the north Indian Himalayan plant *Valeriana wallichii* DC (Valerianaceae) (maaliol chemotype), was used in applications similar to "elemi" but with a greater focus on psychotropic activity, in the treatment of a broad range of psychological disorders including stress, epilepsy and "insanity" [124]; and also in the treatment of a range of skin disorders. The study by Sah *et al.* [124] demonstrated activity from the maaliol (**55**) rich essential oil consistent with sedation or analgaesia via inhibition of the opioidergic pathway; or consistent with a peripheral antinociceptive effect via inhibition of prostaglandin synthesis. Maaliol (**55**) may be involved in these activities. It would be interesting to know if Australian Aboriginal people were aware of similar effects following the use of maaliol (**55**) rich *Prostanthera* species. Possible anti-inflammatory and analgaesic activity has also been demonstrated using 1,8-cineole (**29**) [129], indicating that oils containing both maaliol (**55**) and 1,8-cineole (**29**) are good candidates for further pharmacological tests. Again, *Prostanthera* essential oils have great potential as novel additions to Australia's aromatherapy and/or natural product industry.

The potential for cultivation of novel essential oil yielding crops utilizing species endemic to the Australian landmass is implicit in the literature. Such endemic species yield appreciable amounts of secondary metabolites with known *in vitro* pharmacological activities, such as antimicrobial activity. Table 3 gives a brief summary of the species broached by this review, but a substantially greater number of species are yet to be fully describes in the literature.

Table 3. Possible commercial scale applications from essential oil yielding flora in Australia.

Species	Chemotype	Use
Geijera parviflora	geijerene (**46**)/pregeijerene (**45**) (and germacrene D)	Commercial plantation: Insect repellent, topical analgaesia (linalool content). "Australian Green Lavender".
Geijera parviflora	osthole (**35**), isopsoralen (**33**), xanthyletine (**34**)	Commercial plantations: therapeutic effects
Zieria floydii	car-3-en-2-one (**48**)	Commercial plantations: Chemical scaffold for further drug development and antimicrobial activities
Prostanthera prunelloides	maaliol (**55**)	Commercial plantations: Medicinal applications consistent with the Indian *Valeriana willichii*
Prostanthera rotundifolia, P. centralis	prostantherol (**54**)	Commercial plantations: Antimicrobial activities
Eremophila dalyana	NA	Essential oil requires characterisation—useful in topical applications to treat fungal or bacterial infections. Also an effective decongestant in coughs and colds.
Eremophila deserti	ngaione	Commercial plantation: antifungal treatment
Eremophila deserti	methoxymyodesert-3-ene	Commercial plantation: chemical scaffold
Eremophila longifolia	isomenthone (**41**)/menthone (**39**)	Commercial plantation: topical, gastrointestinal for antimicrobial activities, topical for muscle aches and pains, active in applications for treatment of thrush (*Candida*)
Eremophila longifolia	fenchyl- (**22**)/bornyl acetate (**20**)	Commerical plantations: possible activity in gastrointestinal disease, possible activity in aromatherapy for headache sufferers
Eremophila longifolia	Limonene (**3**)/sabinene/ α-terpinolene, (−)-genifuranal (**43**)	Commercial plantations: derive (−)-genifuranal for therapeutic effects (*i.e.*, treatment of MRSA)
Callitris glaucophylla	NA	(**1**) Bioactive γ-lactones; ferruginol, pisiferal, pisiferol. (**2**) Occurrence of slightly hydrophilic antibiotic highly active against *S. aureus* (MRSA) and *B. subtilis*—requires purification and structure elucidation. Medicinal applications consistent with the Japanese species *Chamaecyparis pisifera*

4. Conclusions: Suggested Areas for Further Research

The demonstration of multiple chemotypes in *E. longifolia* emphasises the chemical variability expressed by this species, which may be an intrinsic general character of this genus. Thus, it is quite probable that other species from *Eremophila* may demonstrate similar geographical chemical variability. The observed correlation of diploidy with higher abundance of secondary metabolites may have more general implications. Therefore it would be worthwhile examining other species for both chemogeography and karyotype. Perhaps this search should start in *E. deserti*, as this has already been shown to possess an abundance of essential oil chemotypes each with high yields of essential oil. Another species, *E. glabra*, produces no essential oil at all; however NSW specimens are either hexaploid or tetraploid, but a diploid biotype can be found in far western WA. It may be worthwhile seeing if this diploid specimen yields any amount of essential oil. This may be a fruitful area of investigation for all *Eremophila* and its allied genus, *Myoporum*.

With regard to further investigation of species of *Eremophila* for derivatives produced in smoke fumigation rituals, no other species was as frequently used for this purpose as was *E. longifolia*, implying a lower likelihood that volatile therapeutic compounds could be found in other species of *Eremophila*. Perhaps a few exceptions would be *Eremophila freelingii* F.Muell and *E. neglecta* J.M.Black.

Derivatives or larger molecular mass compounds produced/evaporated during smoke fumigation methodology may alternatively be produced from hydrodistillation. However, due to the less destructive nature of conventional hydrodistillation as compared to smoke fumigation methodology, such derivatives or larger molecules could possibly be distilled in higher abundance at shorter time-intervals. This is particularly true if higher temperatures and pressures are employed, not without the risk of decomposition or fragmentation of volatiles. In this regard a modified pressure cooker, with a 5–15 psi pressure release valve positioned for horizontal airflow into an adjacent condenser, could be used to achieve this end. It has already been used successfully at 15 psi to derive genifuranal (**43**) from *E. longifolia* over a very short time span, but with the consequence of complete decomposition of the essential oil and most of the genifuranal (**43**). This effect may be reduced if adjusted to an optimal lower pressure.

An interesting and unexpected consequence of the current review is the "resurrection" of chemotaxonomy, which was utilised in Australia by botanists in the 70's and 80's before molecular fingerprinting became possible and quickly grew in popularity. In this regard, the question still begs an answer "how do you define a species"? Chemotaxonomy is challenged by the divide between "new species" and "new chemotype of the one species". To complicate the matter further, in some cases it has been demonstrated that a correlation could be made between genetics (karyotype) and chemotype such as with *E. longifolia* [68]. This is in stark contrast to the classical view that chemotypes result from differences in soil climate. In the former "genetics" view, seedlings from one chemotype could be transplanted into different soil types and different climates without any serious variation to the chemical character of its essential oil. In the latter more classical view, most certainly there would be a difference.

The view that chemotype derives from soil type is borrowed from Europe and Great Britain, where cultivar selection over thousands of years has caused a kind of genetic uniformity across many species used in cultivation. However, because this cultivar selection was not a practice employed by Australian Aboriginal people it is more likely that unique soil types and various climates favour certain biotypes—meaning the plant itself is different and better suited to that environment.

Over long stretches of time, geographically isolated chemotypes may diverge into new species, but again, the challenge lies in deciding exactly what amount of divergence warrants delimitation of a new species. Because of the inherent ambiguity in answering such a question, the best resolution for now is that consistent morphological differences should stand alone in defining a new species, but chemotaxonomy and phylogenetics may be utilised to demonstrate that such morphological variability is not merely a consequence of naturally occurring variability within the one species.

Acknowledgments

The authors would like to acknowledge advice provided by both Aboriginal and non-Aboriginal people in the preparation of this manuscript, particularly Don Murray and James Tribe from the Kamilaroi (Gomeroi) Tribe, to Kenneth Watson for overseeing the draft and the University of New England for providing reference materials wherever appropriate.

Author Contributions

This manuscript was compiled by Nicholas Sadgrove with extensive English editing, comments and suggestions provided by Graham Jones. This work summarises research with emphasis on collaborations between the two authors.

Appendix

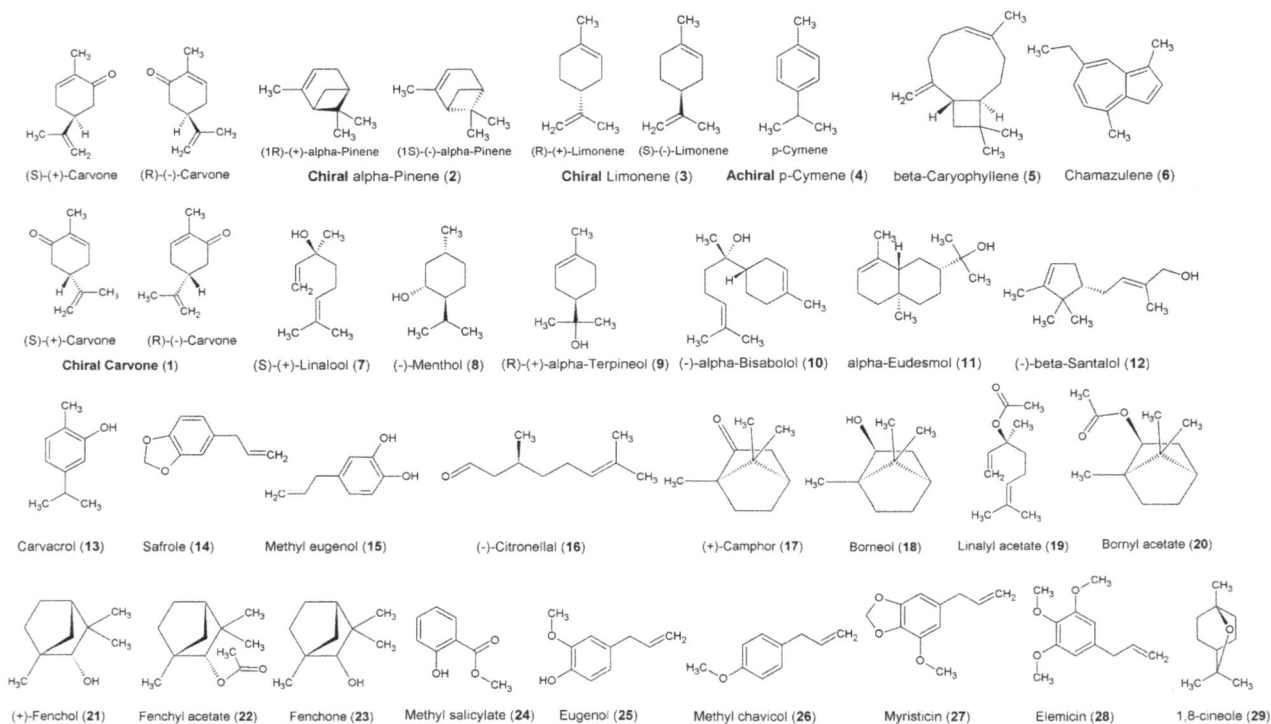

Figure A1. *Cont.*

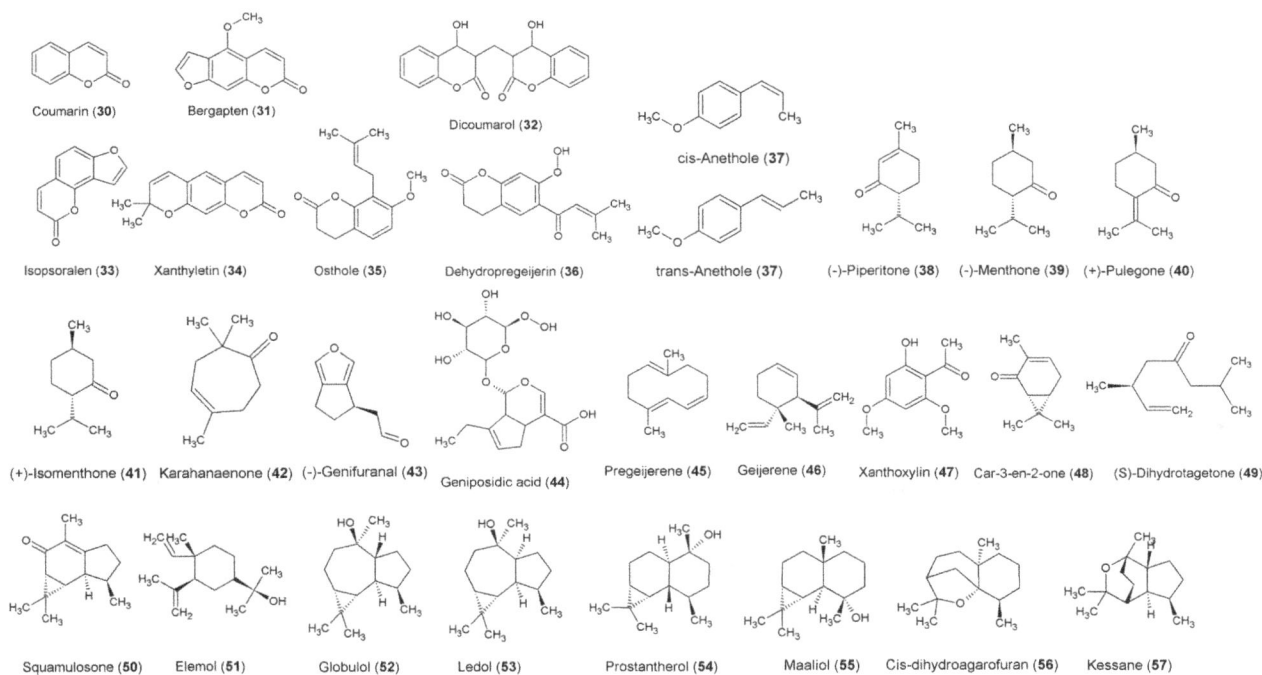

Figure A1. Chemical structure.

Introduction to Line Structures and Chiral Concepts Used in Organic Chemistry

Briefly, the most common atoms occurring in a volatile organic molecule, such as an essential oil molecule, are carbon, hydrogen and oxygen. Throughout this review the molecular structures are represented by line structures, combining 3-D structural formula where necessary (Figure A2, Lesson B) and incorporating chemical formula on the lines where methyl groups (CH_3) or other groups are specified (Figure A2, Lesson C). In a line structure diagram, the lines represent connections between carbons (one line connects two carbons) with hydrogen atoms bonded to them. For convenience (tidiness) the chemical symbols for both carbon and hydrogen (C—carbon, H—hydrogen) are generally replaced with lines, where single lines represent single bonds between carbons and double lines represent double bonds (Figure A2, Lesson D). Because carbons usually only have a maximum of four bonds, hydrogens occupy the remaining (invisible) bonds (Figure A2, Lessons C and E). Where another atomic element is present, such as oxygen (O—oxygen), the chemical symbol is always included. Some chemists like to show methyl groups (CH_3), and that is what we have done here.

In chemical identification the 3D spatial constitution, or stereochemistry, of connective parts of a molecule, as well as the position of a double bond, can significantly influence the chemical behaviour and pharmacological activity of the compound. Usually small differences in the spatial configuration (not to be confused with conformation) result in detectable differences in chemical analysis, such as in gas chromatography (GC) or nuclear magnetic resonance spectroscopy (NMR). However, often a single change in the spatial configuration of one molecule can produce another compound that is its exact mirror image, called an enantiomer. When a molecule is chiral this means it has an enantiomer or a mirror image of itself. Figure 2 depicts the two enantiomers of carvone (**1**), which is a chiral molecule. Although there appears to be four molecules in Figure 2, there are actually only two, with each enantiomer (either S+ or R−) depicted from both front and back.

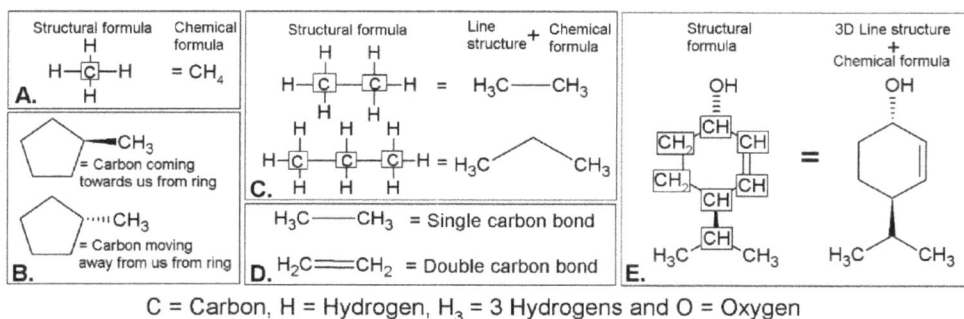

C = Carbon, H = Hydrogen, H₃ = 3 Hydrogens and O = Oxygen

Figure A2. Lessons **A–E** demonstrating how to interpret a line structure representation of an organic molecule. (**A**) The difference between a structural formula and a chemical formula; (**B**) How spatial distribution (stereochemistry) is conveyed; (**C**) The difference between structural formula and line structures that use chemical formula for methyl groups; (**D**) Single and double bonds; (**E**) The structural formula compared to its equivalent line structural diagram utilizing a 3D effect and chemical formula for methyl groups.

Briefly, a chiral centre is identified by a central carbon that is bonded to four different groups (Figure 3A). Often one of those bonds is to a hydrogen atom, but generally not shown in the line structure. Although ρ-cymene (**4**) does not have a chiral centre, one of the two hypothetical compounds depicted in Figure 3 (**B**) does. The compound on the left appears to have a chiral centre, but it does not because two of the bonds are identical and the compound is symmetrical. This means that although there is a 3D spatial constitution, it does not create a new molecule because it is superimposable over its mirror image. However, the compound on the right does have a chiral centre on the same carbon, but with the double bond in the molecule it means that it does not have a plane of symmetry. Therefore the compound on the left is chiral and the other is not (it is achiral).

In the unlikely event that a molecule has both a chiral centre and a plane of symmetry, it is called a "meso" compound, but this can only occur if two chiral centres are in the one molecule, each cancelling the other out by rotating plane polarised light in equal and opposite directions. However, unlike an achiral compound, which rotates 180° about its plane of symmetry, parallel to its mirror image to realise their synonymy, meso compounds rotate about their plane of symmetry 180° perpendicular to their mirror image. Generally meso compounds are not discussed in essential oil chemistry. In a meso compound the two chiral centres must have opposite configurations (*i.e.*, both *S* and *R*) and a plane of symmetry. *S* and *R* configurations are demonstrated in Figure 3.

Conflicts of Interest

The authors declare no conflict of interest.

References

1.   Asakawa, Y.; Ludwiczuk, A.; Nagashima, F. *Chemical Constituents of Bryophytes: Bio- and Chemical Diversity, Biological Activity, and Chemosystematics: 95 (Progress in the Chemistry of Organic Natural Products)*; Springer: New York, NY, USA, 2012.

2. Schnaubelt, K. *Medical Aromatherapy: Healing with Essential Oils*, 1st ed.; Frog Books: Berkeley, Canada, 1999.

3. ISO. International Standards Organisation—Home Page. Available online: http://www.iso.org/iso/home.htm (accessed on 12 December 2014).

4. Fadel, O.; Ghazi, Z.; Mouni, L.; Benchat, N.; Ramdani, M.; Amhamdi, H.; Wathelet, J.P.; Asehraou, A.; Charof, R. Comparison of microwave-assisted hydrodistillation and traditional hydrodistillation methods for the *Rosmarinus eriocalyx* essential oils from eastern Morocco. *J. Mater. Environ. Sci.* **2011**, *2*, 112–117.

5. Asghari, J.; Touli, K.C.; Mazaheritehrani, M. Microwave-assisted hydrodistillation of essential oils from *Echinophora platyloba* Dc. *J. Med. Plants Res.* **2012**, *6*, 4475–4480.

6. Mohamadi, M.; Shamspur, T.; Mostafavi, A. Comparison of microwave-assisted distillation and conventional hydrodistillation in the essential oil extraction of flowers of *Rosa damascena* Mill. *J. Essent. Oil Res.* **2013**, *25*, 55–61.

7. Stewart, D. *Chemistry of Essential Oils Made Simple: God's Love Manifest in Molecules*; NAPSAC Reproductions: Marble Hill, MO, USA, 2005.

8. Kostadinovic, S.; Jovanov, D.; Mirhosseini, H. Comparative investigation of cold pressed essential oils from peel of different mandarin varieties. *IIOAB J.* **2011**, *3*, 7–14.

9. Markley, K.S.; Nelson, E.K.; Sherman, S.M. Some wax-like constituents from expressed oil from the peel of florida grapefruit, *Citrus grandis*. Food Research Division and Fertlizer Investigations, Bureau of Chemistry and Soils, United States Department of Agriculture, Washington. *J. Biol. Chem.* **1937**, *118*, 433–441.

10. Sadgrove, N.; Jones, G.L. A possible role of partially pyrolysed essential oils in Australian Aboriginal traditional ceremonial and medicinal smoking applications of *Eremophila longifolia* (R. Br.) F. Muell (*Scrophulariaceae*). *J. Ethnopharmacol.* **2013**, *147*, 638–644.

11. Braithwaite, M.; Vuuren, V.S.F.; Viljoen, A.M. Validation of smoke inhalation therapy to treat microbial infections. *J. Ethnopharmacol.* **2008**, *119*, 501–506.

12. Sadgrove, N.; Jones, G.L.; Greatrex, B.W. Isolation and characterisation of (−)-genifuranal: The principal antimicrobial component in traditional smoking applications of *Eremophila longifolia* (Scrophulariaceae) by Australian Aboriginal peoples. *J. Ethnopharmacol.* **2014**, *154*, 758–766.

13. Guenther, E. *The Essential Oils—vol 1: History—Origin in Plants—Production—Analysis*; Van Nostrand: New York, NY, USA, 1948.

14. Stewart, D. *Healing Oils of the Bible*; Care Publications: Marble Hill, MO, USA, 2003.

15. Burt, S. Essential oils: Their antibacterial properties and potential applications in foods—A review. *Int. J. Food Microbiol.* **2004**, *94*, 223–253.

16. Bauer, K.; Garbe, D. *Common Fragrance and Flavor Materials. Preparation, Properties and Uses*; VCH Verlagsgesellschaft: Weinheim, UK, 1985.

17. Sell, C. Chemistry of essential oils. In *Handbook of Essential Oils: Science, Technology, and Applications*; Başer, K.H.C., Buchbauer, G., Eds.; CRC Press, Taylor and Francis Group: London, UK, 2010.

18. Moussaieff, A.; Rimmerman, N.; Bregman, T.; Straiker, A.; Felder, C.C.; Shoham, S.; Kashman, Y.; Huang, S.M.; Lee, H.; Shohami, E.; *et al.* Incensole acetate, an incense component, elicits psychoactivity by activating TRPV3 channels in the brain. *FASEB J.* **2008**, *22*, 3024–3034.

19. Moussaieff, A.; Shein, N.A.A.; Tsenter, J.; Grigoriadis, S.; Simeonidou, C.; Alexandrovich, A.G.; Trembovler, V.; Ben-Neriah, Y.; Schmitz, M.L.; Fiebich, B.L.; *et al.* Incensole acetate: A novel neuroprotective agent isolated from *Boswellia carterii. J. Cereb. Blood Flow Metab.* **2008**, *28*, 1341–1352.

20. Moussaieff, A.; Shoham, E.; Kashman, Y.; Fride, E.; Schmitz, M.L.; Renner, F.; Fiebich, B.L.; Munoz, E.; Ben-Neriah, Y.; Mechoulam, R. Incensole acetate, a novel anti-inflammatory compound isolated from *Boswellia* resin, inhibits nuclear factor-κB activation. *Mol. Pharm.* **2007**, *72*, 1657–1664.

21. Başer, K.H.C.; Demirci, F. Chemistry of essential oils. In *Fragrance and Flavours: Chemistr, Bioprocessing and Sustainability*, 1st ed.; Berger, R.G., Ed.; Springer: Leipzig, Germany, 2007.

22. Della, E.W.; Jefferies, P.R. The chemistry of *Eremophila Species*. 111. The essential oil of *Eremophila longifolia* F. Muell. *Aust. J. Chem.* **1961**, *14*, 663–664.

23. Sainsbury, M. *Aromatic Chemistry*; Oxford University Press: New York, NY, USA, 1992.

24. Clayden, J.; Greeves, N.; Warren, S. *Organic Chemistry*, 2nd ed.; Oxford University Press Inc.: New York, NY, USA, 2012.

25. Kelvin, W.T. *The Molecular Tactics of a Crystal*; Clarendon Press: London, UK, 1894.

26. König, W.A.; Hochmuth, D.H. Enantioselectie gas chromatography in flavor and fragrant analysis: Strategies for the identification of know and unknown plant volatiles. *J. Chromatogr. Sci.* **2004**, *42*, 423–439.

27. Sadgrove, N.; Telford, I.R.H.; Greatrex, B.W.; Dowell, A.; Jones, G.L. Dihydrotagetone, an unusual fruity ketone, is found in enantiopure and enantioenriched forms in additional australian native taxa of *Phebalium* (Rutaceae: Boronieae). *Nat. Prod. Commun.* **2013**, *8*, 737–740.

28. Leitereg, T.J.; Guadagni, D.G.; Harris, J.; Mon, T.R.; Teranishi, R. Chemical and sensory data supporting the difference between the odors of the enantiomeric carvones. *J. Agric. Food Chem.* **1971**, *19*, 785–787.

29. Zellner, B.D.A.; Dugo, P.; Dugo, G.; Mondello, L. Analysis of Essential Oils. In *Handbook of Essential Oils: Science, Technology and Applications*; Başer, K.H.C., Buchbauer, G., Eds.; CRC Press, Taylor and Francis Group: London, UK, 2010.

30. Reineccius, G.A. Flavour-Isolation Techniques. In *Fragrance and Flavours: Chemistr, Bioprocessing and Sustainability*, 1st ed.; Berger, R.G., Ed.; Springer: Leipzig, Germany, 2007.

31. Mosandl, M. Enantioselective and Isotope Analysis—Key Steps to Flavour Authentication. In *Fragrance and Flavours: Chemistr, Bioprocessing and Sustainability*, 1st ed.; Berger, R.G., Ed.; Springer: Leipzig, Germany, 2007.

32. Australian Standards—Home Page. Available online: http://www.standards.org.au/Pages/default.aspx (accessed on 12 December 2014).

33. Association Française de Normalisation—Home Page. Available online: http://www.afnor.org/en (accessed on 12 December 2014).

34. British Pharmacopoeia—Home Page. Available online: https://www.pharmacopoeia.gov.uk/reference-standards.php (accessed on 12 December 2014).

35. Adams, T.B.; Taylor, S.V. Safety evaluation of essential oils: A constituent-based approach. In *Handbook of Essential Oils: Science, Technology and Applications*; Başer, K.H.C., Buchbauer, G., Eds.; CRC Press, Taylor and Francis Group: London, UK, 2010.

36. Tisserand, R.; Balacs, T. *Essential Oil Safety: A Guide for Health Care Professionals*; Churchill livingstone: New York, NY, USA, 1995.

37. Schnaubelt, K. *Advanced Aromatherapy: The Science of Essential Oil Therapy*; Healing Art Press: Rochester, VT, USA, 1995.

38. Sadgrove, N.; Gonçalves-Martins, M.; Jones, G.L. Chemogeography and antimicrobial activity of essential oils from *Geijera parviflora* and *Geijera salicifolia* (Rutaceae): Two traditional australian medicinal plants. *Phytochemistry* **2014**, *104*, 60–71.

39. Therapeutic Goods Administration—Home Page. Available online: https://www.tga.gov.au/ (accessed on 12 December 2014).

40. Bowles, J.E. *The Chemistry of Aromatherapeutic Oils*; Allen and Unwin: Crows Nest, NSW Australia, 2003.

41. Poucher, W.A. *Poucher's Perfumes, Cosmetics and Soaps*, 9th ed.; Chapman & Hall: London, UK, 1993.

42. Kusari, S.; Spiteller, M. Metabolomics of Endophytic Fungi Producing Associated Plant Secondary Metabolites: Progress, Challenges and Opportunities. In *Metabolomics*; Roessner, U., Ed.; InTech: Rijeka, Croatia, 2012.

43. Emiliani, G.; Mengoni, A.; Maida, I.; Perrin, E.; Chiellini, C.; Fondi, M.; Gallo, E.; Gori, L.; Maggini, V.; Vannacci, A.; *et al.* Linking bacterial endophytic communities to essential oils: Clues from *Lavandula angustifolia* Mill. *Evid. Based Complement. Altern. Med.* **2014**, *2014*, doi:10.1155/2014/650905.

44. Mucciarelli, M.; Camusso, W.; Maffei, M.; Panicco, P.; Bichi, C. Volatile terpenoids of endophyte-free and infected peppermint (*Mentha piperita* L.): Chemical partitioning of a symbiosis. *Microb. Ecol.* **2007**, *54*, 685–696.

45. Gersbach, P.V. The essential oil secretory structures of *Prostanthera ovalifolia* (Lamiaceae). *Ann. Bot.* **2002**, *89*, 255–260.

46. Schrader, J. Microbial flavour production. In *Fragrance and Flavours: Chemistr, Bioprocessing and Sustainability*, 1st ed.; Berger, R.G., Ed.; Springer: Leipzig, Germany, 2007.

47. Noma, Y.; Asakawa, Y. Biotransformation of monoterpenoids by microorganisms, insects, and mammals. In *Handbook of Essential Oils: Science, Technology and Applications*, Başer, K.H.C., Buchbauer, G., Eds.; CRC Press, Taylor and Francis Group: London, UK, 2010.

48. Asakawa, Y.; Noma, Y. Biotransformation of sesquiterpenoids, ionones, damascones, adamantanes, and aromatic compounds by green algae, fungi, and mammals. In *Handbook of Essential Oils: Science, Technology and Applications*; Başer, K.H.C., Buchbauer, G., Eds.; CRC Press, Taylor and Francis Group: London, UK, 2010.

49. Scragg, A.H. The production of flavours by plant cell cultures. In *Fragrance and Flavours: Chemistr, Bioprocessing and Sustainability*, 1st ed.; Berger, R.G., Ed.; Springer: Leipzig, Germany, 2007.

50. Hunter, M. *Essential Oils: Art, Agriculture, Science, Industry and Entrepreneurship (a Focus on the Asia-Pacific Region)*; Nova Science Publishers, Inc.: New York, NY, USA, 2009.

51. Blazquez, M.A. Role of natural essential oils in sustainable agriculture and food preservation. *J. Sci. Res. Rep.* **2014**, *3*, 1843–1860.

52. Başer, K.H.C.; Franz, C. Essential oils used in veterinary medicine. In *Handbook of Essential Oils: Science, Technology and Applications*; Başer, K.H.C., Buchbauer, G., Eds.; CRC Press, Taylor and Francis Group: London, UK, 2010.

53. Kloucek, P.; Frankova, A.; Smid, J. Effect of Warm Air Flow and Reduced Pressure on Antibacterial Activity of Essential Oil Vapors. In Proceedings of the 42th ed International Symposium on Essential Oils, Antalya, Turkey, 11–14 September 2011.

54. Sangwan, N.S.; Farooqi, A.H.A.; Shabih, F.; Sangwan, R.W. Regulation of essential oil production in plants. *Plant Growth Regul.* **2001**, *34*, 3–21.

55. Schwab, W. Genetic engineering of plants and microbial cells for flavour production. In *Fragrance and Flavours: Chemistr, Bioprocessing and Sustainability*, 1st ed.; Berger, R.G., Ed.; Springer: Leipzig, Germany, 2007.

56. Sadgrove, N.J.; Jones, G.L. University of New England, Armidale, NSW, Australia. Unpublished work, 2008.

57. Buchbauer, G. Chapter 9. Biological activities of essential oils. In *Handbook of Essential Oils: Science, Technology and Applications*; Başer, K.H.C., Buchbauer, G., Eds.; CRC Press, Taylor and Francis Group: London, UK, 2010.

58. Koroch, A.R.; Juliani, H.R.; Zygadlo, J.A. Bioactivity of essential oils and their components. In *Fragrance and Flavours: Chemistr, Bioprocessing and Sustainability*, 1st ed.; Berger, R.G., Ed.; Springer: Leipzig, Germany, 2007.

59. CLSI. *Performance Standards for Antimicrobial Disk Susceptibility Tests; Approved Standard—Tenth Edition*; Clinical and Laboratory Standards Institute: Wayne, PA, USA, 2009; Volume 29, pp. 1–54.

60. Mann, C.M.; Markham, J.L. A new method for determining the minimum inhibitory concentration of essential oils. *J. Appl. Microbiol.* **1998**, *84*, 538–544.

61. CLSI. *Methods for Dilution Antimicrobial Susceptibility Testing for Bacteria that Grow Aerobically Approved Standard—Eight Edition*; Clinical and Laboratory Standards Institute: Wayne, PA, USA, 2009; Volume 29, pp. 1–66.

62. Sadgrove, N.; Mijajlovic, S.; Tucker, D.J.; Watson, K.; Jones, G.L. Characterization and bioactivity of essential oils from novel chemotypes of *Eremophila longifolia* (F. Muell) (Myoporaceae): A highly valued traditional Australian medicine. *Flavour Fragr. J.* **2011**, *26*, 341–350.

63. Van Vuuren, S.F.; Viljoen, A.M. Plant-Based antimicrobial studies—Methods and approaches to study the interaction between natural products. *Plant. Med.* **2011**, *77*, 1168–1182.

64. Yi, A.-K.; Yoon, J.-G.; Hong, S.-C.; Redford, T.W.; Krieg, A.M. Lipopolysaccharide and CPG DNA synergise to tumor necrosis factor-alpha production through activation of NF-κB. *Int. Immunol.* **2001**, *13*, 1391–1404.

65. Ashley, J.W.; McCoy, E.M.; Clements, D.A.; Shi, Z.; Chen, T. Development of cell-based high-throughput assays for the identification of inhibitors of receptor activator of nuclear factor-kappa B signalling. *Assay Drug Dev. Technol.* **2011**, *9*, 40–49.

66. Semple, S.J.; Reynolds, G.D.; O'Leary, M.C.; Flower, R.L. Screening of Australian medicinal plants for antiviral activity. *J. Ethnopharmacol.* **1998**, *60*, 163–172.

67. Dong, L.; Schill, H.; Grange, R.L.; Porzelle, A.; Johns, J.P.; Parsons, P.G.; Gordon, V.A.; Reddell, P.W.; Williams, C.M. Anticancer agents from the Australain tropical rainforest: Spiroacetals EBC-23, 24, 25, 72, 73, 75 and 76. *Chem. A Eur. J.* **2009**, *15*, 11307–11318.

68. Brand-Williams, W.; Cuvelier, M.E.; Berset, C. Use of a free radical method to evaluate antioxidant activity. *Lebensm. Wiss. Technol.* **1995**, *28*, 25–30.

69. Benzie, I.F.F.; Strain, J.J. The ferric reducing ability of plasma (FRAP) as a measure of antioxidant power: The frap assay. *Anal. Biochem.* **1996**, *239*, 70–79.

70. Rogers, K.L.; Fong, W.F.; Redburn, J.; Griffiths, L.R. Fluorescence detection of plant extracts that affect neuronal voltage-gated Ca^{2+} channels. *Eur. J. Pharm. Sci.* **2002**, *15*, 321–330.

71. Behrendt, H.-J.; Germann, T.; Gillen, C.; Hatt, H.; Jostock, R. Characterization of the mouse cold-menthol receptor TRPM8 and vanilloid receptor type-1 VR1 using fluorometric imaging plate reader (FLIPR) assay. *Br. J. Pharmacol.* **2004**, *141*, 173–745.

72. Van Vuuren, S.F.; du Toit, L.C.; Parry, A.; Pillay, V.; Choonara, Y.E. Encapsulation of essential oils within a polymeric liposomal formulation for enhancement of antimicrobial efficacy. *Nat. Prod. Commun.* **2010**, *5*, 1401–1408.

73. Karlsen, J. Encapsulation and other programmed release techniques for essential oils and volatile terpenes. In *Handbook of Essential Oils: Science, Technology and Applications*; Başer, K.H.C., Buchbauer, G., Eds.; CRC Press, Taylor and Francis Group: London, UK, 2011.

74. Sadgrove, N.; Jones, G.L. Medicinal compounds, chemically and biologically characterised from extracts of Australian *Callitris endlicheri* and *C. glaucophylla* (Cupressaceae): Used traditionally in Aboriginal and colonial pharmacopoeia. *J. Ethnopharmacol.* **2014**, *153*, 872–883.

75. Elisabetsky, E.; Coelho de Souza, G.P.; Santos, M.A.C.; Siqueira, I.R.; Amador, T.A. Sedative properties of linalool. *Fitoterapia* **1995**, *66*, 407–414.

76. Horak, S.; Koschak, A.; Stuppner, H.; Striessnig, J. Use-dependent block of voltage-gated $Ca_v2.1$ Ca^{2+} channels by petasins and eudesmol isomers. *J. Pharmacol. Exp. Ther.* **2009**, *330*, 220–226.

77. Asakura, K.; Kanemasa, T.; Minagawa, K.; Kagawa, K.; Ninomiya, M. The nonpeptide alpha-eudesmol from *Juniperus virginiana* Linn. (Cupressaceae) inhibits omega-agatoxin IVA-sensitive calcium currents and synaptosomal $^{45}Ca^{2+}$ uptake. *Brain Res.* **1999**, *823*, 169–176.

78. Hongratanaworakit, T.; Heuberger, E.; Buchbauer, G. Evaluation of the effects of east indian sandalwood oil on alpha-santalol on humans after transdermal absorption. *Plant. Med.* **2004**, *70*, 3–7.

79. Sadgrove, N.; Jones, G.L. Antimicrobial activity of essential oils and solvent extracts from *Zieria* species (Rutaceae). *Nat. Prod. Commun.* **2013**, *8*, 741–745.

80. Maia, M.F.; Moore, S.J. Plant-based insect repellents: A review of their efficacy, development and testing. *Malar. J.* **2011**, *10* (Suppl. 1), S11.

81. Sadgrove, N.; Hitchock, M.; Watson, K.; Jones, G.L. Chemical and biological characterization of novel essential oils from *Eremophila bignoniiflora* (F. Muell) (Myoporaceae): A traditional Aboriginal Australian bush medicine. *Phytother. Res.* **2013**, *27*, 1508–1516.

82. Wang, J.; Cai, Y.; Wu, Y. Antiinflammatory and analgesic activity of topical administration of *Siegesbeckia pubescens*. *Pak. J. Pharm. Sci.* **2008**, *21*, 89–91.

83. Semnani-Morteza, K.; Saeedi, M.; Hamidian, M. Anti-inflammatory and analgesic activity of the topical preparation of *Glaucium grandiflorum*. *Fitoterapia* **2004**, *75*, 123–129.

84. Kalbhen, D. Abbo Nutmeg as a narcotic. A contribution to the chemistry of pharmacology of nutmeg (*Myristica fragrans*). *Angew. Chem. Int.* **1971**, *10*, 370–374.

85. Beyer, J.; Ehlers, D.; Maurer, H.H. Abuse of nutmeg (*Myristica fragrans* Houtt.): Studies on the metabolism and the toxicologic detection of its ingredients elemicin, myristicin, and safrole in rat and human urine using gas chromagrography/mass spectrometery. *Ther. Drug Monit.* **2006**, *28*, 568–575.

86. Baker, R.T.; Smith, H.G. On a new species of *Prostanthera* and its essential oil. *J. Proc. R. Soc. NSW* **1912**, *46*, 103–110.

87. Southwell, I.A.; Tucker, D.J. *cis*-Dihydroagarofuran from *Prostanthera* sp. aff. *ovalifolia*. *Phytochemistry* **1993**, *22*, 857–862.

88. Dellar, J.E.; Cole, M.D.; Gray, A.I.; Gibbons, S.; Waterman, P.G. Antimicrobial sesquiterpenes from *Postanthera* aff. *melissifolia* and *P. rotundifolia*. *Phytochemistry* **1994**, *36*, 957–960.

89. Pala-Paul, J.; Copeland, L.M.; Brophy, J.J.; Goldsack, R.J. Essential oil composition of two variants of *Prostanthera lasianthos* Labill. from Australia. *Biochem. Syst. Ecol.* **2006**, *34*, 48–55.

90. Lassak, E.V.; McCarthy, T. *Australian Medicinal Plants*; Methuen Australia Pty Ltd.: North Rhyde, Australia, 2011.

91. Lassak, E.V.; Pinhey, J.T. The constituents of *Eriostemon Trachyphyllus*. The structure of trachyphyllin, a new coumarin. *Aust. J. Chem.* **1969**, *22*, 2175–2185.

92. Lahey, F.N.; MacLeod, J.K. The coumarins of *Geijera parviflora* Lindl. *Aust. J. Chem.* **1967**, *20*, 1943–1955.

93. Carotti, A.; Carrieri, A.; Chimichi, S.; Boccalini, M.; Cosimelli, B.; Gnerre, C.; Carotti, A.; Carrupt, P.-A.; Testa, B. Natural and synthetic geiparvarins are strong and selective MOA-B inhibitors. Synthesis and sar studies. *Bioorg. Med. Chem. Lett.* **2002**, *12*, 3551–3555.

94. Tanaka, K.; Pescitelli, G.; Di Bari, L.; Xiao, T.L.; Nakanishi, K.; Armstrong, D.W.; Berova, N. Absolute stereochemistry of dihydrofuroangelicins bearing C-8 substituted double bonds: A combined chemical/exciton chirality protocol. *Organic Biomol. Chem.* **2004**, *2*, 48–58.

95. Hutt, A.J. Chirality and pharmacokinetics: An area of neglected dimensionality? *Durg Metab. Drug Interact.* **2007**, *22*, 79–112.

96. Maiden, J.H. *The Useful Native Plants of Australia*; Alexander Bros Vic: Sydney, Australia, 1889.

97. Cribb, A.B.; Cribb, J.W. *Wild Medicine in Australia*; William Collins, Pty, Ltd.: Sydney, Austrilia, 1981.

98. Cribb, A.B.; Cribb, J.W. *Useful Wild Plants in Australia*; William Collins Pty Ltd.: Sydney, Australia, 1981.

99. Behnam, S.; Farzaneh, M.; Ahmadzadeh, M.; Tehrani, A.S. Composition and antifungal activity of essential oils of *Mentha piperita* and *Lavandula angustifolia* on post-harvest phytopathogens. *Commun. Agric. Appl. Biol. Sci.* **2006**, *71*, 1321–1326.

100. Barr, A. *Traditional Bush Medicines: An Aboriginal Pharmacopoeia*; Greenhouse publications Pty Ltd.: Richmond Vic, Australia, 1988.

101. Jirovetz, L.; Buchbauer, G.; Denkova, Z.; Stoyanova, A.; Murgov, I.; Gearon, V.; Birkbeck, S.; Schmidt, E.; Geissler, M. Comparative study on the antimicrobial activities of different sandalwood essential oils of various origin. *Flavour Fragr. J.* **2006**, *21*, 465–468.

102. Thompson, J.; Johnson, L.A.S. *Callitris glaucophylla*, Australia's white cypress pine—A new name for an old species. *Telopea* **1986**, *2*, 731, doi:10.7751/telopea19864610.

103. Low, T. *Bush Medicine: A Pharmacopoeia of Natural Remedies*; Greenhouse publications Pty Ltd.: Richmond Vic, Australia, 1990.

104. Oprava, A.; Leach, D.N.; Beattie, K.; Connellan, P.; Forster, P.I.; Leach, G.; Buchbauer, G.; Shepherd, K.; Deseo, M. Chemical composition and biological activity of the essential oils from native Australian *Callitris* species. *Plant. Med.* **2010**, *76*, doi:10.1055/s-0030-1264273.

105. McKern, H.H.G. Arthur de Ramon Penfold. *J. Proc. R. Soc. N.S.W.* **1980**, *113*, 100.

106. McKern, H.H.G. Arthur de Ramon Penfold, 1890–1980. *Chem. Aust.* **1981**, *48*, 327.

107. Guenther, E. *The Essential Oils—Vol 1–6*; Van Nostrand Company, Inc.: New York, NY, USA, 1948; Volume 2.

108. Carson, C.F.; Hammer, K.A.; Riley, T.V. *Melaleuca alternifolia* (tea tree) oil: A review of antimicrobial and other medicinal properties. *Clin. Microbiol. Rev.* **2006**, *19*, 50–62.

109. Plummer, J.A.; Wann, J.M.; Spadek, Z.E. Intraspecific variation in oil components of *Boronia megastigma* Nees. (Rutaceae) flowers. *Ann. Bot.* **1999**, *83*, 253–262.

110. Beattie, K.; Waterman, P.G.; Forster, P.I.; Thompson, D.R.; Leach, D.N. Chemical composition and cytotoxicity of oils and eremophilanes derived from various parts of *Eremophila mitchellii* Benth. (Myoporaceae). *Phytochemistry* **2011**, *71*, 400–408.

111. Thomas, J.; Narkowicz, C.K.; Jacobson, G.A.; Davies, N.W. An examination of the essential oils of tasmanian *Kunzea ambigua*, other *Kunzea* spp. and commercial *Kunzea* oil. *J. Essent. Oil Res.* **2010**, *22*, 381–385.

112. Trevena, G. Essentially Australia. Available online: https://essentiallyaustralia.com.au/about-us/ (accessed on 12 December 2014).

113. Amri, I.; Mancini, E.; de Martino, L.; Marandino, A.; Lamia, H.; Mohsen, H.; Bassem, J.; Scognamiglio, M.; Reverchon, E.; de Feo, V. Chemical composition and biological activities of the essential oils from three *Melaleuca* species grown in Tunisia. *Int. J. Mol. Sci.* **2012**, *13*, 16580–16591.

114. Smith, J.E.; Tucker, D.J.; Alter, D.; Watson, K.; Jones, G.L. Intraspecific variation in essential oil composition of *Eremophila longifolia* F. Muell (Myoporaceae): Evidence for three chemotypes. *Phytochemistry* **2010**, *71*, 1521–1527.

115. Sadgrove, N.; Jones, G.L. Cytogeography of essential oil chemotypes of *Eremophila longifolia* F. Muell (Schrophulariaceae). *Phytochemistry* **2014**, *105*, 43–51.

116. Murray, D. Kamilaroi Tribe, Collarenebri, NSW, Australia. Personal communication, 2014.

117. Sadgrove, N.; Jones, G.L. Chemical and biological characterisation of solvent extracts and essential oils from leaves and fruit of two Australian species of *Pittosporum* (Pittosporaceae) used in Aboriginal medicinal practice. *J. Ethnopharmacol.* **2013**, *145*, 813–821.

118. Bäcker, C.; Jenett-Siems, K.; Bodtke, A.; Lindequist, U. Polyphenolic compounds from the leaves of *Pittosporum angustifolium*. *Biochem. Syst. Ecol.* **2014**, *55*, 101–103.

119. Bäcker, C.; Jenett-Siems, K.; Siems, K.; Wurster, M.; Bodtke, A.; Chamseddin, C.; Crüsemann, M.; Lindequist, U. Triterpene glycosides from the leaves of *Pittosporum angustifolium*. *Plant. Med.* **2013**, *79*, 1461–1469.

120. Bäcker, C.; Jenett-Siems, K.; Siems, K.; Wurster, M.; Bodtke, A.; Lindequist, U. Cytotoxic saponins from the seeds of *Pittosporum angustifolium*. *Zeitzchrift Naturforschung. C J. Biosci.* **2014**, *69*, 191–198.

121. Vesoul, J.; Cock, I.E. An examination of the medicinal potential of *Pittosporum phylliraeoides*: Toxicity, antibacterial and antifungal activities. *Pharmacogn. Commun.* **2011**, *1*, 8–17.

122. Sadgrove, N.J.; Telford, I.R.H.; Greatrex, B.W.; Jones, G.L. Composition and antimicrobial activity of essential oils from the *Phebalium squamulosum* species complex (Rutaceae) in New South Wales, Australia. *Phytochemistry* **2014**, *97*, 38–45.

123. Collins, T.L.; Jones, G.L.; Sadgrove, N. Volatiles from the rare australian desert plant *Prostanthera centralis* B.J.Conn (Lamiaceae): Chemical composition and antimicrobial activity. *Agriculture* **2014**, *4*, 308–316.

124. Sah, S.P.; Mathela, C.S.; Chopra, K. *Valeriana wallichii* DC (maaliol chemotype): Antinociceptive studies on experimental animal models and possible mechanism of action. *Pharmacologia* **2012**, *3*, 432–437.

125. Lassak, E.V. New essential oils from the Australian flora (October 1980)—Perfumes and flavours symphony of nature. In Proceedings of the 8th International Congress of Essential Oils—Paper No. 120, Fedarom Grasse, France, 1980; pp. 409–415.

126. Hellyer, R.O. Occurence of maaliol, elemol, and globulol in some australian essential oils. *Aust. J. Chem.* **1962**, *15*, 157–157.

127. Lassak, E.V.; Southwell, I.A. Essential oils isolates from the australian flora. *Int. Flavours Food Addit.* **1977**, *8*, 126–132.

128. Buchi, G.; Wittenau, S.V.; White, D.M.; Terpenes, X. The constitution of maaliol. *J. Am. Chem. Soc.* **1959**, *81*, 1968–1980.

129. Santos, F.A.; Rao, V.S.N. Antiinflammatory and antinociceptive effects of 1,8-cineole a terpenoid oxide present in many plant essential oils. *Phytother. Res.* **2000**, *14*, 240–244.

7

National Livestock Policy of Nepal: Needs and Opportunities

Upendra B. Pradhanang [1], Soni M. Pradhanang [2,*], Arhan Sthapit [3], Nir Y. Krakauer [4],
Ajay Jha [5] and Tarendra Lakhankar [6]

[1] Shankar Dev Campus, Faculty of Management, Tribhuvan University, Kathmandu 44600, Nepal;
    E-Mail: upradhananga@gmail.com
[2] Department of Geosciences, University of Rhode Island, Kingston, RI 02881, USA;
    E-Mail: spradhanang@uri.edu
[3] Department of Management, Public Youth Campus, Tribhuvan University, Kathmandu 44600,
    Nepal; E-Mail: arhansthapit@gmail.com
[4] Department of Civil Engineering, The City College of New York, City University of New York,
    New York, NY 10031, USA; E-Mail: nkrakauer@ccny.cuny.edu
[5] Department of Horticulture and Landscape Architecture, College of Agricultural Sciences,
    Horticulture and Landscape Architecture, Colorado State University, Fort Collins, CO 80523, USA;
    E-Mail: ajay.jha@colostate.edu
[6] NOAA-Cooperative Remote Sensing Science & Technology (CREST) Center,
    The City College of New York, City University of New York, New York, NY 10031 USA;
    E-Mail: tlakhankar@ccny.cuny.edu

* Author to whom correspondence should be addressed; E-Mail: spradhanang@uri.edu

Academic Editor: Milan Shipka

Abstract: This paper describes Nepal's national livestock policies and considers how they can be improved to help meet the pressing national challenges of economic development, equity, poverty alleviation, gender mainstreaming, inclusion of marginalized and underprivileged communities, and climate vulnerability. Nepal is in the process of transforming its government from a unitary system to a federal democratic structure through the new constitution expected by 2015, offering the opportunity to bring a new set of priorities and stakeholders to policymaking. Nepal's livestock subsector comes most directly within the purview of the National Agricultural Policy 2004, Agro-Business Policy, 2006 and Agricultural Sectoral Operating Policies of the Approach Paper to 13th Plan,

2012/13–2015/16 policy instruments. We systematically review these and other livestock-related national policies through analysis of their Strengths, Weaknesses, Opportunities and Threats (SWOT). We conclude with the need to formulate a separate, integrated national livestock policy so that Nepal can sustainably increase livestock productivity and achieve diversification, commercialization and competitiveness of the livestock subsector within the changing national and international contexts.

Keywords: livestock policy; federal structure; SWOT; Nepal; livelihood; climate change

1. Introduction

Nepal remains a predominantly agrarian economy. About 66 percent of its population is involved in agriculture, which accounts for 35 percent of the gross domestic product or GDP [1,2]. The livestock subsector of agriculture contributes 24 percent of the total agricultural GDP [3], and also plays important roles in human food and nutritional security, livelihood, regional balance, gender mainstreaming, and rural poverty alleviation [4]. Yet, there is no separate national livestock policy in Nepal, and instead, its national livestock-related policies are spread across agriculture and other sectors. The three most crucial relevant policies are National Agricultural Policy 2004 [5], Agro-Business Promotion Policy 2006 [6], and Agricultural Sectoral Operating Policies (ASOPs) [6] of the Approach Paper to Thirteenth Plan, 2012/13–2015/16. These three policy documents and a number of other livestock-related policies are reviewed in this paper with a view to building the groundwork for a new, integrated national livestock policy.

Today's changed context demands that national policies toward the livestock subsector (LSS) of the agricultural sector should be dealt with anew. While (I) unstable and increasing world food prices; (II) climate change; (III) rapidly growing regional markets; and (IV) globalized marketplaces have changed international contexts, new dimensions in the national context have resulted from (I) better connectivity (roads, internet, mobile phones); (II) outmigration and remittances; (III) movement towards decentralization and community participation; and (IV) new political developments [3].

Nepal is in the process of transforming itself from a unitary system to a federal democratic structure through the New Constitution originally expected by 22 January 2015 as per the Constituent Assembly's work schedule [3]. According to the current Interim Constitution-2007, Article 138 and 139, the organizational structure of the New Nepal—which the upcoming New Constitution will underpin—will most likely entail the democratic, inclusive and progressive restructuring of the state by replacing the centralized and unitary form of the state with a local self-governance system based on decentralization and devolution of power. This changing political context is an opportune moment to review the development of policies affecting livestock development and suggest directions for revision.

Nepal is one of the least developed countries (LDCs) in the world. In fiscal year (FY) 2012/13, its per capita income was US$ 721, with an annual economic growth rate estimated at 3.56 percent [7]. Nepal has 23.8 percent of its people below the national poverty line (NPC, 2013). There are sharp regional disparities in economic opportunity, so that poverty is more rampant and severe in the

Mountain and Hill regions compared to the Terai (plains) in the south and in rural areas compared to in urban areas.

Nepal is committed to alleviate extreme forms of poverty and hunger in line with the United Nations Millennium Development Goals (MDGs). The 13th Plan sets the goal to reduce poverty to 18 percent within the plan period, and to achieve 6 percent annual economic growth [6]. Nepal also has the goal of graduating to the status of a developing country from its current least-developed status by 2022. Nepal aims to achieve inclusive, wider and sustainable economic growth by integrating the contributions of the private, public and cooperative sectors in the development process.

In Nepal, the LSS has been formally recognized as an integral component of agriculture sector of the national economy at least since the inception of the First Plan in 1956. The National Agriculture Policy, 2004 [5] states that the term "Agriculture" includes LSS-related production, industry, and business. According to the National Sample Census of Agriculture-2011/12 (December 2013) [1], Nepal's livestock population stood at 22,135,058, including cattle (cows), mountain cows (yak, nak and chauri), buffaloes, goats, sheep, pigs, horses, mules, asses, rabbits and others, while domestic birds numbered 26,267,815, which encompassed poultry, ducks, pigeons and others.

Livestock farming prevails in all regions of the country, including the Mountain, Hill and Terai belts, with variations based on climate, topography, and socio-economic factors. Nepal has largely a smallholder livestock system under which farmers raise small numbers of livestock in small land holdings. Many farmers with livestock smallholdings are marginalized, close to the survival threshold and driven by subsistence needs rather than market demand. Such farmers are characterized by socio-economic vulnerability due to their inability to withstand adverse impacts from multiple stressors and risks. A common stressor is livestock mortality and morbidity due to poor nutrition and disease. The LSS has a strong gender dimension in that women contribute some 70 percent of the livestock farming work but usually have no significant role in livestock marketing and finance [3]. The LSS contributes to human food security and nutrition, livelihood of farmers, employment and income generation, inputs for farm operation (such as draft and manure), industrial production, and rural transportation.

Given the vital role of the LSS in the country's economy, the Agricultural Perspective Plan, 1995–2015 (APP) targeted growth of livestock share in AGDP from 31 percent in the pre-APP period to 45 percent, driven by livestock growth rates of 2.9 percent to 6.1 percent during the Plan period from 1995 to 2015. However, this targeted livestock growth was not achieved.

Promotion of the LSS is administered and facilitated by the Department of Livestock Services (DoLS) under the Ministry of Agricultural Development (MoAD) of the Government of Nepal (GoN) which assumes ultimate responsibility for total agricultural development including livestock development in the country, although the LSS also relates to other development and infrastructure sectors falling under different ministries and agencies. The organizational structure of DoLS includes Directorates of Animal Health, Livestock Production, Animal Services Training and Extension and Market Promotion, 5 regional directorates and 75 District Livestock Services Offices (DLSOs) with several livestock service centers and sub-centers. Other government organizations related to livestock include units of the DoLS Central Veterinary Hospital (CVH), and Veterinary Epidemiology Unit (VEU) under the Central Lab of Animal Health Directorate, as well as different laboratories. Nepal Veterinary

Council (NVC) and Nepal Agriculture Research Council (NARC) are other, autonomous livestock-related bodies.

2. Study Methodology

This paper adopts a desk-research and policy-review approach aiming at a Strengths, Weaknesses, Opportunities and Threats (SWOT) analysis of the existing livestock-related national policies. SWOT is a strategic analysis technique widely used for getting an overview of the situation of any policy or organization [8]. It provides a logical framework guiding discussion and reflection about a sectoral policy via its positives (strengths and opportunities) and negatives (weaknesses and threats). We attempt to identify the factors needed for the formulation of a new national livestock policy responsive to the needs of previously underserved stakeholders, particularly the poor, and to the challenges of increasing globalization.

With these goals in mind, we first present an overview of existing livestock-related national policies of Nepal as they relate to critical issues concerning the LSS (including breeding, livestock insurance, agri-business promotion, animal health and other socioeconomic issues, Figure 1). The key national laws and planning documents from recent decades are then reviewed for their impacts on LSS using a SWOT analysis. The effects of cross-sectoral policies and interrelated macro-policies on the LSS are examined more briefly. Nepal's LSS policy regime is then discussed from the point of view of system vulnerability resulting from livelihood vulnerability and climate change vulnerability. Finally, we present the rationale for formulating a separate national livestock policy by identifying categorical premises for such a new policy and specific policy suggestions for making a new separate national livestock policy effective (Figure 1).

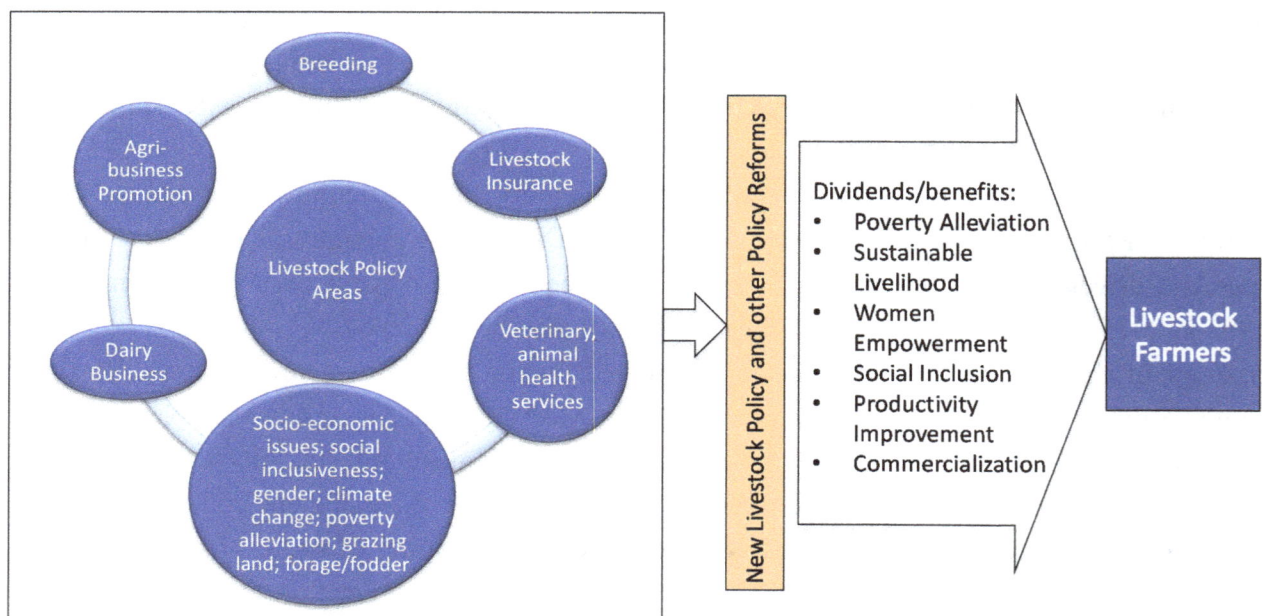

Figure 1. Model of study: interrelationships of livestock policy areas and expected outcomes (dividends) of the new livestock policy to livestock farmers and country.

3. Review and Analysis of Livestock-Related National Policies

The Interim Constitution of Nepal 2007 directs Nepal as an independent, sovereign and democratic state to "develop agriculture as an industry to improve economic conditions of majority of the people dependent on agriculture" [9]. A national-level priority has been accorded to agricultural development since Nepal's planned development initiative began in 1956 with the introduction of the periodic plans. A 10-year long-term agriculture development plan was pushed forward in 1975 between the five-year periodic plans. Livestock development is subsumed under this stress on agricultural development. The Asian Development Bank (ADB) has been the major development partner supporting the LSS. The ADB assisted in the formulation of Livestock Master Plan, 1993, Agriculture Perspective Plan (APP), 1995–2015, Policies such as National Agricultural Policy, 2004 (NAP), and Agriculture Business Promotion Policy, 2007 (ABPP), and Dairy Development Policy, 2007 (DDP).

We proceed to evaluate the existing livestock-related national policies of Nepal by analyzing their *strengths* and *weaknesses* as well as their industry-specific *opportunities* and *threats* (SWOT; Figure 2) *vis-à-vis* the changing context of external (national and international) environments. For each policy, we contrast internal *strengths* and external (macro-environmental) *opportunities* (*i.e.*, the SO combination) against internal *weaknesses* and macro-environmental *threats* (*i.e.*, the WT combination). Such a policy analysis highlights positive (SO) and negative (WT) aspects that should be considered in a new LSS policy.

Internal Aspects → External Aspects ↓	Strengths	Weaknesses
Opportunities	SO	WO
Threats	ST	WT

Figure 2. The SWOT Matrix/Grid.

Currently, the relevant major policy documents that guide livestock development are NAP, 2004; NBPP, 2007; and the Agricultural Sectoral Policy (ASP), 2013–2016 contained in the Approach Paper to the 13th Plan, 2013–2016 (TY, 2013–2016). These policy documents, analyzed in detail below using the SWOT framework, all explicitly aim at increasing livestock production and productivity and achieving diversification, commercialization and competitiveness of livestock products. Many other national policies that include livestock-related policy provisions in explicit or implicit forms (Table 1) are also discussed more briefly as they relate to the needs of the LSS.

3.1. Agriculture Perspective Plan, 1995–2015

The 20-year APP is a strategic document formulated in the mid-1990s and reviewed in 2001–2005 and 2006. The APP recognized livestock as one of its four priority outputs, and saw livestock as contributors to regional balance and gender mainstreaming.

Table 1. List of main and secondary national livestock policies/plans [1] reviewed.

Main Policies/Plans	Secondary Policies (Livestock-Related Policies in Other National Policies)
- Agricultural Perspective Plan (APP), 1995–2015	- Animal Feed Act, 1976
	- Animal Health and Livestock Services Act, 1999 and Animal Health Program Implementation Procedure, 2013
	- Animal Slaughterhouse and Meat Inspection Act, 1999
	- Labor Policy, 1999
- National Agriculture Policy (NAP), 2004	- Child Labor Act, 2000
	- Forestry Sector Policy, 2000 (Forest Policy, 2000)
	- National Micro-Finance Policy, 2005
- Agro-Business Promotion Policy (ABPP), 2006	- Dairy Development Policy, 2007
	- Agriculture Bio-diversity Policy, 2007
	- Trade Policy, 2009
- Approach Paper to the 13th Plan (2013/14–2015/16) and Agriculture/ Livestock Development Policies	- Climate Change Policy, 2011
	- Breeding Policy, 2011
	- Birds Rearing Policy, 2011
	- Rangeland Policy, 2012
	- Livestock Insurance Policy and Agriculture and Livestock Insurance Regulation
	- National Land Use Policy, 2012

[1] All the laws, policies and rules are under the Interim Constitution of Nepal, 2007, the fundamental law of land.

3.1.1. Strengths and Opportunities

The APP identified livestock as one of the four priority outputs, and planned to raise its share in the AGDP to 45 percent, through a targeted livestock annual growth rate increasing to 6.1 percent for the plan period from 2.9 percent previously.

(a) The APP introduced the vision of commercialization of agricultural sector, including the LSS [10].

(b) The APP emphasized the promotion of private sector investment in livestock, as livestock output is assumed to be largely demand-driven and dependent on private sector activities.

(c) With regard to public sector investment in the livestock sector, APP accorded first-order priority to transportation followed by irrigation and research and extension. The APP estimated public sector investment in the livestock subsector over 7 years (1997–2004) at 5 billion Nepalese rupees (NPR), which was almost fully met by the government's budget allocation.

(d) The APP planned to use the livestock sector to aid regional balance and gender mainstreaming.

(e) The APP recognized livestock as a high-value product and acknowledged the need for import substitution.

(f) To achieve growth in livestock business, the APP recommended several policy reforms, including the removal of subsidies in livestock processing and marketing, resolution on the privatization of the Dairy Development Corporation (DDCN), enforcement of standards, establishment of a market information system, introduction of seasonal pricing of milk, and removal of restrictions on the movement of livestock commodities [3].

(g) In the poultry business, an annual growth of 5.37 percent was achieved against the APP target of 5.1 percent [3].

(h) The APP exempted fresh milk and yoghurt from the Value Added Tax (VAT), making them cheaper to consumers and helping the dairy industry expand.

(i) The APP's long-term vision recognizes the need of formulating policy in view of recent developments including the liberalized economic environment, increased role of private sector, Nepal's Millennium Development Goal commitment, and Nepal's commitment to the WTO and regional trading arrangements.

3.1.2. Weaknesses and Threats

(a) In most cases agriculture and livestock related policies have not been implemented effectively because of various factors, including limited human resources and implementation capacity, lack of supportive legislation and monitoring and evaluation, poor coordination and weak planning [3].

(b) The APP's livestock sector strategy laid emphasis on milk and meat production, animal nutrition (specifically, nutritious fodder supply), and health and marketing, while it did not prioritize wool and pashmina, despite the large export manufacturing industries based on these two products. The threats of market domination from Tibetan wool producers that supply wool to the Nepali pashmina and carpet industry continue to exist, as the current amount of pashmina and sheep wool produced in the country is not sufficient or of acceptable quality for producing exportable carpet and pashmina to even meet the 8 percent minimum local wool requirement stipulated by the WTO's rules of origin (RoO). Scattered herders in the mountains produce wool, but there is no aggregation into viable-scale lots, and there is little or no primary processing to connect the value chain. As a result, raw wool is sold into Tibet, where primary processors raise the value significantly through simple removal of coarse hair, washing and bulk-packing, and the Tibetan traders subsequently sell it back to Nepal at about 10 times the price. To address this problem, the new policy should promote processing pashmina and sheep wool within Nepal by collecting it from small, scattered herders.

(c) The APP's actual achievement of targeted annual growth rates in the milk and meat areas fell short, as up to 2003/04 they amounted to 2.77 percent and 2.87 percent, respectively against the targets of 4.2 percent and 4.5 percent [3].

(d) The APP did not address the impact of the existing legislation on the APP implementation, nor did it propose specific legislative changes.

(e) APP did not take into consideration climate change and its effect on livestock activities.

(f) There is a lack of policy consistency and clarity. Not laying a substantial emphasis on the privatization of the DDCN implies that the APP recognized the importance of promoting a level playing field between the DDCN and the private sector. Hence, in view of the privatization of the Pokhara Milk Supply Scheme, a project under the DDCN, in June 2004 in order to comply with the conditions associated with the ADB Community Livestock Development Project loan, new livestock-related policy should make a clear policy decision on whether to privatize DDCN.

(g) The DDCN's role as the price-setter and controller in the dairy business instead of market mechanisms goes against the APP spirit. Set prices do not reflect geographic location differences, cross-border prices, general business cycles, and the impact of rising costs, wages, utilities, and taxes. Donor agencies and analysts have concluded that the government fixing of producer and retail prices is a major detriment to dairy development in Nepal. This reiterates the need for policy reforms as part of the new national livestock policy.

(h) Although APP's vision of commercializing the livestock sub-sector is consistent with the Ninth Plan, Tenth Plan and NAP-2004, little has been done, except for the implementation of the Pocket Package Strategy (PPS) approach. The concept of contract farming proposed in these five-year plans can be applicable to livestock subsector also, and can be proposed in the new livestock policy.

(i) The APP posted hardly any progress in implementing programs on slaughterhouse and meat inspection. The House and Meat Inspection Act, 1999 and Slaughter House and Meat Inspection Rule, 2001 are yet to be implemented at the national level.

(j) APP policy to exempt livestock and livestock products (including poultry products, feed and feed ingredients) from local taxes as well as to remove barriers to livestock commodity movements was not fully implemented. Local bodies levying taxes on livestock products have often inhibited such products' free movement. Therefore, the new livestock policy should incorporate a categorical policy on implementing such tax exemption provisions.

(k) The APP did not exempt dairy products such as cheese, paneer and ice cream from the value-added tax. To promote the commercial potential of dairy products, the new livestock policy should consider VAT exemption on them.

In the nutshell, the performance of the APP was mixed. Combined with the changed internal and external context, this led the Nepal Government to formulate a new long-term strategy for Nepal, namely the Agriculture Development Strategy (ADS) [3].

3.2. National Agriculture Policy (NAP), 2004

While the APP was being implemented, National Agriculture Policy, 2004 was introduced. The NAP, 2004 adopts a long-term vision oriented towards transforming the current subsistence-oriented farming system into a commercial and competitive one. The NAP aims to contribute to ensuring food security and poverty alleviation. Its objectives are:

(a) To increase agricultural production and productivity,

(b) To develop the basis of a commercial farming system and make it competitive in the regional and world markets, and

(c) To conserve, promote and properly utilize natural resources, as well as the environment and bio-diversity.

The policies of the NAP provided for achieving its objectives include:

(a) to ensure the needs of farmers (I) with access to resources; and (II) with comparatively less access to resources,

(b) to provide special facilities by classifying farmers into (I) those having less than half a hectare of land and lacking irrigation facilities; and (II) those belonging to *dalit* (so-called untouchable) and *utpidit* (downtrodden, underprivileged) classes and other marginal farmers and agricultural workers.

The policy area coverage of the NAP is comprehensive, and it provides a participatory method to ensure the involvement of the stakeholders at the concerned level (village, district, region or nation) in the process of formulating, monitoring and evaluating plans connected with the agricultural sector.

The NAP provides for the formation of a National Agricultural Development Board at the national level, and Agricultural Development committees at national, regional, district and VDC levels. The NAP aims to make Village Development Committees (VDCs) and District Development Committees (DDCs) responsible for the formulation, implementation, monitoring and evaluation of plans in accordance with the Local Self-governance Act, 1997. The Policy gives special priority to a set of high-value agricultural products, and seeks to develop commercial and competitive farming systems by a gradual extension of livestock insurance programs and organic farming.

Overall, the NAP, 2004 has the merit of being decentralization-based, friendly to small-holder farmers of livestock, and inclusive of untouchables, marginalized groups and poor communities. Our specific policy suggestions *vis-a-vis* the existing strengths of the NAP, 2004 from the livestock point of view are given in Table 2.

Table 2. Outcomes of SWOT analysis of National Agricultural Policy.

Strengths (Positive Provisions, Potentials)	Weaknesses (Gaps, Shortcomings): Areas of Improvement
- Identification of priority areas of agriculture in Nepal	- Identification of too many areas as priority impedes effective implementation
- Developing integrated National Agricultural Resource Centers capable of operating survey/surveillance and laboratory services for diagnosis of livestock disease, soil analysis, seed certification, and crop protection, and providing capacity development training to entrepreneurs, business persons, cooperative workers and agriculture activity workers.	- The institution responsible for developing National Agricultural Resource Centers and procedure for them should be specified
	- Capacity development training authority and systems should be specified and put in place.
- Provisions on livestock and crop insurance and extension of livestock insurance program	- Legislation, responsible institutions and implementation modality need to be developed for effective livestock and crop insurance.
- Policy on commercializing different agro-products and attracting investors in agriculture	- Agro-products to be prioritized for commercialization and investment should be identified based on market potential and specific strategies for them should be developed, augmenting the agro-based commodities listed in the National Trade Integration Strategy (NTIS)-2010.
	- There should be policy incentives and programs to promote forage crops to ensure uninterrupted availability of feed and fodder to livestock. The forage-manure-crop nutrient cycle should be promoted to supply manure essential for agricultural crops.
- Policy to systematize and strengthen livestock quarantine services to raise the quality of livestock products and market confidence in them	- A system and institutional arrangement for livestock quarantine services should be put in place to limit disease outbreaks and market impacts.

3.3. Agro-Business Promotion Policy (ABPP), 2006

The ABPP, 2006 aims at diversification, service delivery and private sector involvement to transform agriculture from subsistence to commercial farming. The policy aims to reduce poverty by encouraging production of market-oriented and competitive agro-products and promoting internal and external markets. The policy's specific measures include public-private partnership in services delivery and in infrastructure for storage, marketing and processing. Such measures have been successful, and could be replicated with further investments (Table 3).

Table 3. Outcomes of SWOT analysis of Agro-Business Promotion Policy.

Strengths (Positive Provisions, Potentials)	Weaknesses (Gaps): Areas of Improvement
- The Policy highlights diversification, commercialization and promotion of agriculture sector with private sector involvement in commercial farming, for transforming agriculture from the subsistence to the commercial form	- Specific policies on livestock farming and livestock-business promotion are essential to address issues of livestock farming which is more subsistence in nature than crop and cereal farming in Nepal.
- This policy considers infrastructure development as a cornerstone for commercialization, and emphasizes establishing business service centers for quality agriculture inputs and services.	- The Policy's initiatives on diversification and commercialization of livestock farming and promotion of livestock products should specifically include meat products, dairy products, hide and other animal-based products for both domestic and international markets.
- Emphasis on promoting a partnership approach between Government and the private sector for export of quality goods.	- Livestock farming areas should be linked with special economic zones (SEZs) including commercial production areas and export-oriented production areas.
- Provision on providing 25 percent discount on electricity charge for the first ten years from industry establishment.	- The provision should be translated into actually providing agro-entrepreneurs with this discount. The government should annually budget for assistance/subsidy/rebate/tax structuring to carry out this commitment.
- Provisions to treat agri-business projects as collateral for loans.	- Credit providing institutions including banks and micro-finance institutes should be instructed through the central bank's directives to treat livestock and associated equipment and facilities as collateral.
- Development of market and processing facilities under public-private partnerships (PPP)	- Need institutional linkages to support commercialization of agri-business cooperatives
- Policy to strengthen agricultural information (statistics and market information)	- Mechanism to support commercialization by accelerating and strengthening development of agribusiness and commercial farm management capacity in the Ministry of Agriculture Development (MoAD) research and extension should be implemented.

3.4. Forestry Sector Policy, 2000 (Forest Policy, 2000)

The Forestry Sector Policy, 2000 is relevant to the LSS, as farmers use forests to graze livestock and to collect fodder to feed livestock.

3.4.1. Strengths and Opportunities (Positive Provisions, Potentials)

I   The policy simplified the process of handover of institutional as well as group leasehold forestry to Community Forestry User Groups (CFUGs) and has stressed integration of the leasehold forestry program to local community development.

II   The Policy recommended commercial management for forests in larger blocks in Terai and Inner Terai districts. It provided for the Operational Forest Management Plan (OFMP), a plan consistent with the priority objectives of the erstwhile APP.

III   The Policy aims to base livestock quantities on the amount of fodder production and highland pasture so as to improve forest management and increase the production of fodder by community efforts. The Policy calls to immediately design an integrated national forage development program and an appropriate institutional arrangement for its implementation in order to complement the Master Plan for the Forestry Sector and the APP.

3.4.2. Weaknesses and Threats (Gaps, Shortcomings)

I   Disputes on use of forest resources including the fodder and grazing land (range-lands) should be properly resolved through collaborative forest management mechanisms so that livestock farming can make use of appropriate grazing land.

II   Integrating livestock rearing into community forestry is needed to share forest resources and address the shrinking size of grazing lands.

III   In the livestock, agriculture and forestry sectors, efforts should be made to bridge the gap in national support for Clean Development Mechanism (CDM) projects developed to address climate change.

IV   Livestock management should be initiated in close coordination with the mechanism of handing over forests to community forest user groups (CFUGs) and using community forest resources.

V   The proposed integrated national forage development program and institutional arrangement for its implementation should now be aligned not with the APP but with the forthcoming Agriculture Development Strategy, which has identified feed shortage as one of the main constraints for increasing livestock production.

VI   The livestock projects should be strengthened to create awareness of the farmers in cultivating forage and pasture.

3.5. *National Micro-Finance Policy, 2005*

The National Micro-Finance Policy, 2005 mainly aims at helping to alleviate poverty through micro-financial services, and targets the agricultural and livestock sectors, among others.

3.5.1. Strengths

I   Provides for micro-finance facilities to agriculture sector and allows for group guarantees

II   Sees agriculture as a priority sector for micro-finance

3.5.2. Weaknesses

I   It is highly essential to have separate policies governing micro-finance for the livestock subsector instead of a blanket agriculture-related one, as this subsector has its own salient features that vary substantially from other subsectors of agriculture

II   There is the need for according specific priority to livestock subsector micro-finance by making it mandatory not only for micro-finance institutes but also for commercial banks to allocate/ earmark at least a specific fraction, such as one third (1/3), of the total agricultural loans/ investments in the livestock business.

3.6. Dairy Development Policy, 2007 (2064 BS)

The policy (DDP) envisions investment in the income and employment generating and poverty-alleviating dairy business. To achieve this vision, it has adopted the policy of providing pasture (grazing land) and cattle feed year-round.

3.6.1. Strengths

The DDP encourages concerned organizations to provide collateral-free soft (concessional) loans, group loans and technical assistance to farmers, particularly women and underprivileged communities, with a view to promoting livestock farming. The DDP seeks to mobilize farmers' cooperatives to promote livestock insurance service extension, and provides that the Nepal government may subsidize the premium on livestock insurance obtained by farmers through their cooperatives and groups. Accordingly, the Nepal Government, in its national budget speech-2014/15, has announced a 75 percent subsidy on the livestock insurance premium [11].

I   To ensure quality dairy production, the DDP has entrusted the Department of Livestock Service with the responsibility to provide technical services, manage cattle-feed and livestock health training, and minimize costs.

II   The Policy is based on a long-term vision to encourage participation of public, private and cooperative sectors in dairy production.

III A DDP objective is to increase production and productivity of milk in rural areas which helps alleviate rural poverty.

3.6.2. Weaknesses

I   To assure micro finance lenders and banks of the security of collateral-free loans, there should be provisions for community-based group-guarantee and group-monitoring of dairy borrowers.

II   Incentives to disadvantaged communities will not materialize until social laws and practices effectively erase the old social stigma against using milk and dairy items produced by so-called low-caste people. This effort should coordinate with other social laws and law-enforcing agencies including police and civil servants.

III A large-scale livestock insurance system is yet to be developed. Subsidies should be coordinated with the Livestock Insurance Policy, and there is a need for a setting and monitoring mechanism in this regard.

IV   Resource centers of improved livestock (dairy animal) breeds should be developed.

V   A mechanism to ensure participation of smallholder farmers (backward linkage) at all stages of the value chain to retail products is essential.

VI   Access of rural farmers to livestock support services and loans should be increased through district livestock offices and bank and micro-finance institutes.

3.7. Agriculture Bio-Diversity Policy, 2007

The Agriculture Bio-diversity Policy resulted from Nepal's accession in 1993 to the International Convention on Biological Diversity (CBD)—informally known as Convention on Biodiversity—adopted at the Earth Summit in Rio de Janeiro, Brazil in 1992. This policy is in accordance with the objectives of NAP-2004 to protect, promote and utilize bio-diversity and to promote ecological balance. It intends to lead to benefit from protection and utilization of genetic resources for food security, livelihood security and poverty reduction. Livestock genetic improvements enabled by diversity are expected to contribute to increased productivity.

3.7.1. Strengths

The Policy has the overriding objective to protect, promote and utilize genetic resources and protect biodiversity for sustainable agricultural development coupled with food and nutritional security.

3.7.2. Weaknesses

I   Regulation for research and experimentation on Nepalese bio-diversity and genetic resources of livestock is yet to be developed and implemented.

II   A system for registration and allocation of agro- and livestock biodiversity should be developed.

3.8. Trade Policy, 2009

The Trade Policy is formulated to address issues of international trade dynamics such as affiliation with the regional and multilateral trading system, expansion of bilateral free trade areas, simplification of trade procedures, development of new border transit system, sanitary and phyto-sanitary (SPS) measures, and managing technical barriers to trade (TBT), which could contribute towards sustaining the export trade. This policy aims to support economic development and poverty alleviation through enhanced development of the trade sector. Nepal is a chronic trade-deficit country, having a trade imbalance of more than national budget in the fiscal year 2013/14.

3.8.1. Strengths

The Policy lays emphasis on commercial livestock farming and the promotion and supply of improved breeds. The Policy offers capital and technical assistance for the commercial farming of animals to ensure supply and export of high-quality rawhide and skins.

3.8.2. Weaknesses

I   Nepal's international (export) trade of livestock products and animal-based goods makes it mandatory to develop specific programs and infrastructure/facilities on sanitary and phyto-sanitary measures in compliance with WTO's Agreement on Trade Related Aspects of Intellectual Property Rights (TRIPS) and World Intellectual Property Organization (WIPO)'s Conventions, to which Nepal is a party [12–15].

II   National policies and programs to develop an export led production zone or processing zone with market link are essential.

III National policies and programs on initiating effective international marketing and competitiveness-enhancement for Nepali livestock-related products should be specifically developed to address the export requirements specific to livestock, including proper packaging and advertising policies to address the sanitary and hygienic concerns of foreign consumers.

3.9. Climate Change Policy, 2011

The Climate Change Policy, 2011 is based on Nepal's ratification on November 1, 1993 of the United Nations Framework Convention on Climate Change (UNFCCC) negotiated at the UN Conference on Environment and Development (UNCED) or the Earth Summit held at Rio de Janeiro, Brazil in June 1992 [16]. Nepal also acceded to the Kyoto Protocol (KP) which is an international agreement linked to the UNFCCC, on 19 September 2005. Nepal also adopted a National Adaptation Program of Action (NAPA), 2010 and Local Adaptation Program of Action (LAPA), 2011 for climate change adaptation [10,17,18].

The main goal of the Climate Change Policy, 2011 is to improve livelihoods by mitigating and adapting to the adverse impacts of climate change, adopting a low carbon emissions socio-economic development path, and meeting the spirit of the country's national and international agreements related to climate change.

3.9.1. Strengths

Sustainable management of forests, agro-forestry, pasture, rangeland, and soil conservation that can address the impacts of climate change are in urgent need and are identified as potentials to be prioritized and implemented. Climate adaptation implementation needs to be linked with socio-economic development and income-generating activities to the extent possible.

3.9.2. Weaknesses

Despite advocating sustainable management of pasture and rangeland, the Climate Change Policy-2011 lacks specific strategies and policies in this area. Livestock-specific policies and programs are essential, as intensification of climate change effects adversely affects biodiversity, diminishes and damages grazing lands, and puts livestock at great risk. Policy should also specifically address the needs of livestock development and income-generating activities relating to livestock.

3.10. Rangeland Policy, 2012

The Policy defines rangeland as natural pasture land, grassland and shrub-land. It aims to increase productivity by improving forage/grass productivity, to protect livestock farmers' traditional rights for pasturing livestock in community rangeland and forest, and to determine stocking density to minimize competition between grazing domestic and wild animals.

3.10.1. Strengths and Opportunities

The Policy seeks to secure the facilities traditionally enjoyed by livestock farmers using range-lands located within community forests. The Policy identifies provisions to collect and conserve the green forage (grass) during the rainy season and winter and dry seasons in order to ensure continuous supply of cattle feed round the year. The Policy seeks to determine livestock density on the basis of capacity of the rangelands for minimizing the grazing competition and pressure of both domesticated and wild animals, and imposes charges or penalties on cattle for using rangeland with the goal of limiting unproductive cattle on the rangeland.

3.10.2. Weaknesses and Threats

I   There is a need for integrating livestock rearing into community forestry to share forest-resources and meet the threat of shrinking size of grazing lands.

II Institutional arrangements and collaboration of government with livestock farmers and community forest users at the local level are essential to ensure a continuous year-round supply of cattle feed.

III The rangeland charge—important as an income-generating source for the rangeland management—should be well streamlined and managed.

IV The rangeland charge should serve as a sufficient control measure to discourage use of rangeland by unproductive cattle, as the owners of such cattle would not let them graze in the rangeland; it is impractical and often cumbersome for the rangeland officials to directly identify productive and unproductive cattle.

3.11. Livestock Insurance Policy and Agriculture and Livestock Insurance Regulation (2013)

Livestock insurance is extremely important, as livestock husbandry is risky, particularly for small and low-income farmers who face financial ruin in case of theft, injury, illness or death of an animal. According to DOLS, premature mortality is about 2 percent to 3 percent per annum for cattle and buffalo and considerably higher for small ruminants and pigs. Livestock insurance helps livestock farmers to cope with such risks, and facilitates farmers' access to finance by increasing their creditworthiness.

Although general insurance was introduced in Nepal in 1937 after the establishment of Nepal Bank Ltd, the country's first commercial bank, and the National Insurance Corporation was established in 1967 [13], livestock insurance began only in 1987 in form of livestock credit or micro-finance guarantee insurance against animal mortality and loss. In Nepal, many organizations provide livestock insurance services on a limited scale; they include the Small Farmers' Development Bank (SFDB), Micro-Finance Institutions (MFI), Community Livestock Development Projects (CLDPs) sponsored community-based organization (CBOs) and Financial Intermediary Non-Governmental Organizations (FI-NGOs) which are not regulated by the Insurance Board (IB), the national-level regulating body.

This gap should be addressed by a proper policy mechanism. In recognition of the need of systematizing livestock insurance, Nepal introduced Livestock Insurance Regulation and a Livestock Insurance Policy. The Livestock Insurance Regulation under the Insurance Board aims at encouraging financial institutions to finance more agricultural projects, as most financial institutions abstain from extending loans and advances to livestock and agricultural projects in the absence of proper insurance coverage. The Agriculture and Livestock Insurance Directive makes it obligatory for non-life insurance companies to issue insurance policies on livestock, crops and poultry. Table 4 outlines the SWOT analysis of National Livestock Insurance Policy.

Table 4. SWOT Analysis of Livestock Insurance Policy.

Strengths	*Weaknesses*
Policy to promote livestock and crop insurance for encouraging financial institutes to invest more on livestock and crop projects.	- There should be specialized insurance companies licensed by the Insurance Board to provide crop and livestock insurance policies. - The number of microfinance institutions, agriculture cooperatives and financial Non Government Organizations (NGOs) that provide livestock and crop insurance as part of their credit-plus program to their beneficiaries should be increased. The ceiling on the amount of insurance they may provide should be raised from the current limit of Nepalese Rs 100,000.

3.12. National Land Use Policy, 2012

The Policy aims to encourage optimal use of land for agriculture by classifying the country's land territory into seven land use categories—agricultural, forest, residential, commercial, public, industrial, and others. Land in the agriculture category is for agricultural cultivation, livestock farming, and tree plantation. The Policy also aims to increase agricultural productivity by systematizing land fragmentation and by adopting a land pooling system. The goal is to encourage commercial, cooperative and contractual farming. The SWOT analysis of National Land Use Policy is outlined in Table 5.

Table 5. SWOT Analysis of National Land Use Policy.

Strengths	*Weaknesses*
- Policy to allocate land for agricultural purposes including livestock farming - The Policy also aims to increase agricultural productivity by controlling land fragmentation, systematizing land-pooling activities, and encouraging commercial, cooperative and contractual farming	For all these policies to be effectively implemented there is an imperative need of a separate Land Use Act that provides adequate legislative backing.

3.13. Breeding Policy, 2011 (2068)

The Policy aims at increasing productivity of milk, meat and eggs and hence increasing farmers' income through improvement of livestock and poultry.

3.13.1. Strengths and Opportunities

The policy has set goals of utilization, conservation and improvement of genetic resources and capabilities of livestock and poultry and hence achievement of increasing productivity.

3.13.2. Weaknesses

I   There is a need to exploit potentials for breed improvement and increasing livestock productivity.

II There should be a policy and implementation mechanism on development of livestock (and poultry) resource centers through public-private partnership.

III Programs should be urgently initiated to conserve indigenous breeds which are in danger of extinction.

IV Policy and programs to promote breeding of productive livestock (genetic resources) are needed as a part of implementation of agreements from the New Earth Summit 1992 (on livestock genetic resource conservation and their improvement).

3.14. Animal Health Program Implementation Procedure, 2013 and Animal Health and Livestock Services Act, 1999

The Animal Health Program Implementation Procedure, 2013 was introduced in line with the Animal Health and Animal Service Act, 1999 and its related Regulation.

3.14.1. Strengths and Opportunities

The policy rightly aims at promoting production, distribution, consumption and export of healthy livestock and making animal-health related programs more effective, as these functional areas are crucial in livestock management.

3.14.2. Weaknesses

There is a need to develop policy and procedures to protect livestock from emerging internationally endemic diseases such as swine flu and foot-and-mouth disease, which should be explicitly tied to implementing the WTO Agreement on Sanitary and Phyto-sanitary (SPS) Measures [12,13,15].

3.15. Labor Policy, 1999, and Child Labor Act, 2000

Nepal, a member of the International Labor Organization (ILO), has so far signed 11 ILO conventions [4,19]. Hence, it should make its labor policy and practices fully compliant with its commitment to international labor standards and practices. The Labor Policy, 1999 in compliance with the Child Labor Act, 2000, bans use of child labor in economic activities. The policies and provisions also have bearing on livestock management.

3.15.1. Strengths and Opportunities

Policies to ban forced or voluntary use of child laborers in economic or business activities are identified. By implication, the use of children is restricted in the risky work of herding big ruminants.

3.15.2. Weaknesses

The Labor Policy needs to specifically address the problem of the widespread use of child labor in livestock farming.

3.16. Birds Rearing Policy, 2011

The policy was issued within the framework of National Agriculture Policy (NAP), 2004 and Agri-business Promotion Policy, 2006. The policy covers the poultry business, encompassing chickens, cocks, hens, ducks, turkeys, quails and other local bird species.

3.16.1. Strengths and Opportunities

I   The policy is compliant with key agricultural national documents including the supplementary to the National Agriculture Policy (NAP), 2004 and Agri-business Promotion Policy, 2006.

II   It plans to make the poultry business more productive, competitive and sustainable by improved quality of chicks through well-managed hatchery and rearing as well as by systematizing distribution of poultry products.

III   The policy envisages programs to base bird rearing and poultry businesses on comparative cost advantages and production potentials.

IV   The policy is consistent with the national policy thrusts of public-private partnership and environment protection as far as its implementation plans are concerned.

3.16.2. Weaknesses and Threats

The policy has not spelled out plans and mechanisms to fight the consequences of sudden outbreaks of bird flu and other diseases that have devastated Nepalese poultry in recent years. Such policies and functional strategies should be in accordance with the WTO Agreement on Sanitary and Phyto-sanitary Measures.

3.17. Approach Paper to 13th Plan and Agriculture/Livestock Development Policies

After 2007, different stakeholders in the Government of Nepal (GoN) and society at large increasingly perceived that the APP—viewed in a new national and international context—had not been successful in achieving its main targets and that there was the need of a new long-term strategy, which in 2012 resulted in the formulation of the Agricultural Development Strategy [3]. The agricultural sector development policy in the ASOPs of the Approach Paper to 13th plan (2013/14–2015/16) has made provisions for the livestock sub-sectoral development by including it in objectives, strategies and operating policies. The agricultural sector objectives set in the Approach Paper to the 13th Plan are

I   To increase the production and productivity of crops and livestock products,

II   To make crops and livestock products competitive and commercial,

III To develop and disseminate environment-friendly agro-technologies to minimize the adverse impacts of climate change, and

IV   To conserve, promote and utilize agro-biodiversity

The sectoral strategies for achieving the objectives are directed towards

I   Promoting commercialization and diversification of agriculture and livestock

II   Developing crop and livestock industries and enhancing their product quality

III Encouraging youths to take up commercial farming as a prestigious profession

IV   Promoting agricultural and livestock marketing, and

V   Promoting the results-oriented application of technologies in the sector; while many of the 46 operating policies set in line with the eight priorities or strategies are generally related to the overall agricultural sector, some others are specific to the livestock subsector:

I   to expand promoting campaigns regarding artificial insemination and fodder and forage plantation,

II   to develop rural infrastructures such as agro-roads, electricity, and communications,

III   to develop agricultural marketing network including livestock wholesale markets and hat bazaars (open-air retail markets), and expand access of livestock information at local levels,

IV   to develop technical manpower for agricultural sector and provide entrepreneurship and skill development training required for agro-business,

V   to encourage production of high quality seeds, high-yielding breeds and vaccination, and to develop bio-pesticides to treat animal for parasites,

VI   to make provisions for livestock insurance, concessional agricultural loans, subsidy on livestock related industrial equipment and tax rebate on trade to small and marginalized farmers, entrepreneurs and business people,

VII   to promote contract and cooperative farming with involvement of private entrepreneurs and cooperative sectors,

VIII to establish agriculture and livestock extension centers under the local bodies at each VDC,

IX   to strengthen livestock related laboratories, and

X   to provide integrated agricultural and livestock services and make effective involvement of national and international non-governmental organizations, universities and local bodies in providing such services.

The TP Approach Paper—although the most recent of the three major policies and policy-documents we have been considering—failed to make any mention of either of the other two, the NAP, 2004 and ABPP, 2007. No plan can be implemented in isolation without coordinating with other existing policy frameworks and implementation mechanisms. Hence, it is desirable that the 13th Plan formulates and executes plans and policies in pursuance and compliance with the previously issued and/or existing national policies as starting points for revision. Even though the Approach Paper envisages commercializing the livestock business and making this subsector competitive, lack of coordination and collaboration with other subsectors of agriculture as well as with existing agriculture-related policies is likely to handicap the accomplishment of the purpose. For instance, the National Agriculture Policy, 2004 has the policy to promote programs on improved livestock production and productivity, controlling livestock-related diseases and systematizing livestock quarantine services; but the Approach Paper has established no explicit linkage with such existing frameworks.

4. Critical Cross-Cutting Concerns Facing the Livestock Subsector

The ASOPs in the Approach Paper to the 13th Plan (FY 2013/14–2015/16) also recognized the national policies on poverty alleviation, human resource development, labor and employment, sustainable and balanced development, environment and climate change, gender equality and inclusion and disaster management as cross-sectoral policies. Nepal's livestock subsector relates closely to several of these critical issues, which demand serious consideration in LSS policy-making initiatives.

4.1. Sectoral Contribution to Regional Balance and National Trade Balance

The LSS contributes to within-country regional balance in Nepal [10]. Compared to the southern belt of Terai and the valleys and towns in the Hill region, the potential of industrial businesses is almost non-existent in the Mountain region located in the country's north, where livestock businesses display the most potential to contribute to economic activity. The LSS is also immensely important for its tremendous potential to contribute to national goals of export promotion, trade deficit reduction, poverty reduction and import substitution.

4.2. Gender Mainstreaming and Child Labor

The Platform for Action of the World Conference on Women held in Beijing, 1995 provided an impetus to address gender inequality, and the Government of Nepal formulated a national plan of action to implement 12 critical areas of concern in line with the UN Millennium Development Goals (MDGs), including women's poverty, access to education, health services, participation in decision making, and vulnerability to violence [20,21]. The Interim Constitution of Nepal, 2007 provides that the state shall not discriminate in any form against women and shall pursue a policy of encouraging maximum participation of women in national development [9]. The ADB-assisted Third Livestock Development Project (TLDP), 1997 and Community Livestock Development Project (CLDP), 2004, for Nepal also accorded high priority to gender aspects in livestock. Nepal has ratified two relevant international conventions, Elimination of All Forms of Discrimination against Women and UN Declaration against Discrimination of Women and Men. Nepal's commitment to non-discrimination, gender equality, and social justice at national and international levels should be reflected in its national policies and practices, and hence gender is a critical factor in the formulation of the new national livestock policy.

The LSS plays a predominant livelihood role in the hills and mountains, and given that some 70 percent of the work in livestock farming is performed by women [9], the livestock business possesses tremendous potential to promote gender equality via empowerment and inclusivity of women and can contribute to national gender mainstreaming initiatives. However, rural women's dominant role in livestock activities is so far limited to livestock rearing, including fodder/ forage collection, feeding, herding, breeding and animal-health care. Women are largely denied any role in and access to economic returns from the livestock. Instead, men in the family solely handle the sales of livestock outputs including milk, wool, manure, and the livestock themselves and hold the proceeds. Gender balance is therefore a critical issue which should be addressed in the National Livestock Policy. Migration of men from rural areas has resulted in growing feminization of rural households, which increases the burden to women who also face vulnerability due to lack of access to financial and educational resources and restriction on their social roles and mobility.

Use of child labor is common and widespread in livestock farming, particularly in collection of feed and fodder and in grazing. Use of children in livestock farming—whether voluntary or imposed as their familial obligation—deprives them of the opportunity to go to school, and also puts them at physical danger in handling and herding large cattle.

4.3. Gaps between Policy and Implementation

National policy is a broad course of action adopted by the government in pursuit of its objectives. Nepal has already a rich body of policies in favor of agriculture [3]. The National Agriculture Policy and the Approach Paper to the 13th Plan emphasize the central role of agriculture. Nevertheless, formulation of some important policies has been excessively delayed. For example, breeding policy remains under review though its improvement was initiated more than 50 years ago. The gaps in policy and in the implementation of existing policies are outcomes of

I   Lack of supportive adequate legislation (acts), rules and regulations for credible enforcement,

II   Inadequate resource allocation,

III   Ineffective coordination,

IV   Irregular and weak policy and program monitoring and evaluation,

V   Lack of climate change monitoring,

VI   Limited human resources and implementation capacity, and

VII Lack of continuity in leadership (short tenures of ministers and secretaries).

4.4. Gaps in Planning Process

The ADS (2012) identified the following gaps in the planning process [3]:

I   Poor data base for agriculture sector, especially in the areas of productivity, inputs, trade, seeds, improved breeds and agribusiness.

II   The periodic plans do not cover programs/projects to the implemented through private sector, community-based organizations (CBOs) and non-government organizations (NGOs); the plans very much concentrate on programs to the implemented by the government only.

III   There is no system of output and impact monitoring and evaluation.

4.5. Pro-Poor Policy and Poverty Reduction

Proper LSS development is critical to poverty reduction, the overriding goal of all national plans and policies in Nepal. With almost three out of every four Nepali households and almost every household in Mountain and Hill rural areas involved in livestock rearing [22], improved performance in livestock activities could be instrumental in Nepal's national bid to reduce poverty. At the same time, livestock vulnerability resulting from poverty, climate hazards, declining ecosystem services and socioeconomic inequality leads to low yields, indebtedness, and worsening poverty among poor livestock farmers with small holdings and no other socio-economic backups. There is an imperative need for making the national livestock policy pro-poor. The livestock subsector holds tremendous potential for bringing communities of marginalized, underprivileged and indigenous groups as well as so-called untouchables into the national mainstream, since many of these groups are heavily reliant on livestock.

4.6. Political Restructuring

The Approach Paper to the 13th Plan aims at promoting good local governance and empowering local bodies politically, financially, administrative and judicially in federal structure. Obviously, the livestock subsector and policy cannot function and thrive unless they are matched with Nepal's state structure. Nepal is committed to restructure its unitary, highly centralized government to form a federal, decentralized democratic system. Policies governing the livestock subsector are to be enforced at both federal (central) and state levels, for which coordination and consistency are indispensable. The federal or central government will have responsibilities for designing and implementing national policies, and the state or regional, district and city level governments have to manage local affairs concerning administration, law enforcement, and development under the federal national policies, regulations and mandates [23].

5. Livestock Policy and Vulnerability

The vulnerability of Nepal's livestock subsector includes livelihood vulnerability and climate-change vulnerability.

5.1. Livelihood Vulnerability

Livestock is an important resource of livelihood in Nepal [22]. The threats of livelihood vulnerability to livestock farmers in Nepal are mainly due to small holdings, poverty, and socioeconomic marginality. A majority of farmers in Nepal have poor resource endowments, small land holdings and lack of access to adequate land, low bargaining power, and weak risk bearing capacity. For smallholders, losing livestock has a great impact and lasting effect on livelihood so that livestock sickness and mortality could even trigger chronic poverty. High vulnerability and reduced livelihood options has increased off-season migration to India and more distant countries, which has increased the risks of indebtedness of poor families and put additional burden on women, children and elder population to cope [24].

Despite its crucial importance, Nepal's policies did not pay adequate attention to livelihood vulnerability in agriculture sector. For instance, Cameron (1998) found that the 20-year APP failed to grasp the nature of livelihood inequalities and there was no targeting of livelihood vulnerable people by economic, social, geographical or age factors [25,26].

5.2. Climate Change Vulnerability

Climate change impacts on the overall livestock system are mainly due to the changed water resource supply, forest health, soil health, land use, and human settlement and migration patterns. Climate change impacts in Nepal have added new dimensions of challenges to many sectors of natural resource management. More severe impacts have been observed in the rural and remote areas where the livelihoods of people are based on subsistence agriculture with limited livelihood options. People are vulnerable to extreme weather events, have poor access to information and lack resources to cope with and recover from climate-related disasters.

The impacts of climate change are not evenly distributed between different communities. Poor and marginalized communities, who often live in disaster-vulnerable areas with limited information, limited livelihood options and low adaptive capacity, are most vulnerable to climate change [25,27]. Similarly, women are on the front line of climate change due to their multiple burdens to obtain livelihoods. The predicted impacts of climate change will heighten existing vulnerabilities, inequalities and exposure to hazards [28,29]. Effects of climate change tend to be more severe where people rely on weather-dependent rain-fed agriculture for their livelihoods. In rural mountain communities with limited livelihood options, adaptive capacity is low due to limited information, poor access to services, and inequitable access to productive assets. Few studies have reported on the status of rural and remote mountain areas in Nepal and on adaptation strategies in use.

Therefore, to address the climate-change vulnerability associated with the livestock subsector, there is a great need of reorientation of the livestock related national policy, restructuring of the national organizational system, enlargement of strong infrastructures and support services and promotion of gender equality with increased inclusivity and empowerment of women. The national livestock policy should address the needs of lasting sustainability, increased productivity and profitability, commercialization, expanded markets, and diversification. Nepal's National Adaptation Program of Action (NAPA), 2010 recognized agriculture and food security as one of six thematic areas [17]. Although the APP did not consider climate change issues, the 13th Plan (2013/14–2015/16) in its Approach Paper has accorded 6th priority for the promotion of mitigating and adaptive techniques and practices to minimize the adverse impacts of climate change on agricultural sector. But the same plan accorded the first (top) priority to the implementation of the NAPA, 2010 and LAPA, 2011 for poverty alleviation. And, only 6th priority has been given to designing and implementing programs related to climate change adaptation by local bodies.

6. Rationale and Categorical Premises for a New National Livestock Policy

Since the livestock subsector is an extremely important component of agriculture and has a nature and characteristics substantially different from other subsectors of agriculture, there is an imperative need for a separate National Livestock Policy that caters to specific needs of the livestock subsector at national, regional and local levels, so that sustainable livestock development can be fully integrated with related national policies, laws (acts) and other national rules and international conventions and be truly pro-poor and growth-driven.

At present, Nepal's national level policies related to the livestock subsector are scattered across different national policies. These policies should be integrated into a new National Livestock Policy. Furthermore, the policy and programs provisions in different national policies broadly relate to agriculture as a whole, whereas there is a need to have such policies and programs specifically for the livestock subsector.

There is no demarcation of objectives and strategies specific to the livestock subsector in the Plan Approach Paper. The Approach Paper in Chapter 3 deals with the sectoral development policies, of which livestock area is one of the subsectors, but makes no reference to existing agriculture-related policies. The approach paper lists 46 operating policies, but these policies are not explicitly related to the specific needs of the livestock subsector. On the other hand, as covered above, APP (1995) and the Agro-Business Promotion Plan have brought forward many pertinent issues but have not fully and adequately covered and addressed them. These problems can be resolved by formulating a new, separate national livestock policy that links the livestock subsector with other subsectors of agriculture as well as with existing related national policies.

At a time when Nepal is in the process of federating its state structures, another important issue concerning the policy on livestock subsector development is to decide whether to bring all responsibilities to livestock stakeholders (including public and private sector organizations, NGOs, livestock-farmers and service-providers) into the fold of central or federal government to avoid duplication, or to decentralize to the local levels. In view of the direction of the larger process of state restructuring, the national livestock policy should match the new federal dispensation, where states or provinces will have their own state/provincial livestock policies according to their requirements within a national policy which plays more of a coordinating role. The national livestock policy should guide and assist the local governments in framing their livestock development policies according to the location-specific needs.

7. Policy Recommendations

In view of the livestock-related national policies being scattered across different national policies of Nepal, there is an imperative need to formulate and implement an integrated National Livestock Policy, so that sustainable livestock development can be achieved integrated with National Agriculture Policy and other related national policies, laws, rules and international conventions. On the basis of the gaps discussed, the following general policy suggestions in regard to formulating a new National Livestock Policy may also be put forth in addition to the specific ones elucidated earlier in the review of individual policies.

Livestock policy should consider livelihood vulnerability which is potentially caused by livestock vulnerability and climate-change vulnerability. General suggestions and conclusions related to policy orientation are as follows:

I   Policies on livestock insurance and corresponding institutional arrangements are essential.

II   Policies and programs should provide soft loans for livestock farmers and other livestock workers including livestock health services providers.

III   National Livestock Policy should be in compliance with measures on climate change, as biodiversity loss results from intensification of climate change effects leading to degradation of pasture lands that has put livestock at great risk. There is need for capacity building on climate science and policy to inform livestock policymakers, implementers, and stakeholders.

IV   The new policy should incorporate priorities specific to geographic locations (like Terai, Hill, Mountain, Valley and Siwalik/Inner Terai) and take into account the differing density of livestock to result in regional balance within Nepal.

V   There should be appropriate legislative backing to the national livestock policy to support and ensure its effective implementation and adjudication. Also in other countries where mixed crop-livestock production systems exist, policies have not successfully accommodated realities of a mountain environment where livestock are key to smallholder livelihood [30] .

VI   An integrated approach to research and development (R&D) should be developed for fostering new technologies and policies

VII   A comprehensive and forward-looking policy environment should provide for policies to be updated and improved regularly [31].

VIII   Livestock policy studies should be extended to accommodate the emerging concerns of socio-economic and national resources management and climate change mitigation and adaptation.

IX   Political parties and mass organizations should be involved in livestock development programs.

X   Participation of villagers, local government bodies (VDCs and DDCs), and NGOs as collaborators in livestock development activities must be encouraged.

XI   The central government should bear the costs of developing livestock technologies that are in the national interest.

XII   Each state (or province) in the new federal structure should identify livestock growth-axes and growth-centers for promoting commercial production.

XIII   Public private partnerships (PPP) along with the appropriate involvement of cooperatives should be promoted in livestock development activities. As suggestions on policy content and planning:

I   There should be adequate and well-placed investments in the livestock subsector from both public and private sectors to allow policy implementation. Nepal's government budget allocation for the entire agriculture sector has been meager and dwindling; in FY 2009/10, it figured only 2.75 percent of the total national outlay against 6.2 and 6 percent in Bangladesh and India respectively, and more than 4 percent even in conflict-hit African nations [32].

II   The new policy should identify livestock projects that can be developed as strategic business units (SBUs), particularly for foreign-assisted projects. To identify such projects, past success stories and experiences in the livestock business can be used as benchmarks. There should be institutional arrangements and mechanisms for developing human resources required for farmer education, business development, and effective policy implementation.

(a)   The newly established Agriculture and Forestry University (AFU) should be strengthened with a focus on developing human resources for livestock-promotion. There should also be roles for other existing universities, including Nepal's pioneer Tribhuvan University, and for private sector institutes working in the area.

(b) Organizational resources at the Centre for Technical Education and Vocational Training (CTEVT) should be effectively managed and mobilized to develop low-level extension workers and technicians for livestock activities.

III   Commercialization and diversification need to be addressed in a broader context that incorporates the emerging challenges and opportunities offered by regional trade agreements like South Asian Free Trade Area (SAFTA) and Bay of Bengal Initiative for Multi-sectoral Technical and Economic Cooperation (BIMSTEC).

IV   There should be policy thrusts and programs for promoting livestock marketing activities to establish stronger forward-linkages that provide farmers better access to market, so that livestock farmers can gain adequate returns from their outputs. Promote a marketing network, including livestock wholesale markets and haat bazaars (open-air retail markets), and centers for livestock information at central, provincial and local levels.

V   In the new policy, there should be a well-functioning mechanism to channelize the livestock-related benefits envisaged directly to marginalized, disadvantaged communities

VI   Socioeconomic issues related to gender, child-labor, markets, community and farmers groups all have direct implications on livestock activities and should be considered in developing the national livestock policy. There is a need to ensure women farmers' participation at all stages of livestock development planning. Women's role should not only be limited to livestock rearing and care, but extended to include marketing and finance. Policy on women's empowerment and participation should comply with Nepal's commitment to the Beyond Beijing Conference and other international conventions including the UN Declaration against Discrimination of Women and Men and Elimination of All Forms of Discrimination against Women. Livestock policy should address the problems arising from use of children in livestock farming.

VIII The new policy should promote projects to set up rainwater harvesting facilities that increase water and fodder for livestock.

IX   The livestock policy should be developed and reviewed from the perspectives of (a) Improved cattle breeds; (b) Improved forage crops and modern varieties; and (c) animal health, including shelter from extreme weather and control of pathogens.

X   It is important to make provisions to delegate responsibilities to local and community levels in appropriate areas:

(a) At the community levels, policies should make adequate provisions for backward linkages to promote livestock activities; they include, if are not limited to, pasture land management, animal health services, scientific breeding, and feed and forage management.

(b) Poor data in the LSS, especially in the areas of productivity, inputs, trade, improved breeds and agri-business, has been a hurdle in livestock development at the local and community level. A new livestock policy should establish an improved database and provide for constant updating and upgrading.

(c) A comprehensive strategy that coordinates various stakeholders in livestock development should be developed. Promotion of the free-market mechanism in the dairy business is deemed necessary and therefore the DDCN's current role as the price setter and controller need to be revisited and redefined. A market mechanism should ensure that dairy prices reflect geographic location differences, cross-border prices, general business cycles and the impact of rising costs, wages, utilities, and taxes.

(d) There can be provisions for subsidizing livestock-related technologies and also for exempting livestock products of small-holder farmers from central and state government taxes as well as local taxes. Nepalese who have returned from foreign employment should be encouraged to take up livestock raising, processing and marketing by providing entrepreneurial support and micro-credit facilities to promote local ecomomy.

The new policy should provide for proper institutional arrangements to perform regular evaluations and controlling functions, and obligate the line agencies and stakeholders to comply and correct the lapses or gaps discovered.

(a) A monitoring and evaluation system should be established and include the impact of livestock programs on national priorities like poverty alleviation and climate change monitoring.

(b) There is a need to review and strengthen the organizational framework and increase the capability of human resources for implementing policies.

(c) Technical and logistical assistance is needed to grassroots and village organizations for implementing resilience-building measures that promote long-term livestock development at specific locations.

8. Conclusions

There is an urgent need for formulating a separate, integrated national livestock policy and for implementing it with adequate institutional support and resources so that Nepal can sustainably increase livestock production and productivity and achieve diversification, commercialization and competitiveness of the livestock subsector to match the changing national and international contexts. The new policy needs to be pro-poor and inclusive to properly address the urgent national agendas of gender mainstreaming, livelihood vulnerability management, climate-change vulnerability management, and the protection and promotion of interests of underprivileged and indigenous communities and of economically underprivileged areas.

Acknowledgements

This work is part of the project "Adaptation for climate change by livestock smallholders in Gandaki river basin", supported by the USAID Feed the Future Innovation Lab for Collaborative Research for Adapting Livestock Systems to Climate Change at Colorado State University under subaward 9650-32. The authors are also thankful to The Small Earth Nepal. All statements made are the views of the authors and not the opinions of the granting agencies or the U.S. government.

Author Contributions

Upendra Pradhanang and Soni M. Pradhanang reviewed, analyzed and conducted SWOT analysis of the National Livestock Policies at various levels. Arhan Sthapit, Nir Krakauer, Ajay Jha, and

Tarendra Lakhankar are involved in providing guidance and evaluating appropriateness of SWOT analysis presented in this paper.

Conflicts of Interest

The authors declare no conflict of interest.

References

1. NSCoA. *National Sample Census of Agriculture-2011/12*; Central Bureau of Statistics, National Planning Commission Secretariat, Government of Nepal: Kathmandu, Nepal, 2013.
2. Department of Agriculture, Ministry of Agricultural Development, Government of Nepal. Available online: http://www.doanepal.gov.np/ (accessed on 20 June 2014).
3. ADS. *Ads Assessment Report, Agricultural Development Strategy Assessment*; Government of Nepal, ADB, IFAD, EU, FAO, SDC, JICA, WFP, USAID, DANIDA, DfID and World Bank: Kathmandu, Nepal, 2012.
4. ILO. *A Fair Globalization: Creating Opportunities for All*; Report of the World Commission on the Social Dimension of Globalization: Geneva, Switzerland, 2004; p. 143.
5. MoAC. *National Agriculture Policy*; Ministry of Agriculture and Cooperatives, Government of Nepal: Kathmandu, Nepal, 2004.
6. NPC. Approach Paper to the Thirteenth Plan (Fiscal Year 2012/13–2015/16). National Planning Commission, Government of Nepal: Singh Durbar, Kathmandu, Nepal, 2012.
7. Shrestha, S. Contraction in Key Sector Hurt Growth Rate. *The Himalayan Times*, 8 April 2013.
8. Pearce, J.A., II; Robinson, R.B.; Mital, A. *Strategic Management: Formulation, Implementation and Control*, 12 ed. McGraw Hill: New Delhi, India, 2012.
9. NLC-Nepal. Interim Constitution of Nepal Nepal Law Commission, Kathmandu, Nepal, 2007.
10. MoAC. Agriculture Perspective Plan (App), 1995–2015; Government of Nepal: Kathmandu, Nepal, 1995.
11. MoF. Budget Speech of 2014/015; Ministry of Finance, Government of Nepal: Kathmandu, Nepal, 2013.
12. Sthapit, A. Boost Investment in Agriculture. *The Rising Nepal*, 10 November 2010.
13. Sthapit, A. *Essentials of International Business & Environment (with Case Studies)*, 2nd ed.; Asmita Books Publishers: Kathmandu, Nepal, 2013.
14. WIPO. *International Classification of Goods and Services for the Purposes of the Registration of Marks*; World Intellectual Property Organization: CH-1211 Geneva 20, Switzerland, 2014.
15. WTO. *Standards and Trade Development Facility, STDF Newsletter*; World Trade Organization, Official Website. Available Online: Http://Www.Wto.Org (accessed on 20 March 2014).
16. UNFCCC. *United Nations Framework Convention on Climate Change*; UNFCCC Secretariat: Bonn, Germany, 1992.
17. NAPA. *National Adaptation Program of Action*; Government of Nepal: Kathmandu, Nepal, 2010.
18. LAPA. *Local Adaptation Program of Action*; Government of Nepal: Kathmandu, Nepal, 2011.
19. Singh, T.M.; Sthapit, A. *Human Resource Management: Text and Cases*; Taleju Prakashan: Kathmandu, Nepal, 2008.

20. ADB. *Overview of Gender Equality and Social Inclusion in Nepal*; Asian Development Bank: Kathmandu, Nepal, 2010.

21. UN-Women. *Review of the Implementation of the Beijing Platform for Action (the Outcome Document of the Twenty-Third Special Session of the General Assembly)*; United Nations Women E/CN.6/2005/2; UN-Women: New York, NY, USA, 2004.

22. Maltsoglou, I.; Taniguchi, K. *Poverty, Livestock and Household Typologies in Nepal*; ESA Working Paper No. 04–15; Food and Agriculture Organization (FAO): Rome, Italy, 2004; pp. 1–48.

23. Bom, P. Federal Governance Structure for Nepal. In *Federal Structure Is Mandatory for Proportional Electoral System*. Government of Nepal: Kathmandu, Nepal, 2007.

24. Piya, L.; Maharjan, K.L.; Joshi, N.P. Vulnerability of Rural Households to Climate Change and Extremes: Analysis of Chepang Households in the Mid-Hills of Nepal. In *28th International Conference of Agricultural Economists: The Global Bioeconomy*; International Association of Agricultural Economists: Foz Do Iguacu, Brazil, 2012; No. 126191.

25. Pettengell, C. Climate Change Adaptation: Enabling People Living in Poverty to Adapt. OXFAM: Carlisle, UK, 2010.

26. Cameron, J. The Agricultural Perspective Plan: The Need for Debate. *Himalaya. J. Assoc. Nepal Himal. Stud.* **1988**, *18*, 8.

27. Ospina, A.V.H.R. *Linking ICTs and Climate Change Adaptation*; University of Manchester: Manchester, UK, 2010.

28. Panthi, J.; Dahal, P.; Shrestha, M.L.; Aryal, S.; Krakauer, N.Y.; Pradhanang, S.M.; Lakhankar, T.; Jha, A.K.; Sharma, M.; Karki, R. Spatial and Temporal Variability of Rainfall in the Gandaki River Basin of Nepal Himalaya. *Climate* **2015**, *3*, 210–226.

29. Parikh, J. Gender and Climate Change Framework for Analysis, Policy & Action; IRADe and UNDP India: New Delhi, India, 2007.

30. Tulachan, P.M.; Partap, T. Development Experiences of Livestock Production Systems in Hindu Kush Himalayan Region. In Proceedings of An IDRC-ILRI International Workshop held at ILRI, Addis Ababa, Ethiopia, 11–15 May 1998; pp. 149–163.

31. PPTA Consultants. *Preparation of the Nepal Agricultural Development Strategy*; ADB project 43447-022; PPTA Consultants: Kathmandu, Nepal, 2012.

32. Sthapit, A. *International Business: Text and Cases*, 1st ed.; Taleju Prakashan: Kathmandu, Nepal, 2005.

8

Food Safety Information Processing and Teaching Behavior of Dietitians: A Mental Model Approach

Lydia C. Medeiros * and Jeffrey T. LeJeune [†]

Food Animal Health Research Program, Ohio Agricultural Research and Development Center,
The Ohio State University, 1680 Madison Ave, Wooster, OH 44691, USA; E-Mail: LeJeune.3@osu.edu

[†] These authors contributed equally to this work.

* Author to whom correspondence should be addressed; E-Mail: Medeiros.1@osu.edu

Academic Editor: Pascal Delaquis

Abstract: Health professionals play an important role in educating the public about food safety risks. However, the ways this important group of educators remains up-to-date on these topics are not well defined. In this study, a national sample of dietitians employed in direct teaching of patients (n = 327) were recruited to complete a web-delivered survey designed to develop a model of factors that promote information processing and teaching in practice about food safety related to fresh vegetables. The resulting mental model demonstrates that dietitians teach fresh vegetable safety using systematic information processing to intellectually understand new information, but this is also associated with a gap in the dietitian's knowledge of food safety. The juxtaposition of an information processing model with a behavioral model provides valuable new insights about how dietitians seek, acquire and translate/transfer important information to move patients toward a higher goal of food safety. The study also informs food safety educators as they formulate teaching strategies that are more effective than other approaches at promoting behavior change.

Keywords: foodborne illness; dietitians; mental models; information processing behavior

1. Introduction

In Europe and the United States (US), the public ranks trust in health professionals, followed by food safety authorities and university scientists, as important when evaluating sources of food safety information [1–3]. Trust in health professionals is particularly important among consumers who are at increased risk of opportunistic infections as a result of immune suppression, medical therapy, life stage, or pharmaceutical use [4]. A series of qualitative studies with highly-susceptible patient groups in the US queried where patients wanted to find food safety information that related to them and their condition [5–8]. Participants in these studies wanted credible information from trusted sources, and health care providers were the information source they preferred. Each group named their physician as their preferred primary source and a member of the health care team, such as a nurse, dietitian, or social worker, as their preferred secondary source. The participants only occasionally mentioned other sources, such as web sites that reported medical information (for example, WebMD [9]).

Registered dietitians were named as a credible source of food safety information by two groups highly-susceptible to opportunistic infections: cancer [7] and transplant [8] patients. Relying on the dietitian as an information source is appropriate, as food safety competence is required for registration as a dietitian in the US [10]. Additionally, post-graduate continuing education is required to maintain competence in topics of importance to dietetic practice. Dietitians self-select the topics they wish to pursue in continuing education; it is not required that dietitians maintain post-graduate competence in food safety or any other specific topic.

Although dietitians are charged with the career-long responsibility to monitor and update their knowledge base, personal preferences, biases, and professional experiences have the potential to influence the choices they make with respect to continuing educational content. This study was designed to better understand factors that influence how dietitians process information as they use their knowledge of food safety for their personal benefit, and to teach their patients. The goal was to elucidate information processing clues, or the mental model used by dietitians when they seek out new information about the safety of food. This information could aid educational providers as they develop risk communication strategies and continuing education opportunities for dietitians. This study focused on foodborne illnesses (FBI) and information processing behavior associated with teaching fresh vegetable safety because dietitians prize fresh vegetables as a major source of fiber, vitamins and minerals, and because they are traditionally strong advocates for consuming fresh vegetables [11]. Nevertheless, *Salmonella* outbreaks associated with sprouts, tomatoes and lettuce has cast doubt on the safety of these foods, causing many consumers to avoid the products [12]. Furthermore, therapeutic low-microbial diets that limit the consumption of uncooked foods, especially vegetables and fruits are often prescribed in clinical chemotherapy or transplant medical units to minimize opportunistic infections; however, such diets also limit intake from sources of vital nutrients [13].

1.1. The Risk Information Seeking and Processing (RISP) Communication Theory

Prerequisite to trust and credibility is the depth and quality of the teacher's knowledge of their subject. Thus, how a teacher seeks information and then internally processes that information to gain understanding are important characteristics that informs how the teacher progresses to sharing the information with others, the actionable behavior of interest in this study. The RISP theory was designed to assess information-related behavior and subsequent personal behavior toward a variety of environmental hazards. Two sub-models compose the overall theory: Heuristic-Systematic Model of Information Processing [14] and the Theory of Planned Behavior [15]. A portion of the RISP theory predicts that previous personal experiences with a health-threatening condition may influence a person's perception of the need to seek education, or in the case of the dietitian, to provide educational information to others they see as being susceptible to hazardous conditions they too may have experienced. A mental model approach is appropriate because the underlying perceptions and beliefs structure can either stimulate or obstruct the intellectual processing of new scientific information. The Risk Information Seeking and Processing communication theory (RISP theory) [16] was used as the theoretical base for developing the mental model of registered dietitians who teach patients about fresh vegetable safety, that is microbial or food safety.

1.2. Mental Models

All of life's behavioral decisions are based on a personal or mental model that uses past experiences, knowledge, affect, and logic to formulate a future behavioral course of action. To the researcher and policy maker, mental models are useful to study because they show how influential factors may interrelate, they reveal salient critical issues, and can show causal links, depending on the methodology used to describe a mental model [17]. Even though a mental model is typically unique to the individual, group or collective mental models are useful to educators and academics because they can reveal shared or commonly-held thought processes and a consensus that is helpful in developing therapeutic or educational strategies. Likewise, a company mental model of individuals with dissimilar values or backgrounds who are charged with the completion of a common goal, could be useful to public policy development because divergent-stakeholder viewpoints can be identified that can simplify implementation and promote acceptance of policy [18].

All mental models contain core elements, such as relatively stable, long-term knowledge about a topic, and they contain peripheral elements that change depending on situations that can affect cognitive processes, which provides for the dynamic dimensions of a mental model [17]. Another aspect of a mental model relates to the attributes of the individual, or a person's need for new knowledge, their capacity to grasp new information, or methods for acquiring new information. Thus, a mental model is dependent on pre-existing knowledge, but is compiled as needed to respond to situations, both anticipated situations and actual situations [19].

The development of a personal, collective, or company mental model requires both a foundational discovery process that should be rooted in social science and risk communication theory [20], and an elucidation process that directly queries the target stakeholder. This way, the causal elements of the mental model are more clearly understood, as well as the dynamic elements that will necessitate periodic review and revision [18]. Frewer *et al.* [20] also concluded that the mental model discovery process should be extended to consider subsequent actionable behavior, as that is essential to the educational and public policy application of the mental model.

1.3. Study Objectives

This study is third in a series to explore if health professionals are influenced by personal beliefs and experience with topics such as foodborne illness, or if they are independently influenced by professional concern for the health status of their patients? We have previously reported knowledge, information sources and training used by dietitians [21,22] that most registered dietitians use to teach patients about the safety of fresh vegetables as potential sources of invasive foodborne pathogens [12]. When food safety was not taught, important barriers to dietitians' provision of information included time constraints, having incorrect information about food safety, and having a lack of confidence in personal knowledge of food safety [21]. Therefore, the first objective of the present study was to identify information processing behavior of registered dietitians, specifically information about fresh vegetable safety, in the discovery of the causal factors used by dietitians in their decision to gain new food safety information and that leads to teaching food safety to patients. The second objective was to create the mental model of information processing and teaching behavior used by the dietitians in this study, based on statistical discovery of salient model elements that are used by dietitians when they teach food safety to patients.

2. Results and Discussion

The priority of dietetics is to provide nutritional information to patients and to educate about high-nutrient food sources; for example, fresh vegetables are highly desirable and nutritious foods [11]. However, vegetables are also frequently cited as the food attribute associated with foodborne illness outbreaks [12]. Where and how dietitians inform themselves about the food safety risks associated with consumption of fresh vegetables is essential to the quality of the education they provide patients.

There were two types of causal models developed in this study, the information processing models and the actionable behavior models. These are the theoretical components of the RISP communication theory but were approached separately in this study because of the different construction of the outcome variables. Together and supported by statistical analyses (means separation, discriminant analysis, and hierarchical linear regression), these causal models informed the dietitian's mental model of food safety information processing; and, most important to their health and the health of their patients, the actionable behavior of teaching food safety to others.

2.1. The RISP Communication Theory

Individual characteristics, previous experience with the hazard of interests and political philosophy are the control variables in the RISP theory. Gender, Ethnic group, Income, Political philosophy and Education level did not differ between Teaching-behavior groups. Age differed with a significant number of younger dietitians in the Do not currently teach group ($P < 0.001$) and more dietitians in the oldest Age category in the Currently-teach group ($P < 0.001$). To construct the RISP theory, Education is entered into hierarchical linear regression (HLM) prior to the variables measuring Information sufficiency to account for variation in Current knowledge independent of Education level. In this study, Education failed tests of multicollinearity and was removed from the final models. Dietitians had little experience with foodborne illness as few had been medically treated for FBI ($n = 22$, 6.7%). Patients counseled by the dietitians were more likely than dietitians to have had FBI as indicated by responding that they counseled patients who had been medically treated for FBI (overall $n = 140$, 42.8%), with a difference found between Teaching-behavior groups ($n = 56$, 40%, Do not currently teach; $n = 84$, 60%, Currently-teach; $P < 0.001$). Political philosophy was not a significant variable between Teaching groups, nor was it a significant contributing factor in either the HLM analysis for Systematic information processing (Table 1) or for Heuristic information processing (Table 2). Less diversity among the findings for the control variables was consistent with the characteristics of the population of dietitians in the US [23]. The educational and post-graduate training requirements for credentialing inadvertently selects for greater homogeneity among dietitians' characteristics.

Table 1. Hierarchical linear model of the Risk Information Seeking and Processing (RISP) model variables for Systematic information processing (standardized β-coefficients for final model).

Variables	Systematic information processing		
	Do not Currently Teach ($n = 144$)	Currently Teach ($n = 155$)	Overall ($n = 299$)
Individual characteristics			
Caucasian ethnic group	0.223	0.314	0.265 *
Other ethnic group	0.220	0.279	0.224 *
Income $64K or less	0.159	−0.004	0.020
Income $65K to $ 99K	0.169	−0.071	0.008
Income $100K or more	0.055	−0.033	−0.066
Age 18–29 years	-------	−0.142	−0.077
Age 30–44 years	0.096	−0.145	-------
Age 45 years and older	0.158	-------	0.072
Liberal political philosophy	−0.008	−0.027	−0.013
Neutral political philosophy	−0.053	−0.010	−0.027
Conservative political philosophy	−0.058	0.153	0.035
FBI-self, medical	0.111	0.021	0.039
FBI-self, no medical	−0.097	0.101	0.039
FBI-others, medical	0.129	−0.062	0.009
FBI-others, no medical	0.057	−0.135	−0.052
FBI-patients, medical	−0.035	−0.071	−0.060
FBI-patients, no medical	0.080	0.003	0.028
Perceived hazard characteristics			
Risk judgment	0.001	0.119	0.091
Institutional trust	0.055	−0.037	0.017
Personal control-self	0.155	0.125	0.141 *
Personal control-patients	0.094	−0.123	−0.009
Informational subjective norms			
People	0.022	−0.016	0.030
Patients	0.145	0.049	0.063
Affect, worry			
Self	0.121	−0.175	−0.083
Patient	−0.032	0.149	0.013

Table 1. *Cont.*

Variables	Systematic information processing		
	Do not Currently Teach (*n* = 144)	Currently Teach (*n* = 155)	Overall (*n* = 299)
Channel beliefs			
Media bias beliefs	−0.106	−0.189 *	−0.131 *
Validity cues beliefs	0.077	0.189 *	0.131 *
Information source credibility	0.248 **	0.079	0.169 **
Information gathering capacity			
Capacity	0.038	−0.043	0.002
Current knowledge			
General food safety knowledge	−0.005	−0.001	−0.027
Pathogen awareness	0.060	−0.008	0.036
Pathogen understanding	−0.124	0.206	0.056
Information insufficiency			
Knowledge threshold	−0.093	0.158	0.070
Final model statistics	$P = 0.019$ $R^2 = 0.126$	$P = 0.023$ $R^2 = 0.114$	$P < 0.000$ $R^2 = 0.139$

* $P \leq 0.05$; ** $P \leq 0.01$; *** $P \leq 0.001$ Standardized β-coefficients.

Table 2. Hierarchical linear model of RISP variables for Heuristic information processing (standardized β-coefficients for final model).

Variables	Heuristic information processing		
	Do not Currently Teach (*n* = 144)	Currently Teach (*n* = 156)	Overall (*n* = 300)
Individual Characteristics			
Caucasian ethnic group	0.167	−0.083	−0.005
Other ethnic group	0.038	−0.085	−0.022
Income $64K or less	0.105	−0.064	0.005
Income $65K to $ 99K	0.082	0.054	0.065
Income $100K or more	0.054	−0.142	−0.037
Age 18–29 years	-------	0.104	0.044
Age 30–44 years	−0.018	0.127	-------
Age 45 years and older	−0.153	-------	−0.105
Liberal political philosophy	0.140	0.166	0.118
Neutral political philosophy	0.018	0.174	0.084
Conservative political philosophy	0.070	0.210	0.102
FBI-self, medical	−0.040	−0.067	−0.055
FBI-self, no medical	0.089	−0.086	0.004
FBI-others, medical	−0.124	0.037	−0.035
FBI-others, no medical	−0.054	−0.055	−0.073
FBI-patients, medical	0.000	−0.060	−0.037
FBI-patients, no medical	0.084	0.204 **	0.152 **

Table 2. *Cont.*

Variables	Heuristic information processing		
	Do not Currently Teach ($n = 144$)	Currently Teach ($n = 156$)	Overall ($n = 300$)
Perceived hazard characteristics			
Risk judgment	0.087	0.067	0.056
Institutional trust	−0.062	0.145	0.032
Personal control-self	−0.043	0.162	0.102
Personal control-patients	0.012	−0.156	−0.081
Informational subjective norms			
People	−0.066	-0.248 *	−0.156 *
Patients	−0.014	0.088	0.049
Affect, worry			
Self	−0.239	0.038	−0.034
Patient	0.055	0.019	0.000
Channel beliefs			
Media bias beliefs	0.062	0.193 *	0.143 **
Validity cues beliefs	−0.033	0.047	0.011
Information source credibility	−0.288 **	−0.365 ***	−0.330 ***
Information gathering capacity			
Capacity	0.178 *	0.082	0.133 *
Current knowledge			
General food safety knowledge	−0.117	−0.053	−0.122 *
Pathogen awareness	−0.079	−0.132	−0.082
Pathogen understanding	0.039	0.096	0.039
Information insufficiency			
Knowledge threshold	−0.035	−0.080	−0.088
Final model statistics	$P < 0.000$ $R^2 = 0.232$	$P < 0.000$ $R^2 = 0.311$	$P < 0.000$ $R^2 = 0.286$

* $P \leq 0.05$; ** $P \leq 0.01$; *** $P \leq 0.001$ Standardized β-coefficients.

The type of information processing that a person engages in when seeking new information is a function of influencing factors, such as their perception of trust in institutions, their judgment about the seriousness and their personal susceptibility to FBI, or their sense of control over the hazard, either personally or for the sake of their patients. Since sanitary handling of foods in retail markets and institutional food preparation are public health-regulated in the US [24], trust in persons and institutions that are responsible for ensuring the safety of food, and the ability to protect the personal and public health can be highly predictive of a person's subsequent attempt to acquire information about the risk hazard and how that external influence affects their actionable behavior regarding the hazard. Slovic *et al.* [25] found that perception of a risk hazard is a function of how well the risk is understood and how much feelings of dread are invoked by thoughts of the health risk. A person's reaction to risk hazards depends on their feeling of personal control and the likelihood that they will become ill from eating a contaminated food, which could lead to a reactionary emotional response. The motivation to begin the information acquisition process that could alleviate negative feelings has been shown to be the degree of worry felt by the individual [26].

For dietitians in this study, however, none of the measures of Perceived hazard characteristics were influencing factors contributing to the RISP theory statistical analysis. Means for the four variables measuring the construct were not different between Teaching-behavior groups. In addition, none of the variables were notable in either HLM or discriminant analysis, with two exceptions. Personal control–self was significant ($P = 0.054$) in the Systematic information processing-overall HLM model (Table 1), and Personal Control–Patients was a minor classifying variable in discriminant analysis (standardized canonical coefficient, −0.263).

The influence of Affect was measured as the degree of Worry respondents felt over possible health risks associated with eating fresh vegetables for either their self or for their patients. There was no difference between Teaching-behavior groups for Affect, Worry–self. The variable was a minor contributor in discriminant analysis (standardized canonical coefficient, −0.257). Those who currently taught fresh vegetable safety to their patients were more concerned for their patients' health risks than were dietitians who did not teach (Affect, Worry–patients, $P = 0.044$). Affect, Worry–patient was also a major classifying variable in discriminant analysis (standardized canonical coefficient, 0.441). Neither of the two Affect measures were significant contributors to the overall variation accounted for in the HLM models (Tables 1 and 2).

In the RISP theory, a person's Current knowledge of a risk hazard and antecedent factors that influence knowledge are, in turn, antecedent to that person's search and understanding of risk-related information. If a person perceives they have sufficient knowledge (as a function of their education and the normative influence of referent others), they may or may not seek and process additional information. Because dietitians are so similarly educated, that effect was removed from the statistical analysis due to multicollinearity. Perception of the influence of people (Informational subjective norms–people) who were important to the respondent, or referent others, was not different between groups. For Informational subjective norms-patients, respondents in the Currently-teach group were highly influenced by the health and food safety learning needs of their patients, compared to the Do not currently teach group ($P < 0.001$). The strength of these variables was reflected in discriminant analysis as each were major variables on which the Teaching-behavior groups were classified (standardized canonical coefficients, Informational subjective norms–patients, 0.641; Informational subjective norms–people, −0.353). The variable, Informational subjective norm–people was significant in the HLM models for Heuristic information processing (Table 2).

For dietitians who are counseling patients at high-risk for FBI, both broad and deep understanding of food safety principles is critical. We have previously reported that teaching food safety in this sample of dietitians was a function of their greater general knowledge of food safety [21]. Food safety knowledge and information sources were also explored with the finding that knowledge credibility is compromised by the depth of factual understanding and the type of information sources selected [22]. Previously reported measures of General food safety knowledge were also used in this study of information processing and actionable behavior. Additionally, food safety knowledge as measured by Pathogen awareness and Pathogen understanding are reported.

General food safety knowledge was not significant ($P > 0.05$) between Teaching groups, but scores were higher for the Currently-teach group. Mean differences for both Pathogen awareness and Pathogen understanding were highly significant ($P < 0.001$), with the Currently-teach group having higher scores or greater awareness and understanding of four common FBI-causing pathogens (*Listeria monocytogenes*, *Salmonella* spp., *E. coli* O157:H7, or *Campylobacter jejuni*). This was a positive finding because of the unique learning needs of high-risk patients and the need for their health providers to have in-depth and mechanistic knowledge when devising medical nutritional therapies. In contrast, all of the measures of Current knowledge were associated with lower Systematic information processing (Table 1) and Heuristic information processing scores (Table 2).

Griffin *et al.* [27] argued that knowledge alone is not a strong predictor of information processing, but that perception of the need for new or additional knowledge to meet a personal or professional goal is the primary driver toward active information gathering and processing. A variable, Knowledge threshold, was added to the RISP theory as a variance-adjusted score independent of a person's current knowledge. It is a representation of an individual's perception of their need for additional information, a construct called Information sufficiency. The construct represents the gap between what a person already knows about the risk hazard and the additional information they perceive they need for them to progress toward actionable behavior. We found that dietitians who Currently-teach food safety of fresh vegetables to their patients also had higher scores for Knowledge threshold ($P < 0.001$) than those who did not teach food safety. Knowledge threshold was also a major classifying variable in discriminant analysis (standardized canonical coefficient, 0.433). Knowledge threshold was not significant in any of the HLM models tested (Tables 1 and 2).

The construct, Information sufficiency, becomes the main predictor of the type of information processing utilized. A large gap reflects greater use of Systematic information processing and smaller difference is indicative of greater use of Heuristic information processing. More recently, the name, Information sufficiency, has evolved in the RISP theory literature into a construct called Information insufficiency [27]. The change from the original iteration of the theory [16] occurred as a truer interpretation of the data analysis outcome. The relationship between Information insufficiency and method of information processing is attenuated by antecedent constructs that measure a person's perception of their ability to locate and understand risk information and beliefs about information sources [27].

A successful search for information may be related to a person's belief in their capacity to locate the information and to understand the complexity of the information (e.g., the variable, Information gathering capacity) [15]. Furthermore, beliefs associated with the source of information can influence the credibility that the person gives to the information coming from that source. Neither Media bias beliefs nor Validity cues beliefs differed between Teaching-behavior groups. Information source credibility ($P = 0.021$) and Information gathering capacity ($P = 0.001$) differed between groups, with the Currently-teach group having the higher mean scores. Media bias beliefs was a major classifying variable (standardized canonical coefficient, 0.369), and Information source credibility (standardized canonical coefficient, 0.221) and Validity cues beliefs (standardized canonical coefficient, −0.204) were minor classifying variables in discriminant analysis. Information gathering capacity was not found to contribute to discrimination of the two Teaching-behavior groups. All of these variables contributed significantly to the HLM analyses shown in Tables 1 and 2. The variables entered into the HLM models were major contributors to the total variance accounted for in the six models. The higher the use of Heuristic information processing, the greater the belief that media are bias information sources (Table 2). Information source credibility, which measured the degree of attention paid to various media, was negatively associated to Heuristic information processing for all HLM analyses with this behavioral outcome. The less attention paid to media, the more Heuristic information processing was used. Similar effects were evident in the three HLM analyses for Systematic information processing, except the direction of the effects were opposite and complimented findings for these variables in the Heuristic information processing models (Table 1).

The information processing variables are the outcome behavioral variables in the Heuristic-Systematic Model of Information Processing portion of the RISP theory [16]. Mean difference between Teaching-behavior groups was significant for Systematic information processing ($P = 0.005$), and the variable was a minor contributor in discriminant analysis (standardized canonical coefficient, 0.200). Heuristic information processing did not differ between groups, nor did the variable contribute to discriminant analysis. For the HLM models with Systematic information processing as the dependent variable, a modest amount of variation was accounted for in the analysis (R^2 below 14% for each of the three models) (Table 1). Models where Heuristic information processing was the dependent variable were stronger, based on R^2 ranging from 23.2% to 31.1% (Table 2). As with Education and depending on the model, various categories of Age were excluded from the analysis because of multicollinearity.

From the three types of statistical analyses completed, we can conclude that in this sample of dietitians, Systematic information processing was the predominant style of information processing used by dietitians who Currently-teach safety of vegetables to their patients as evidenced by their belief that they have the capacity to find information they need and that they will process the information with a critical eye for potential bias toward a particular position regarding the risk hazard.

2.2. The Theory of Planned Behavior

To proceed to the causal model predicting actionable behavior, constructs of the Theory of Planned Behavior were included in the RISP theory. According to Ajzen [15], Indirect attitude is correlated to other Theory of Planned Behavior constructs, and is the most predictive variable of behavioral intent, and in turn actionable behavior. Ajzen [15] describes measures of attitude as the product of a belief and the companion evaluation of that belief; the product of the two measures is a surrogate of attitude or Indirect attitude. Principle component analysis was used to identify factors to serve as predictive variables in the statistical analysis for the Theory of Planned Behavior portion of the causal models (Table 3). As with the information-processing portion of the RISP theory, influence of referent others attenuates progression to behavior. This construct is similar to the Informational subjective norms assessed in the RISP theory portion of the data analysis. Perceived barriers that control behavior differ for each behavior and population and must be accounted for in the study of actionable behavior. Perceived barriers are measures of self-efficacy, as in other behavior decision-making models [15], and they assess whether or not a person believes they have the actual or perceived means to accomplish a desirable behavior. In the RISP theory this is similar in concept to Information gathering capacity.

Table 3. Principle components within Indirect attitude variables [a].

Indirect Attitude Measures	Factor Loading [b]			
	Factor 1	Factor 2	Factor 3	Factor 4
	Local Produce	Benefits Teaching	Organic Produce	Nutrition Quality
Total nutrition content important				0.723
Total fiber content important				0.777
Vitamin content important				0.727
Choose easy to prepare vegetables				0.549
Fresh vegetable cost important				0.540
Less foodborne illness among patients		0.640		
More thankful patients		0.737		
Teaching food safety good for dietitian's reputation		0.732		
More awareness about food safety in community		0.681		
Food safety necessary for highly susceptible patients		0.503		
Local produce has less bacterial contamination	0.583			
Local produce tastes better	0.904			
Local produce is better quality	0.914			
Local produce helps the local economy	0.876			
Being organically grown is important			0.583	
Organic produce tastes better			0.833	
Organic produce is more nutritious			0.855	
Organic produce has less bacterial contamination			0.780	
Organic produce has less pesticide contamination			0.548	
Cronbach alpha	0.89	0.79	0.79	0.76
Variance explain by factor (%)	11.62	11.31	10.14	9.11

[a] Measures of indirect attitude computed as the product of paired survey items assessing behavioral beliefs and evaluation of that belief. Measures with factor loading >0.5 retained; [b] Rotation method = Varimax with Kaiser normalization (KMO = 0.79).

The binary structure of the Teaching behavior outcome variable in the Theory of Planned Behavior portion of the RISP theory necessitated a switch to a non-parametric statistical procedure, but was judged necessary to answer the question about the predictive strength of information processing style on actionable behavior as part of the mental model developed in this study. The significance of this question is that design of educational strategies in food safety will depend on a thorough understanding of the dietitian as a life-long learner. How do they approach information gathering and processing? What attitudes do they have about the teaching of food safety as a risk hazard? What barriers do they perceive to this effort? Furthermore, who influences dietitians suggests whether or not they will be easily persuaded to the desirable goal of teaching patients about fresh vegetable safety, or if they will persist with their prior beliefs, either pro or con, about new information that can give them the broadest and most in-depth knowledge base possible.

The binary logistic models for both Systematic information processing (Figure 1) and Heuristic information processing (Figure 2) included similar significant predictors of teaching behavior. In common, patients were the primary referent other used by the Currently-teach group, and this group felt they had control over whether or not they had the freedom to teach this topic.

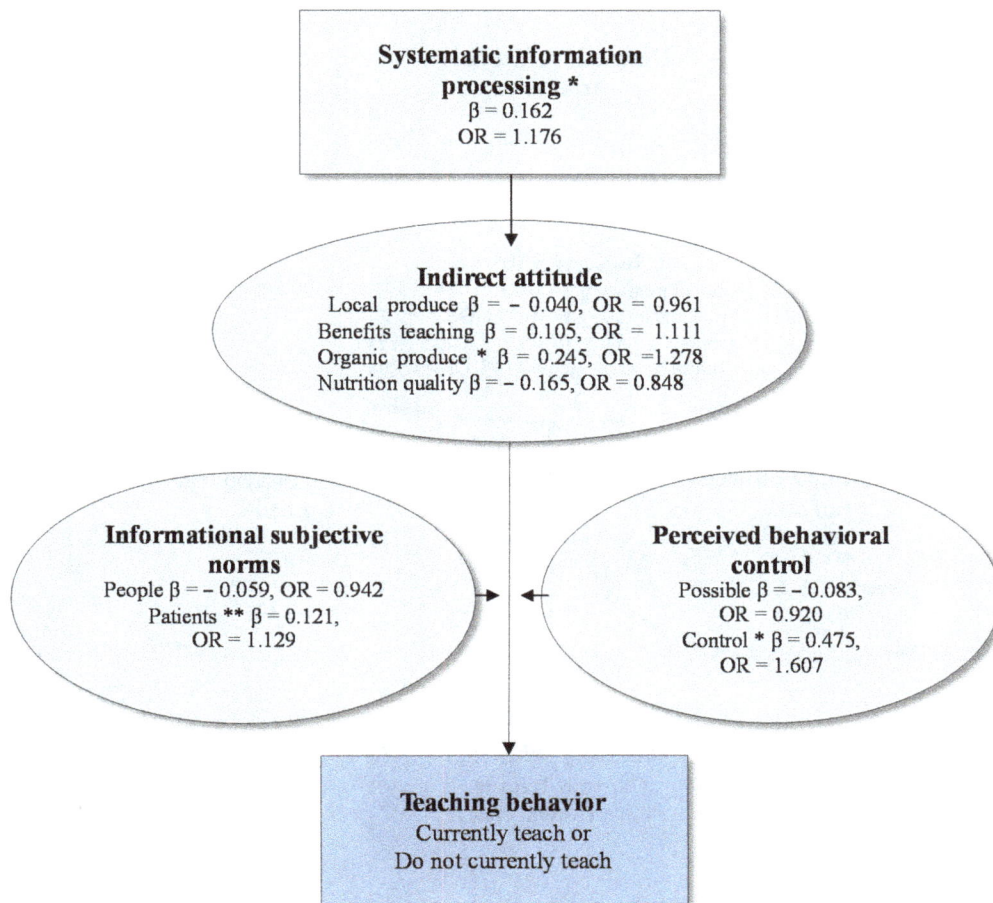

Figure 1. The binary logistic regression model for Systematic information processing and Teaching behavior of dietitians who either Currently teach or Do not currently teach fresh vegetable food safety to their patients. Model, $P < 0.001$, -2 Log Likelihood $= 388.1$, Cox and Snell $R^2 = 0.13$, Nagelkerke $R^2 = 0.17$, Correct classification $= 64.0\%$. Odds ratio (OR). * $P \leq 0.05$; ** $P \leq 0.01$.

Unfortunately, both models had as a significant predictor of Current Teaching behavior the belief in the inherent safety of organic vegetables. The safety of organic vegetables as superior to non-organic vegetables has not been unequivocally confirmed [28], and could represent a strongly-held belief of those in the Currently-teach group who also have high use of the Systematic information processing style. Systematic information processing was a positive and significant predictor of Teaching-behavior (Figure 1), but Heuristic information processing was not (Figure 2). The finding that a significant portion of the dietitians surveyed believed that organic vegetables are the safer choice for selecting vegetables could be problematic for educators and risk communicators who are trying to persuade dietitians to incorporate food safety best practices into their patient education (Figures 1 and 2).

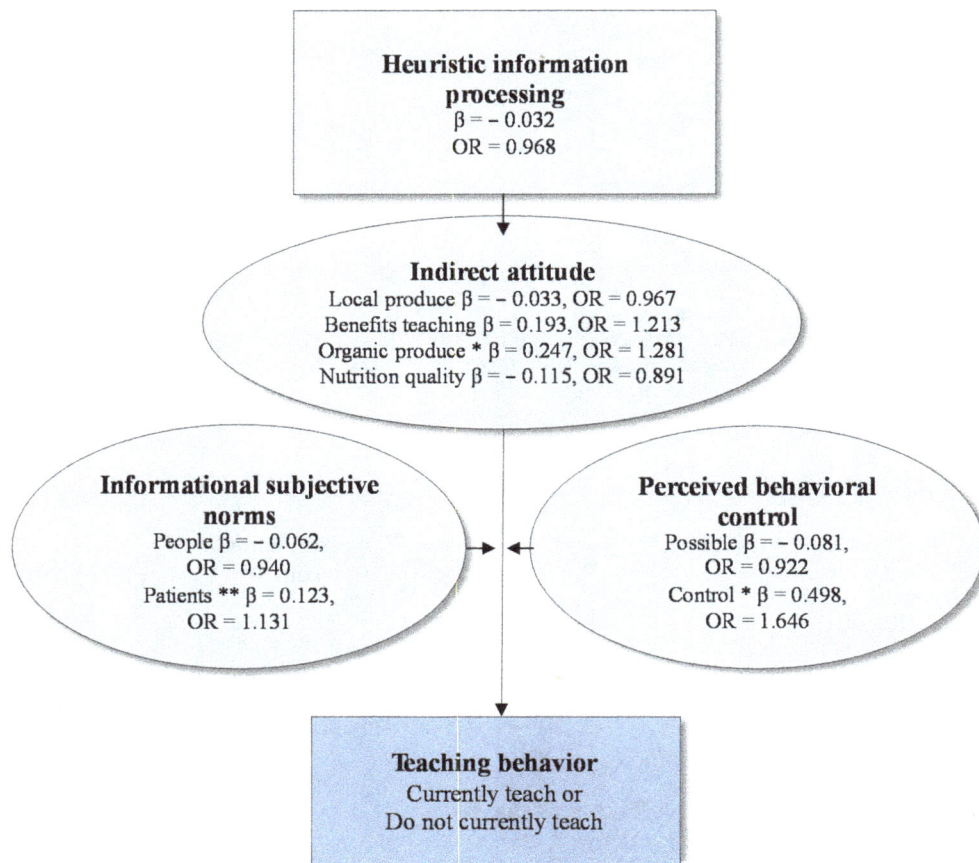

Figure 2. The binary logistic regression model for Heuristic information processing and Teaching behavior of dietitians who either Currently teach or Do not currently teach fresh vegetable food safety to their patients. Model, $P < 0.001$, -2 Log Likelihood $= 392.6$, Cox and Snell $R^2 = 0.12$, Nagelkerke $R^2 = 0.16$, Correct classification $= 63.8\%$. Odds ratio (OR). $* P \leq 0.05$, $** P \leq 0.01$.

In the model with Systematic information processing as a dependent variable (Figure 1), Systematic information processing behavior ($P = 0.047$), the Indirect attitude factor, Organic produce ($P = 0.053$), Informational subjective norms–patients ($P = 0.001$), and Perceived behavioral control–control ($P = 0.014$) accounted for a significant portion of the variation in the binary logistic regression model. The model was significant ($P < 0.001$) and accounted for 13% to 17% of the model variation.

For the binary model with Heuristic information processing as a dependent variable (Figure 2), the Indirect attitude factor, Organic produce ($P = 0.052$), Informational subjective norms–patients ($P = 0.001$), and Perceived behavioral control–control ($P = 0.01$) were significant and accounted for 12% to 16% of the model variation depending on the statistical measure ($P < 0.001$).

2.3. The Mental Model

A mental model is useful to understand how pieces of information are used and intellectually processed when individuals are making a decision toward actionable behavior. The mental model can represent an individual or a group of individuals who operate as a collective or who share common characteristics. The dietitians who Currently-teach food safety of fresh vegetables to their high-risk patients are a collective who are representative of dietitians in the US, thus the mental models proposed in Figure 3 have application to populations of dietitians with similar characteristics.

Frewer *et al.* [20] pointed out that the most representative mental model is one in which the target population is studied both quantitatively and qualitatively. Multiple statistical and modeling approaches are useful to support the findings of the mental model process. There is always, however, a high degree of subjective selection of the various elements that a researcher uses to formulate a mental model. The statistical data shown in the causal models supports the subjective selection of variables to inform the mental model.

Figure 3 is the mental model that we have devised based on the statistical analyses and causal models computed for this study. The model supports the teaching of fresh vegetable safety with the use of Systematic information processing style associated with a large information gap in the dietitian's knowledge of food safety, attenuated by patients as the social or referent influence, worry about health risk associated with FBI in the patient population as the affect influence, and a belief that the dietitian has the control to proceed to actionable behavior; but, factual knowledge of food safety is marginal. This mental model is powerful information that will support and perhaps change the way in which dietitians in the US are prepared to educate others about food safety, especially regarding fresh vegetables, and has the potential to influence both didactic educational preparation and life-long continuing education. There in an inherent challenge exposed from this study for providers of advanced food safety education. Factual food safety information is marginal, is deeply believed and not easily influenced, and highly influenced by others beyond formal educational channels. If the educational approach for dietitians is modified to consider these findings, the ultimate outcome from this study could be fewer cases and outbreaks of FBI, fewer deaths due to FBI, and an improved public health.

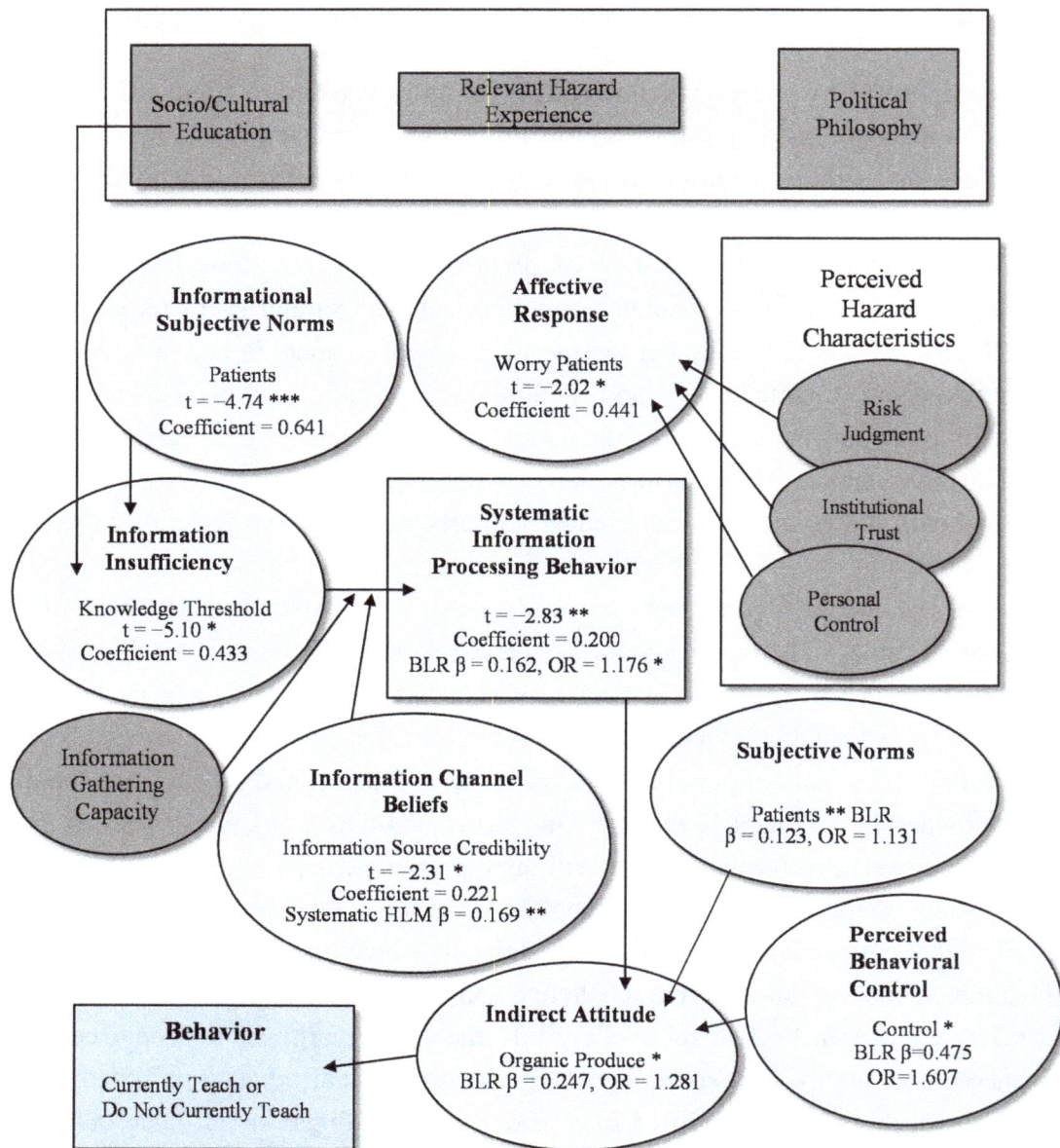

Figure 3. The mental model for teaching behavior of dietitians who either Currently teach or Do not currently teach fresh vegetable food safety to their patients. Grayed objects are the least influential factors in the mental model based on means testing, discriminant analysis, and hierarchical linear modeling. Coefficients are standardized canonical discriminant function coefficients. Beta coefficients (β) = HLM (hierarchical linear models) or BLR (binary logistic regression). Odds ratio (OR). $* P \leq 0.05$; $** P \leq 0.01$; $*** P < 0.0$.

3. Experimental Section

3.1. Respondent Recruitment

The recruitment process for this study has been previously described [21]. In brief, registered dietitians were recruited via advertisements and emails to listservs sponsored by state affiliates of the Academy of Nutrition and Dietetics. The majority of responses (74.3%) originated from Colorado,

Connecticut, Florida, Georgia, Maine, Nevada, North Dakota, Ohio, and Texas. Individual characteristics of the respondents were measured for Age, Gender, Ethnicity, Income, and Education. Although the sample was a non-probabilistic sample of volunteers who responded to the advertisements, the sample was a near representation of the population of registered dietitians in the United States, based on a comparison of the descriptive statistics with demographic statistics of dietitians in the year the study was conducted [23]. Respondents to the survey agreed to a waiver of written informed consent prior to completing the survey, as per human subjects' protocol approved by The Ohio State University Institutional Review Board for Social and Behavioral Sciences (Protocol #2008B0345, approved 2009). Survey respondents were financially compensated for participation.

3.2. The Survey

The survey was designed for implementation on the Internet and was based on the Risk Information Seeking and Processing (RISP) theory of risk communication [16]. Survey items originated from previous studies [29–32] and were tested with five key informants for content and face validity. The online version of the survey was developed using SelectSurvey.NET (ClassApps, Release 2.0, Overland Park, KS, USA). Registered dietitians ($n = 16$) were recruited to pilot test the web survey prototype; and, to beta test functionality, provide further face and content validity, and generate data for item analysis (reported elsewhere, [21]). Revisions to the survey were completed prior to study implementation. The survey was opened for participants for approximately two weeks, or until a minimum of 300 responses was received. Sample size was determined using pilot study data. If participants agreed to waive written consent, the following survey page asked questions to determine if they met inclusion criteria: registered dietitian status, age 18 years or older, and employed in a position requiring direct client/patient education. If inclusion criteria were met, the participant proceeded to the survey. Each item required a response to advance to subsequent items; thus, an option was included in all items to check if they voluntarily choose to not answer the item. Items that were not answered were considered missing data. Only surveys from respondents who advanced through all items were included in the data analysis. There were 349 partial or completed surveys submitted to the survey database, and 327 completed surveys were included in the data analysis.

3.3. Control Variables

Control variables included individual characteristics, previous experience with the risk hazard, foodborne illness, and Political philosophy. All control variable items were constructed as binomial yes/no response choices.

Individual characteristics were asked for Gender, Age, Education, Income, and Ethnicity. Five Age categories were collapsed into three categories and then converted to dummy variables for regression analysis. Since inclusion criteria specified credentialed dietitian status (registration in the US), none of the respondents had less than a college education, as required for registration. Two education categories were created as dummy variables, either college graduate or post-graduate education (including both academic and professional degrees). Income control variables were created from seven original categories. Over 90% of participants were Caucasian and the remainder were a mixtures of other ethnicities. Data for all ethnicities other than Caucasian were collapsed into one variable (named,

Other ethnicities). Six items asked dietitians about their previous experience with foodborne illness (FBI). One item referred to the dietitian's personal experience and another item if they had received medical care. Similar items asked if close friends or relatives, or if the dietitian's patients had experienced a FBI and did they seek medical care. Political philosophy was collapsed from five original categories into three dummy variables.

3.4. RISP Communication Theory Variables

Respondents' perception of the risk hazard posed by FBI was measured by three constructs: Risk judgment (Perceived likelihood and Perceived severity), Personal control, and Institutional trust. Perceived likelihood and Perceived severity were constructed in the survey as 11-point semantic differential scale items with 0 (not very likely) and 10 (very likely) as the discriminators for perceived likelihood and 0 (not very serious) and 10 (very serious) as discriminators for perceived severity. Personal control was measured by two items (5-point Likert scale). One item measured the respondents' sense of personal control over avoiding FBI, and another item asked if they perceived that their patients had control over avoiding FBI. Survey items were constructed as 5-point Likert scales from 1 (strongly disagree) to 5 (strongly agree). A summed scale was computed from four items that measured the respondent's trust in institutions charged with responsibility to maintain the safety of the food supply (Institutional trust, $\alpha = 0.68$, minimum score = 4, maximum score = 20).

Two variables were computed as the product of an item that measured the belief that referent others expected the participant to be informed about food safety (Normative belief) and an item that measured the degree to which the participant believed that referent others' opinions were important to them (Motivation to comply) [15]. One variable addressed personal influences (variable name, Informational subjective norm-people) and the second variable addressed the dietitian's patients (variable name, Informational subjective norm-patients) as their referent others and the influence these external forces exerted on the sufficiency of the participants' food safety knowledge. (Minimum score response = 5, maximum score = 25 for both subjective norm variables).

The Affect construct was constructed from two items asking respondents about how much worry they felt about the possible health risk posed by eating fresh vegetables. One item measured worry according to the health risks to the respondent (variable name, Worry-self) and the other item measured worry according to the health risk of the dietitian's patients (variable name, Worry-patient). Items were constructed as 11-point semantic differential scales from 0 (no worry) to 10 (worry a lot).

Six items measured participant's beliefs about various media they used to gain information about food safety (five-point Likert scale). Principal component analysis was used to identify two factors that addressed beliefs about media bias (Media bias beliefs, $\alpha = 0.62$) and validity of food safety information (Validity cues beliefs, $\alpha = 0.58$). The strength of the principle component analysis was acceptable (KMO = 0.67) and 56.0% of the variance in the six items was accounted for in the two factors. Four items queried the amount of attention or credibility paid to various information sources (television, newspaper, personal discussion, and radio) (0, No attention; 10, A lot of attention). Items were summed to form a computed variable, Information source credibility ($\alpha = 0.89$).

Two items were summed to form the variable, Information gathering capacity. One item addressed how easily the participant could get needed information about the safety of fresh vegetables (five-point

Likert scale), and the other addressed the degree to which the participant agreed that useful information about the safety of fresh vegetables was hard to get (five-point Likert scale, reverse coded). Items were summed to form the variable, Information gathering capacity ($\alpha = 0.72$).

The development of the construct, Current (food safety) knowledge, has been previously reported [22] and consisted of three variables. Four items on the survey addressed self-reported awareness (4-point Likert scale) of *Salmonella* spp., *E. coli* O157:H7, *Campylobacter jejuni*, and *Listeria monocytogenes*. Data were summed to form the variable, Pathogen awareness ($\alpha = 0.79$). Four additional items measured understanding of the same pathogens (5-point Likert scale), and were summed to form the variable, Pathogen understanding ($\alpha = 0.83$). General food safety knowledge items were scored as either correct or incorrect. Of the 17 original items included on the survey, eight were retained for data analysis after item analysis with this data set ($\alpha = 0.51$). Scores for Pathogen awareness, Pathogen understanding, and General food safety knowledge were converted to percentiles for data analysis.

The variable, Knowledge threshold, was computed from two items that measured the degree to which dietitians agreed that they had a sufficient amount of food safety knowledge about fresh vegetables for either their own use, or for their ability to adequately teach their patients. Items were five-point Likert scales and were summed to form the variable ($\alpha = 0.91$). Data were converted to percentiles for data analysis.

Eight items measured the participants' information processing characteristics, either heuristic or systematic processing. Four items were summed to measure Systematic information processing (five-point Likert scale, $\alpha = 0.53$). Four items measured Heuristic information processing on the survey, but two were eliminated from data analysis to improve internal consistency of the summed scale (five-point Likert scale, $\alpha = 0.52$).

3.5. Theory of Planned Behavior Variables

Indirect attitude items were computed as the product of scores from survey items assessing behavioral beliefs and a paired item assessing the perceived evaluation of that belief [15,21]. Item products were considered to be indirect measures of attitude. Thirty attitude products had internal consistency ($\alpha = 0.72$) and were thus used in principle component analysis to identify factors. Nineteen attitude products were reduced to four factors, each with improved internal consistency (Table 3). Factor scores were computed and used in logistic regression as independent variables. The strength of the principle component analysis was strong (KMO = 0.79) and 42.2% of the variance in the 30 original Indirect attitude measures was accounted for in the four factors.

Two items measured barriers to Teaching behavior as indicators of Perceived behavioral control. One item measured the possibility of including information about the food safety of fresh vegetables in their educational sessions (variable name, PBC-possible, five-point sematic differential, definitely impossible to definitely possible). The other item measured the perception of control that the dietitian has over teaching patients about fresh vegetable safety (PBC-control, five-point Likert scale).

An item with three response options queried the participants' food safety Teaching behavior. Response options were: I currently teach my patients about fresh vegetable safety ($N = 159$); Even though I don't currently teach my patients about fresh vegetable safety I plan to in the future ($N = 143$); or, I don't and never plan to teach my patients about fresh vegetable safety ($N = 25$). The Do not currently-teach behavior response option was computed from the summed responses of those who are not teaching food safety to their patients. The new variable name was Teaching behavior, with yes, Currently-teach (n = 159, 48.6%) or no, Do not currently teach (n = 168, 51.4%) as response levels.

3.6. Data Analyses

The study was designed as a cross-sectional, descriptive study. The Statistical Package for the Social Sciences software (SPSS Version 19.0, Chicago IL) was used for all data analyses. Descriptive statistics were calculated for all control variables, and differences by Teaching-behavior group determined with Chi-Square analysis. For summed-scale variables, internal consistency was assessed with Cronbach α and accepted if α was near or exceeded 0.60 [33]. Discriminant analysis was used to identify RISP theory variables that were the most influential in characterizing Teaching-behavior groups. Continuous scaled variables were selected for analysis, which were: Risk judgment, Personal control–self, Personal control–patient, Institutional trust, Informational subjective norm–people, Informational subjective norm–patients, Worry–self, Worry–patient, Media bias beliefs, Validity cues beliefs, Information source credibility, Information gathering capacity, Pathogen awareness, Pathogen understanding, General food safety knowledge, Knowledge threshold, Heuristic information processing, and Systematic information processing. Standardized canonical discriminant function coefficients ±0.30 or above were considered as major classifying variables, and coefficients between ±0.30 and ±0.19 were considered as minor classifying variables. Student's t-test was used to identify mean differences with the same RISP variables that were used in discriminant analysis as the dependent variables and Teaching behavior as the categorical variable. Hierarchical linear modeling (HLM) was used to test the predictive strength of the above RISP theory variables for six models (Heuristic or Systematic information processing; and, Overall; Currently teach, and Do not currently teach). For regression analysis, control variables were recoded to dummy variables. For all tests of significance, differences were declared if probability was at or less than the 5% significance level. Binary logistic regression analysis was used to determine the strength of the regression model for two levels of the dependent variable, Teaching behavior. Independent variables used in binary logistic regression analysis were Systematic information processing, Heuristic information processing, Indirect attitude (Local produce factor, Benefits teaching factor, Organic produce factor, and Nutrition quality factor), Informational subjective norms-people, Informational subjective norms-patients, Perceived behavioral control–possible, and Perceived behavioral control-control).

For objective 2 (the mental model), statistical results from all analyses were inspected and triangulated to qualitatively determine the most salient factors that explained the mental model of dietitians' information processing and teaching behavior.

4. Conclusions

Eight different causal models were generated to understand the relationships between RISP theory variables and Information processing behavior (Tables 1 and 2) and the ultimate outcome of interest in this study, Teaching Behavior (Figures 1 and 2). These models were developed using the theoretical framework of the Heuristic-Systematic Model of Information Processing [14] and the Theory of Planned Behavior [15], as previously proposed by Griffin *et al.* [16]. The juxtaposition of two behavioral models provides valuable new insight about how dietitians seek, acquire and translate/transfer important information about knowledge to move patients toward a higher goal of food safety.

Of interest is this study's finding that dietitians who have significantly higher knowledge and awareness of foodborne illness pathogens use Systematic information processing when learning new information about food safety. Their higher Knowledge threshold scores may suggest that they were aware of their knowledge gap on information that could be critically needed by their patients, prompting them to seek and learn new, and more in-depth information. Supporting this insight is the significantly higher concern for patients associated with the teaching group (Affect), and that patients are normative (Subjective Norms) for those dietitians who teach food safety. Additionally, dietitians who teach food safety were significantly more likely to view information sources critically (Information Source Credibility).

In contrast, General food safety knowledge did not differ between groups but was overall low in comparison to other groups surveyed when the survey items were first developed [31]. Incorrect knowledge may have contributed to the finding that dietitians who believed that organic vegetables are the safer choice for patients was also predictive of who will teach fresh vegetable safety. This suggests that food safety educators should seek ways to place in-depth food safety information in locations that are readily assessable to health professionals who have the critical skills needed to evaluate the information. Suggested locations for placing food safety information are peer-reviewed journal articles, trade-journals for specific professions, and webinars that provide continuing education credit. We have previously reported that this sample of dietitians frequently uses the internet to locate food safety information [22]. This is an opportunity for educators to combine quality food safety instruction with a popular means to access new information. To attract dietitians to the in-depth information locations, educators are advised to use motivational factors, specifically the health of their patients that were shown in this study to be predictors of teaching behavior.

Acknowledgments

We thank Gina Casagrande, Janet Buffer-Pealer, and Eric Lauterbach for their technological expertise in the design, function and implementation of the web survey. We are also grateful for the statistical guidance and advice provided by Dr. Steven Naber.

Author Contributions

Author, LeJeune was the principal investigator for the funding grant from the USDA, National Integrated Food Safety Initiative, Grant # 2007-51110-03817. There was no involvement from this funding source in the design, implementation, interpretation or preparation of this manuscript. LeJeune was responsible for the initial studies on mental model development, supervising experimental work, providing key discussion on the results of the studies and co-writing the manuscript.

Author, Medeiros designed the survey, completed all data analyses, designed the mental model graphics, provided key discussion on the results, and co-writing the manuscript.

Conflicts of Interest

The authors declare no conflict of interest.

References

1. Mazzocchi, M.; Lobb, A.; Bruce Traill, W.; Cavicchi, A. Food scares and trust: A European study. *J. Agric. Econ.* **2008**, *59*, 2–24.

2. Shepherd, J.; Saghaian, S. Consumer response to and trust of information about food-safety events in the chicken and beef markets in Kentucky. *J. Food Distri. Res.* **2008**, *39*, 123–129.

3. Kersting, A.L.; Medeiros, L.C.; LeJeune, J.T. Differences in *Listeria monocytogenes* contamination of rural Ohio residences with and without livestock. *Foodborne Pathog. Dis.* **2010**, *7*, 57–62.

4. Kendall, P.; Medeiros, L.C.; Hillers, V.; Chen, G.; DiMascola, S. Food handling behaviors of special importance for pregnant women, infants and young children, the elderly and immune compromised people. *J. Am. Diet. Assoc.* **2003**, *103*, 1646–1649.

5. Athearn, P.N.; Kendall, P.A.; Hillers, V.; Schroeder, M.; Bergman, V.; Chen, G.; Medeiros, L. Awareness and acceptance of current food safety recommendations during pregnancy. *Matern. Child Hlth. J.* **2004**, *8*, 149–162.

6. Hoffman, E.W.; Bergmann, V.; Schultz, J.S.; Kendall, P.; Medeiros, L.D.; Hillers, V.N. Application of a five-step message development model for food safety education materials targeting people with HIV/AIDS. *J. Am. Diet. Assoc.* **2005**, *105*, 1597–1604.

7. Medeiros, L.C.; Chen, G.; Hillers, V.N.; Kendall, P.A. Discovery and development of educational strategies to encourage safe food handling behaviors in cancer patients. *J. Food Protect.* **2008**, *71*, 1666–1672.

8. Chen, G.; Kendall, P.A.; Hillers V.N.; Medeiros, L.C. Qualitative studies of the food safety knowledge and perceptions of transplant patients. *J. Food Protect.* **2010**, *73*, 327–335.

9. WebMD. Available online: http://www.webmd.com/ (accessed on 13 March 2015).

10. Accreditation Council for Education in Nutrition and Dietetics, Academy of Nutrition and Dietetics. ACEND Accreditation Standards for Dietitian Education Programs Leading to the RD Credential. Available online: http://www.eatrightacend.org/WorkArea/DownloadAsset.aspx?id= 6442468848 (accessed on 15 March 2015).

11. Van Duyn, M.A.S.; Pivonka, E. Overview of the health benefits of fruit and vegetable consumption for the dietetics professional: Selected literature. *J. Am. Diet. Assoc.* **2000**, *100*, 1511–1521.

12. Hanning, I.B.; Nutt, J.D.; Ricke, S.C. Salmonellosis outbreaks in the United States due to fresh produce: Sources and potential intervention measures. *Foodborne Pathog. Dis.* **2009**, *6*, 635–648.

13. French, M.R.; Levy-Milne, R.; Zibrik, D. A survey of the use of low microbial diets in pediatric bone marrow transplant programs. *J. Am. Diet. Assoc.* **2001**, *101*, 1194–1198.

14. Eagly, A.H.; Chaiken, S. *The Psychology of Attitudes*; Harcourt Brace: San Diego, CA, USA, 1993.

15. Ajzen, I. The theory of planned behavior. *Organ. Behav. Hum. Dec.* **1991**, *50*, 179–211.

16. Griffin, R.J.; Dunwoody, S.; Neuwirth, K. Proposed model of the relationship of risk information seeking and processing to the development of preventive behaviors. *Environ. Res. Sec. A* **1999**, *80*, S230–S245.

17. Lynam, T.; Brown, K. Mental models in human-environment interactions: Theory, policy implications, and methodological explorations. *Ecol. Soc.* **2011**, *17*, 24. Available online: http://dx.doi.org/10.5751/ES-04257-170324 (accessed on 13 March 2015).

18. Jones, J.A.; Ross, H.; Lynam, T.; Perez, P.; Leitch, A. Mental models: An interdisciplinary synthesis of theory and methods. *Ecol. Soc.* **2011**, *16*, 46. Available online: http://www.ecologyandsociety.org/vol16/iss1/art46/ (accessed on 13 March 2015).

19. Lynam, T.; Mathevet, R.; Etienne, M.; Stone-Jovicich, S.; Leitch, A.; Jones, N.; Ross, H.; Du Toit, D.; Pollard, S.; Biggs, H.; *et al.* Waypoints on a journey of discovery: Mental models in human-environment interactions. *Ecol. Soc.* **2012**, *17*, 23. Available online: http://www.ecologyandsociety.org/vol17/iss3/art23/ (accessed on 13 March 2015).

20. Frewer, L.J.; Fischer, A.R.H.; Brennan, M.; Bánáti, D.; Lion, R.; Meertens, R.M.; Rowe, G.; Siegrist, M.; Verbeke, W.; Vereijken, C.M.J.L. Risk/benefit communication about food—A systematic review of the literature. *Crit. Rev. Food Sci. Nutr.* **2013**, in press. Available online: http://www.tandfonline.com/doi/abs/10.1080/10408398.2013.801337#.VQMibeG3qf5 (accessed on 13 March 2015).

21. Casagrande, G.; LeJeune, J.; Belury, M.; Medeiros, L.C. Registered dietitian's personal beliefs and characteristics predict their teaching or intention to teach fresh vegetable food safety. *Appetite* **2011**, *56*, 469–475.

22. Medeiros, L.C.; Buffer, J. Current food safety knowledge of registered dietitians. *Food Protect. Trends* **2012**, *32*, 688–696.

23. Ward, B. Compensation and benefits survey 2009: Despite overall downturn in economy, RD and DTR salaries rise. *J. Am. Diet. Assoc.* **2010**, *110*, 25–36.

24. U.S. Public Health Service Food and Drug Administration. 2013. Food Code 2013 Recommendations of the United States Public Health Service Food and Drug Administration. Report number PB2013–110462. Available online: http://www.fda.gov/downloads/Food/GuidanceRegulation/RetailFoodProtection/FoodCode/UCM374510.pdf (accessed on 6 February 2015).

25. Slovic, P.; Finucane, M.L.; Peters, E.; MacGregor, D.G. Risk as analysis and risk as feelings: Some thoughts about affect, reason, risk, and rationality. *Risk Anal.* **2004**, *24*, 311–322.

26. Peters, E.; Slovic, P.; Hibbard, J.H.; Tusler, M. Why worry? Worry, risk perceptions, and willingness to act to reduce medical errors. *Health Psychol.* **2006**, *25*, 144–152.

27. Griffin, R.J.; Zheng, Y.; ter Huurne, E.; Boerner, F.; Ortiz, S.; Dunwoody S. After the flood: Anger, attribution, and the seeking of information. *Sci. Commun.* **2008**, *29*, 285–315.

28. Dangour, A.D.; Dodhia, S.K.; Hayter, A.; Allen, E.; Lock, K.; Uauy, R. Nutritional quality of organic foods. A systematic review. *Am. J. Clin. Nutr.* **2009**, *90*, 680–685.

29. Griffin, R.J.; Neuwirth, K.; Giese, J.; Dunwoody, S. Linking the heuristic-systematic model and depth of processing. *Commun. Res.* **2002**, *29*, 705–732.

30. Griffin, R.; Neuwirth, K.; Dunwoody, S.; Giese, J. Information sufficiency and risk communication. *Media Psychol.* **2004**, *6*, 23–61.

31. Medeiros, L.; Hillers, V.; Chen, G.; Bergmann, V.; Kendall, P.; Schroeder, M. Design and development of food safety knowledge and attitude scales for consumer food safety education. *J. Am. Diet. Assoc.* **2004**, *104*, 1671–1677.

32. Kendall, P.; Elsbernd, A.; Sinclair, K.; Schroeder, M.; Chen, G.; Bergman, V.; Hillers, V.; Medeiros, L. Observation versus self-report: Validation of a consumer food behavior questionnaire. *J. Food Protect.* **2004**, *67*, 2578–2586.

33. Parmenter, K.; Wardle, J. Evaluation and design of nutrition knowledge measures. *J. Nutr. Educ. Behav.* **2000**, *32*, 269–277.

9

Community Perspectives on the On-Farm Diversity of Six Major Cereals and Climate Change in Bhutan

Tirtha Bdr. Katwal [1,*], Singay Dorji [2], Rinchen Dorji [3], Lhab Tshering [3], Mahesh Ghimiray [4], Ganesh B. Chhetri [5], Tashi Yangzome Dorji [6] and Asta Maya Tamang [7]

[1] Specialist III-Maize, RNR Research and Development Center, Yusipang, Department of Forest and Park Services, Ministry of Agriculture and Forests, Thimphu, P.O. Box 212, Bhutan

[2] National Coordinator, Global Environment Facility-Small Grants Programme, UNDP, Thimphu, P.O. Box 162, Bhutan; E-Mail: sdbhutan@gmail.com

[3] Biodiversity Officers, National Biodiversity Center, Serbithang, Ministry of Agriculture and Forest, Thimphu, P.O. Box 875, Bhutan; E-Mails: rinchend2003@gmail.com (R.D.); lhabtsherin@gmail.com (L.T.)

[4] Rice Specialist III, RNR Research and Development Center, Bajo, Ministry of Agriculture and Forest, Wangduephodrang, P.O. Box 1263, Bhutan; E-Mail: mghimiray@gmail.com

[5] Agriculture Specialist II, Department of Agriculture, Ministry of Agriculture and Forest, Thimphu, P.O. Box 392, Bhutan; E-Mail: gbchettri@gmail.com

[6] Program Director, National Biodiversity Center, Serbithang, Ministry of Agriculture and Forest, Thimphu, P.O. Box 875, Bhutan; E-Mail: yangzome2011@gmail.com

[7] Principal Biodiversity Officer, National Biodiversity Center, Serbithang, Ministry of Agriculture and Forest, Thimphu, P.O. Box 875, Bhutan; E-Mail: asta.nbc@gmail.com

[*] Author to whom correspondence should be addressed; E-Mail: tirthakatwal@gmail.com

Academic Editor: Rainer Hofmann

Abstract: Subsistence Bhutanese farmers spread across different agro-ecological zones maintain large species and varietal diversity of different crops in their farm. However, no studies have been undertaken yet to assess why farmers conserve and maintain large agro-biodiversity, the extent of agro-ecological richness, species richness, estimated loss of traditional varieties and threats to the loss of on-farm agro-biodiversity. Information on the number of varieties cultivated by the farmers for six important staple crops were collected from nine districts and twenty sub-districts spread across six different agro-ecological zones of the country to understand farmers reasons for maintaining on-farm crop diversity, estimate agro-ecological richness, species richness and the overall loss of traditional varieties, to know the famers' level of awareness on climate change and the different threats to crop diversity. The results from this study indicated that an overwhelming 93% of the respondents manage and use agro-biodiversity for household food security and livelihood. The average agro-ecological richness ranged from 1.17 to 2.26 while the average species richness ranged from 0.50 to 2.66. The average agro-ecological richness indicates a large agro-ecological heterogeneity in terms of the different species of staple crops cultivated. The average species richness on the other hand shows that agro-ecological heterogeneity determines the type and extent of the cultivation of the six different staple cereals under consideration. The overall loss of traditional varieties in a time period of 20 years stands at 28.57%. On climate change, 94% of the farmers recognize that local climate is changing while 86% responded that they are aware of the potential impacts of climate change on their livelihoods. Climate change and associated factors was considered the most imminent threat to the management and loss of on-farm agro-biodiversity. The results from this study indicate that on-farm agro-biodiversity conservation, development and utilization programs have to be more specific to the different agro-ecological zones considering the agro-ecological heterogeneity. Attention has to be given to individual crops that have low average species richness and high percentage of loss of traditional varieties. The impact of climate change could offset the traditional seed system which primarily supports the persistence of on-farm agro-biodiversity in several ways.

Keywords: agro-biodiversity; climate change; subsistence farming; average agro-ecological richness; average species richness; threats; traditional seed system

1. Introduction

Bhutan represents a fragile mountainous ecosystem and is a least developed country. The economy of the country is one of the world's smallest and continues to depend substantially on the Renewable Natural Resources (RNR) sector that comprises Forest, Agriculture and Livestock. The RNR sector accounts for about 15.7% of the total GDP [1]. The livelihood of over 69% of the population is dependent on the RNR sector. The country is located in the southern slopes of Eastern Himalayas between latitudes 26°42′ N and 28°14′ N, and longitudes 88°44′ E and 92°07′ E. The country has a total geographical area of 38,394 km² of which about 70.46% is under forest cover with only 2.93% of the total area available for cultivation [2]. Rice, maize, wheat, barley, buckwheat and millets are major staple cereals cultivated by farmers. Bhutanese farmers are largely small holders, marginal and practice a self-sustaining, integrated and subsistence agricultural production system. The average land holding is three acres on which farmers grow a variety of crops under different farming practices and rear livestock to meet their household food security. Despite small farm size, farmers grow many types of crops and varieties where farm level agro-biodiversity is the corner stone for sustainable subsistence agriculture. In Bhutan where subsistence farming is still dominant, agro-biodiversity plays a pivotal role for sustainable agricultural development, food security and poverty alleviation [3]. Bellon [4] has noted that agro-biodiversity is the basis of food security both in subsistence and technologically advanced agriculture production systems. The Bhutanese agricultural production can be classified as a classic "small holder system" because it associates with most of the characteristic of a small holder. A small holder is characterized by small farm size less of than 10 hectares; most of the farming is undertaken using family labor; the major portion of the produce is used for household consumption with small surplus for sale that provide them the cash income [5,6].

Due to the significant influence of the high Himalayas, the Bhutanese agriculture is entirely dependent on the monsoon and prevailing weather conditions where even small variations in the onset and retreat of the monsoon could have considerable impacts on crop production [7]. The increasing frequency of extreme and adverse weather events that are considered as the signals of climate change include the floods from glacial lake outburst, flash floods, insurgence of new pest and disease such as the rice blast of paddy in 1995 and Gray Leaf Spot (GLS) in maize in 1997 indicate that climate change will have a major impact on the Bhutanese farmers [8]. Climate change has been proven to have a pervasive influence on food security and livelihoods of the small holder farmers worldwide [6,9,10]. According to the Sector Adaptation Plan of Action (SAPA) [11] of the Ministry and Agriculture and Forest (MoAF) for climate change, Bhutan is projected to experience a peak warming of about 3.5 °C by the 2050s with the overall significant increase in precipitation but with an appreciable change in the spatial pattern of winter and summer monsoon precipitation. Given this backdrop, the Bhutanese agriculture sector and the farming communities are likely to be most vulnerable.

Bhutan's strategic location as a landlocked country, poor accessibility with mostly rugged mountainous terrain in the high Himalayas and, its relative isolation from other parts of the world until 1960 has made it rely on agro-biodiversity for its domestic food production. In contrast to its geographical size, it has a rich and unique domestic diversity of species and varieties which has been enhanced through natural and human selection and, subsequent conservation by the farmers. The local crops are considered to possess tremendous genetic diversity and are well adapted to the specific requirements of the areas where they are grown. It is estimated that there are about 350 landraces of rice, 47 of maize, 24 of Wheat and 30 of Barley in the country [12]. Generally, agro-biodiversity in the Bhutanese context is understood as the cultivation of different types of crops, their landraces and varieties for direct consumption as food and normally includes cereals, oilseeds, legumes, roots and tuber crops, all types of vegetables and fruits [3]. The National Biodiversity Center (NBC) is the nodal agency of the country, which is mandated to develop programs for the sustainable conservation, development and utilization of the agro-biodiversity resources for food security and poverty alleviation. The conservation and use of agro-biodiversity is fundamental for making the agricultural ecosystems sustainable, productive, and resilient and can contribute to better nutrition and livelihoods of poor farmers throughout the world [13]. Thus, agro-biodiversity has been recognized as one of the potential means of adaptation to climate change mainly through the development and use of biotic and abiotic tolerant varieties, strengthening the traditional seed system and enhancing the on-farm diversity as potential insurance to climate change [3,7,11]. Jarvis *et al.* [14] have established that agro-diversity is maintained by farmers as an insurance to meet future environmental changes and socio-economic needs.

As one of the key strategy for the formulation of the Strategic Action Plan (SAP) for conservation of cereals in the country, the NBC commissioned a rapid baseline research to understand the community perspectives on the on-farm diversity, status and trends of the six major cereals considered in this study.

This study attempted to understand why farmers conserve and maintain agro-biodiversity, the extent of agro-ecological richness, average species richness and the estimated loss of traditional varieties and threats to the loss of on-farm agro-biodiversity.

2. Materials and Methods

This study was undertaken as one of the key process for the formulation of SAP for conservation of cereals. A nationwide community vulnerability assessment study was carried out in 2013 covering the five main agro-ecological zones of the country. This research was undertaken through collaboration between the NBC, the Regional Research and Development Centers (RDC), *Dzongkhag* (district) and *Geog* (Sub-district), agriculture extension staff and farmers. The objectives of this study were to understand why farmers conserve and maintain agro-biodiversity, the extent of agro-ecological richness, average species richness and the estimated loss of traditional varieties and threats to the loss of on-farm crop species diversity in the country.

2.1. Study Sites

This survey covered a total of nine *Dzongkhags* (districts) out of the total 20 *Dzongkhags* and 20 of the total 205 *Geogs* (Sub-districts). The districts and sub-districts were randomly selected to represent the five dominant agro-ecological zones (Table 1). The elevations of the study sites ranged from 200 masl to 2500 masl covering four different types of agro-ecological zones namely the humid subtropical, the dry subtropical, warm temperate and the cool temperate. The dominant production environment in the dry-subtropical and cool temperate agro-ecology was the rainfed dryland farming where crop production entirely depends on monsoon rains. In the warm temperate, humid and wet-subtropical agro-ecological zone, terraced rice paddies which are irrigated and those that depend on rainfed streams for source of irrigation were the prevalent land use practice. The key features of the study sites are summarized in Table 1.

2.2. Data Collection Methodology

In order to start the survey, a core team comprising of biodiversity officers, researchers and policy makers was formed. This team designed a comprehensive survey questionnaire to assess the status, extent and trends of cereals diversity and farmers understanding and perceptions on climate change and, its potential impact on cereals diversity. After the finalization of the questionnaire, the sites were randomly selected to represent the different agro-ecological zones from where the data for this study was gathered. The target crops in this study were rice (*Oryza sativa* L), maize (*Zea mays*), wheat (*Triticum asetivum*), barley (*Hordeum vulgare*), buckwheat (*Fagopyrum esculentum*); millets namely finger millet (*Eleusine coracana*), foxtail millet (*Setaria italaca*) and common millet (*Panicum miliaceum*) that constitute the predominant cereals cultivated by the farmers. Information was collected from 404 respondents (Table 1).

Table 1. Detail of the study sites and respondents.

| Agro-ecology/Altitude Range masl | Dzongkhags | Geogs | No. of Informants | | Total |
			Male	Female	
Cool Temperate 2600–3600	Haa	Isu	8	12	20
		Katsho	7	13	20
Warm Temperate 1800–2600	Paro	Tsento	7	13	20
		Dopshari	8	12	20
	Punakha	Shelngana	10	6	16
		Toewang	9	15	24
	Dagana	Drujegang	13	7	20
		Goshi	18	2	20
Dry Subtropical 1200–1800	Mongar	Chali	13	11	24
		Kengkhar	12	9	21
	Trashiyangtse	Khamdang	4	12	16
		Toetsho	16	9	25
	Pemagatshel	Khar	7	9	16
		Yurung	1	15	16
Humid-Subtropical 600–1200	Samtse	Norgaygang	18	4	22
		Pemaling	15	6	21
		Samtse	14	6	20
Wet-Subtropical 150–600	Sarpang	Gelephu	16	3	19
		Jigmechholing	21	2	23
		Shompangkha	20	1	21
Total	9	20	237	167	404

Source: SAP Survey [3].

During the survey, the survey team collected the data based on the information entirely recalled by the farmers. The surveyors visited the farmers and interviewed them to gather information on the crops, varieties grown and lost, cereals displaced and farmer's perceptions on climate changes. The information on crop varieties is exclusively based on those mentioned by the farmers. The names of the varieties reported by the respondents included those being cultivated now or those that used to be cultivated before 20 years and that they had heard of from the village elders. Information was collected for traditional and improved varieties. Improved varieties in this study refers to those crop varieties that have been introduced from different sources from outside the country and evaluated and adapted to local conditions through adaptive research including those varieties that have been developed through hybridization with local varieties. The improved varieties are formally released for cultivation to the farmers by the Technology Release Committee (TRC) of the Ministry of Agriculture and Forest which is coordinated by the Council for RNR Research of Bhutan (CORRB). The improved varieties are in most cases high yielding compared to the traditional varieties. They also confirm to the three basic characteristics of being distinct, uniform and stable.

The respondents were randomly selected to represent a *Geog*. The survey targeted respondents with age group of 40 years and above in order to get a better perspective on the status and trends of cereals and traditional crop varieties cultivated 20 years ago. The survey also gathered farmer's awareness on the importance of farm level diversity of these cereals, practices and capacity to manage cereals diversity and different challenges faced in maintaining the cereals diversity.

2.3. Data Analysis

The survey data was first compiled using MS Excel and then relevant tables were generated using MS Access. To understand the role of the cereals diversity and the reasons why farmers cultivate different crops and varieties we computed the percentage of respondents assigning different reasons for maintaining agro-biodiversity. The reasons that the highest percentage of respondents assigned was considered the most important factor for maintaining the on-farm agro-biodiversity.

We determined the current status of on-farm agro-biodiversity for individual crop by listing the number of varieties that the surveyed farmers are currently cultivating in their farms (Table 2). In estimating the number of varieties grown for each crop we only used the respondents growing the crops. Individual variety named by the respondents for each crop was used as the basic diversity unit. Wherever we came across with varieties with different local names in the same locality we further probed the farmers and asked them to describe and agree if the variety in question were similar or different. We removed the varieties with similar names to avoid duplicates. Jarvis *et al.* [14] and Kiwuka *et al.* [15] have used variety as a basic diversity and when issue of the same variety being reported with different names occurred they used the farmer's knowledge and description to agree whether a variety in question was actually different. In this study, we also used farmer's knowledge and description as the basis to distinguish whether a variety was similar or different when varieties with different names were reported from the same location.

Table 2. Number of improved and traditional varieties cultivated by the interviewed farmers in the study sites.

Crop	Improved Variety	Traditional Variety	Total Varieties Cultivated
Rice	15	84	99
Maize	5	34	39
Wheat	6	19	25
Barley	0	14	14
Buckwheat	1	13	14
Millet	0	30	30
Total (*n* = 219)	25	194	219

Source: SAP Survey [3].

To estimate the level of current on-farm diversity we took into account number of traditional and improved varieties which are currently grown by the respondents in their farm and listed during the survey. Similarly, we asked the respondents to list of varieties for each crop cultivated 20 years ago (1993) but not found in the respondents farm now (2013). We used the chronological difference between numbers of varieties cultivated now and the number of varieties that used to be cultivated 20 years ago to estimate the percentage of varieties lost for each crop (Table 3). Fowler and Mooney [16] have also used number of varieties in a time series gap of 80 years (1903 and 1983) to estimate the percentage loss of diversity for different vegetables as one of the basis to establish the loss of genetic diversity. Thrupp [17] has used the same basis in his review to indicate the loss of traditional crop varieties over time.

Table 3. Status of traditional varieties of cereals cultivated 20 years ago and now.

Crop	20 Years Ago (1993)	Now (2013)	Loss %
Rice	126	86	31.75
Maize	45	34	24.44
Wheat	31	19	38.71
Barley	14	8	42.86
Buckwheat	15	13	13.33
Millet	35	30	14.29
Total	**266**	**190**	**28.57**

Source: SAP Survey [3].

Richness is one notion of diversity that refers to the number of different kinds of individuals regardless of their frequencies [14]. In this study we have attempted to assess the average species richness for each agro-ecological zone and for six important staple crops (Table 4). We first estimated the average richness per agro-ecological zone and per species, as the average number of varieties of the six staple crops cultivated by farmers in the six different agro-ecological zones. We used all the respondents interviewed in each site to estimate the agro-ecological richness for each crop under consideration (Table 4). We computed the average species richness by taking the mean of the species richness from different agro-ecological zones to compare the richness between the six different crops covered in this study.

Table 4. Average agro-ecological and species richness for six main staples, 2013.

Average Agro-ecological and Species Richness for Six Main Staple Cereals							
Agro-ecological Zones (AEZ)	Rice	Maize	Wheat	Barley	Buckwheat	Millet	Average Agro-ecological Richness
Cool Temperate (2600–3600 masl)	0.13	1.65	3.16	1.37	2.20	1.09	1.60
Warm Temperate (1800–2600 masl)	3.70	1.31	0.73	0.09	0.70	0.50	1.17
Dry Sub-tropical (1200–1800 masl)	4.63	5.63	0.73	0.91	0.90	0.78	2.26
Humid Sub-tropical (600–1200 masl)	3.19	2.23	0.16	0.11	0.92	1.94	1.42
Wet Sub-tropical (150–600 masl)	1.50	2.48	0.31	0.03	1.26	1.67	1.21
Average species richness	2.63	2.66	1.02	0.50	1.20	1.19	1.53

Source: SAP Survey [3].

We used the number of respondents (counts) to estimate the percentage of respondents as a simple measure to assess farmers' level of awareness on climate change, their perceptions of climate change and its impact on agro-biodiversity and household food security. Risks and threats to cereals diversity posed by climate change and farmer's current coping strategies for adaptation to climate change were recorded. The threats mentioned by the respondents were listed for each crop. The frequency of threats mentioned was cumulated and ranked to arrive at the seven most widespread threats to cereal diversity (Figure 1). To know which traditional crops were being displaced, we directly asked the farmers for information and synthesized the information and tabulated by *Dzongkhags* and by agro-ecology (Table 5).

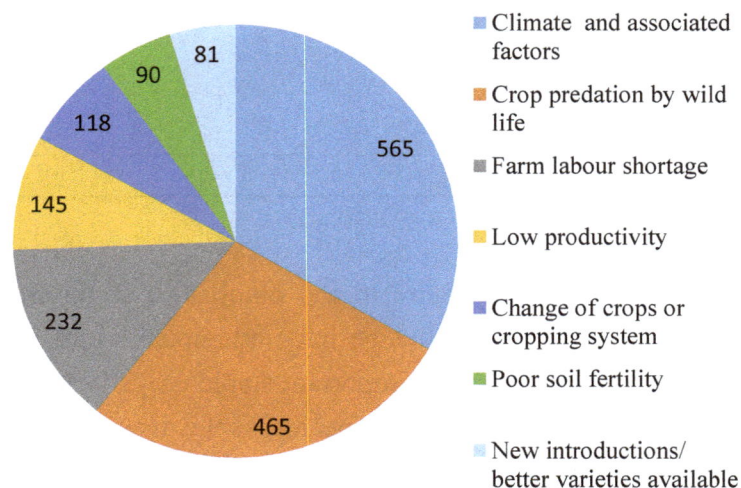

Figure 1. Frequency of different threats to cereals diversity reported by farmers.

Table 5. Information on cereals displaced and crops displacing them in study site.

Dzongkhags	Agro-ecology	Crops Displaced	New Crops
Haa	Cool Temperate 2600–3600	Barley, Buckwheat, Wheat, Millet	Potato, Apple, Vegetable, Pasture
Paro	Warm Temperate 1800–2600	Wheat, Barley, Buckwheat	Potato
Dagana	Warm Temperate 1800–2600	Maize	Citrus
Trashiyangtse	Dry Subtropical 1200–1800	Wheat, Buckwheat	Potato, Vegetables
Samtse	Humid-Subtropical 600–1200	Rice, Maize, Wheat, Barley, Buckwheat, Millet	Ginger, Cardamom, Arecanut, Citrus, Vegetables
Sarpang	Wet-Subtropical 150–600	Rice, Maize, Wheat, Barley, Buckwheat, Millet	Arecanut, Cardamom, Litchi, Vegetables, Citrus, Pasture, Fodder

Source: SAP Survey [3].

3. Results and Discussion

3.1. Importance of On-Farm Agro-Biodiversity

This study indicates that on–farm agro-biodiversity which is dominated by rice, maize, wheat, barley, buckwheat and millet plays a pivotal role in the household food security of the Bhutanese farmers with 93% of the respondents saying that crop and varietal diversity is very important. Of the total respondents

90% ($n = 365$) said that they cultivated six different cereals for household food security and self-sufficiency while 7% ($n = 26$) said that they grow for income generation and only 3% ($n = 6$) mentioned conservation as one of their objective of cultivation. These reasons are highly relevant and valid because at the national level, domestic food production meets 64% of the total cereals requirement [18]. Conservation of diversity of crops is often not the foremost focus of any farmer but Bhutanese farmers primarily perpetuate diversity for household food security and livelihoods. In Ethiopia, small holder farmers in marginal production system use high levels of crop genetic diversity for maintaining food security [19]. Farmers are also known to cultivate traditional varieties because in many situations they still provide the best means of fulfilling their livelihood needs [20]. According to [4], subsistence farmers highly value the diversity of crops mainly for the purpose of spreading the production risks, needs for crops that can adapt to different environments, for protection against pest and diseases, and meeting their social and cultural needs.

3.2. Status of On-Farm Diversity of Six Cereals under Consideration

The information on varieties cultivated by the interviewed farmers in the study sites shows that for each crop the cultivation of traditional variety by far dominates the improved varieties (Table 2). The highest number of variety cultivated was recorded in rice and the least was in barley and buckwheat. No improved varieties are currently cultivated for barley and millets, although two improved millet varieties have been released by the formal breeding program.

The average agro-ecological richness ranged from 1.17 to 2.66 (Table 4) indicating that Bhutanese farmers in the six different agro-ecological zones grow more than one staple crop in their farm. The highest average agro ecological richness of 2.66 was recorded for the dry-subtropical zone. In the dry-subtropical zone farmers have more choice to grow all the six different staples in both terraced wetland paddies and dryland (upland) fields. The low agro-ecological richness in the warm temperate region can be underpinned to rice as the single most dominant crop and fewer farmers growing barley and millet (Table 4). The average species richness for the six crops ranged from 0.50 to 2.66. The highest average richness of 2.66 estimated was for maize and the lowest for barley (Table 4). This can be explained by the fact that maize is a major crop that is widely cultivated by most households across the five different agro-ecological zones as compared to barley which is considered to be a minor cereal and is mostly dominant in the cool temperate zone. Jarvis *et al.* [14] have reported the average farm richness for traditional varieties in the range of 1.38 to 4.25 while Mulumba *et al.* [21] have estimated the household richness for common bean and banana at 2.37 and 8.02 respectively.

In this study the average species richness estimated for some crops is less than one. This can be underpinned to the large altitudinal variation within and among the agro-ecological zones where some of the six staple cereals cannot be cultivated and that some cereals like wheat, barley, buckwheat and millet are cultivated by fewer households as compared to rice and maize. The lower richness of rice than that of maize despite having much higher number of varieties for rice (Table 2) is due to the very low species richness for rice in the cool temperate agro-ecological zone where rice is not cultivated by majority of the respondents. The variation in the agro-ecological richness among the six different agro-ecological zones indicate a much wider agro-ecological heterogeneity in terms of the extent of cultivation of six different cereals under consideration. Some species like barley is more dominant in the cool temperate agro-ecological zone where as in the dry-subtropical zones all the six crops are cultivated. For instance wheat, barley and buckwheat are more predominant in the cool temperate agro-ecological zone whereas millet is more widely cultivated in the humid and wet subtropical agro-ecological zones. The lower average species richness for wheat, barley, buckwheat and millets as compared to rice and maize reveal that households grow much fewer varieties of these crops and these crops are dominant only is a selected agro-ecological zone. This is a cause for concern as growing few varieties entails higher risks of crop failures especially during the outbreak of diseases as in the case of Gray Leaf Spot of maize in 2007 where all the high altitude traditional maize varieties succumbed to the disease and famers lost maize harvest to the tune of 70%–100% [22]. However, Bellon [4] has argued that although it is important to maintain diversity by growing a large number of varieties, it alone does not assure a high level of diversity because the different varieties under cultivation might not be genetically different.

The chronological comparison of the traditional varieties cultivated 20 years before (1993) and now (2013) for each crop shows that quite a substantial percentage of traditional varieties have been lost (Table 3). The overall percentage of traditional varieties lost in all six cereals is estimated at 28.57%. The lowest percentage of traditional varieties lost was 14% for millet and the highest was 43% in barley which is quite alarming (Table 3).

3.3. Farmers Perceptions on Climate Change and Threats to Cereals Diversity

An overwhelming 94% of the respondents in the study sites had the perception that local climate is changing while 86% responded that they are aware of the potential impacts of climate change on their livelihoods. Most farmers' descriptions of climate change interpreted as the fluctuations of weather events and their experiences of droughts, unreliable monsoon, the insurgence of new pest and diseases in crops, rising temperature, prolong occurrence of frost and changing snowfall patterns [23]. A similar study on building farmers' perception and traditional knowledge on climate change undertaken in Peru, Zimbabwe and Vietnam has found that smallholder farmers are aware of the occurrence of climate change and their perceptions of climate change are based on impact on their farming systems, livelihoods and crop performances [24]. As agro-biodiversity is central to the livelihood of the Bhutanese farmers, we asked the farmers to identify and rank potential threats faced in maintaining agro-biodiversity of the staple cereals from their knowledge and experiences. From their experiences farmers assigned climate change and climate associated factors as the most serious threat to the diminishing on-farm cereals diversity in the study sites (Figure 1). The climate associated factors included drought, flood, outbreak of pest and diseases, hail and windstorm, landslide and other forms of natural calamities. The frequency of climate and associated factors was the highest while the threats from the new introduction or from the promotion of high yielding varieties was the lowest. A very unique but a severe threat perceived by farmers to the decline of the on-farm cereal diversity is the crop predation by wildlife. A national review of the status of the human-wildlife conflicts by a task force found that almost 70% of the households attribute losses of crops like maize and paddy to predation by wildlife [25]. The most direct threats to agro-biodiversity that emanates from climate change and increasing human wildlife conflicts could contribute to farmers relinquishing farming, disruption of the traditional seed system, loss of traditional crops, fallowing of productive agricultural lands which ultimately leads to the increasing household food shortages and decrease of the farm level diversity from the extinction of crops and varieties specifically adapted in that area [3].

The Bhutanese subsistence farmers meet 98% of their total cereals seed requirement from the informal traditional seed system managed by themselves [26]. They annually recycle the farm saved seed, select seeds for next season and even exchange seeds from other farmers and communities. Many communities are also engaged in the formal seed production chain as registered seed growers and seed producer groups. Disruption of this established seed system in any form will have a direct implication on the household food security and on-farm agro-biodiversity. The impact of climate change can be very profound on the traditional seed system and can lead to an extreme scenario of no seeds for planting in the next season. The impact on traditional seed system is manifested in the form of entire crop failure and complete loss of source seed for next planting due to drought and untimely monsoon, floods leading to complete crop submergence in low lying areas, early frost and cold temperature affecting seed setting in higher elevations, poor quality seed due to incessant rain during harvesting and curing, higher incidence of storage pest due to inadequate drying resulting from high humidity and poor germination of farm saved seeds. Due to topography that is dominated by high mountains and deep valleys there is a wide variation of micro climate especially rainfall and temperature over short distance and small altitudinal range. Hence Bhutanese farmers often operate on highly location specific crops and varieties that are almost impossible to replace. The diversity and types of agro-ecosystems is known to have a positive influence on the on-farm varietal diversity [27]. The traditional seed system and production of quality seed on-farm are considered the key elements for the perpetuation of the agro-biodiversity [4,28]. In Vietnam, Tin et al. [29] have established that the number of rice varieties managed by seed clubs in the informal system was generally very high which lead to high on-farm diversity. Similarly in Mexico a large number of small-scale maize farmers continue to rely on the traditional seed systems which is fundamental for farmers' livelihoods and conservation of maize land races [30].

We considered the hypothesis that the increasing emphasis of the formal research and development programs for the promotion of improved high yielding varieties as one of the key incitement for the loss of diversity of traditional crops and varieties. Although, the displacement of the traditional crops and varieties by the new introductions was ranked as the least threatening to cereals diversity, it is apparent that there is an actual displacement of traditional crops (Table 5). In all the districts across all agro-ecological zones, farmers have identified different cash crops that are displacing the traditional cereals. In the Dzongkhags representing the cool temperate and warm temperate agro-ecosystem potato, vegetables, apples and pasture grasses are the new crops that are displacing wheat, barley, buckwheat and millet. Maize is displaced by citrus only in Dagana Dzongkhag. In the dry-subtropical agro-ecology represented by Trashiyangtse Dzongkhag, wheat and buckwheat is displaced by potato and vegetables. Interestingly, rice has been displaced only in Samtse and Sarpang which is largely attributable to the increasing cultivation of cardamom an attractive plantation crop that fetches a very high cash returns. The standing government policy that restricts the conversion of irrigated wetland designated for rice cultivation to other forms of land use acts as deterrent to the large scale non-displacement of rice by other crops.

The majority of the farmers (61%) responded that during any natural hazards and crop failures, they usually manage themselves; 31% said they report the problems to the local government and seek government support, while 8% said they mobilize community support. The majority of Bhutanese farmers who operate on diverse and risk prone environment have been traditionally known to make key farming decisions to spread risks at household level. Hansen *et al.* [31] have established that the availability of information on climate fluctuations and its potential impact on agriculture to the farmers well in advance immensely help farmers to prepare better to cope and adapt to such adverse extreme events.

4. Conclusions

This study brings into light the perceptions of the farmers on the status and significance of on-farm varietal diversity of six staple crops and climate change based on their experiences. It is apparent that household food security and livelihood of the subsistence Bhutanese farmers largely hinges on the on-farm agro-biodiversity which provides with crops and varieties that have specific adaptation for the diverse risk prone farming environments spread across five different agro-ecological zones.

This study also shows that subsistence Bhutanese farmers still continue to cultivate different types of staple crops and their varieties in their farms maintaining a rich on-farm agro-biodiversity across different agro-ecological zones. The average agro-ecological richness estimated in this study indicates a wider agro-ecological heterogeneity which determines the types and extent of crops cultivated. Some agro-ecological zones like the dry-subtropical agro-ecological zone grow more crops and their varieties as compared to other agro-ecological zones. The average species richness further indicates that some species like barley is more dominant in the cool temperate agro-ecological zone where as in the dry-subtropical zones all the six crops are cultivated. From the six staple cereals included in this study wheat, barley and buckwheat are more predominant in the cool temperate agro-ecological zone whereas millet is more widely cultivated in the humid and wet subtropical agro-ecological zones. The average species richness estimated for wheat, barley, buckwheat and millets was much lower compared to rice and maize. The lower species richness reveals that households grow much fewer varieties of these crops and these crops are dominant only in a selected agro-ecological zone. Maize and rice had much higher species richness as these two crops are undoubtedly the most popular staple grown across all agro-ecological zones in the country. The findings from this study establish that on-farm conservation programs for the six different staple cereals needs to be more specific and concentrated considering the agro-ecological heterogeneity. The on-farm agro-biodiversity conservation program has to focus on individual crops that have low average species richness and high percentage of loss of traditional varieties.

Climate change is a reality at the household level and farmers perceive climate change including all associated climatic factors as the most impending threat to the loss of on-farm agro-biodiversity. The impact of climate change could directly offset the traditional seed systems in several forms and on farm agro-biodiversity that is vital for household food security. There is abundant existing scientific evidence that agro-biodiversity is fundamental to maintain food production and adaption to climate change for small subsistence farmers. The national research and development programs should start giving more attention towards selection, seed production and dissemination of locally adapted traditional varieties or using the local genes in the future crop breeding programs. The six staple cereals under consideration in this study being displaced by plantation crops rather than the perception that local varieties are being displaced by high yielding improved varieties need more in-depth analysis and attention.

The exploitation of the benefits of agro-biodiversity has been rightly recognized in the National Adaptation Programme of Action (NAPA) and the Sector Adaptation Plan of Action (SAPA) of the MoAF as one of the coping strategies for climate change. To harness the existing potential of agro-biodiversity for adaption to climate change, more dynamic on-farm agro-biodiversity conservation, development and utilization programs that enhance household food security and resilience to climate change will be very important. Farmer's current coping strategies at the household level needs to be strengthened for adaptation to climate change through the provision of information on potential adverse extremes, enhancing farm level diversity through participatory variety selection and diversification of food crops by improving and strengthening the traditional seeds system.

Acknowledgments

We are very grateful to South East Asia Regional Initiative for Community Empowerment (SEARICE), Philippines for encouraging us to take up this research as a baseline for the formulation of the national Strategic Action Plan for cereals. This work was supported from the Benefit Sharing Fund (BSF) allocated for the development of the national SAP cereals by the International Treaty for Plant Genetic Resources for Food and Agriculture (ITPGFRA). The authors are grateful to the Program Director, National Biodiversity Center, Ministry of Agriculture and Forest for approving the study. We also acknowledge the valuable support provided by the Renewable Natural Resources Research and Development Centers, *Dzongkhag* and *Geog* Extension Officers that enabled us to undertake this research. We highly appreciate the whole hearted participation and cooperation of the farmers for sparing their time to provide required information. Finally, we take this opportunity to extend our gratitude to Vernooy Ronnie of Bioversity International, Rome for helping us to source useful and relevant references.

Author Contributions

The major research and the development of this manuscript has been done by Tirtha Bdr. Katwal, who is the main author and designated author for correspondence. Singay Dorji who is the second co-author substantially contributed in conceptualizing the research paper and the final analysis. Lhab Tshering, one of the co-authors assisted in data input and generation of information. All the other co-authors have contributed equally in the development and finalization this manuscript.

Conflicts of Interest

The authors declare no conflict of interests and the views expressed herein are those of the authors, and do not necessarily reflect the views of National Biodiversity Center and the Ministry of Agriculture and Forests.

References

1. Statistical Year Book. *Statistical Year Book of Bhutan*; National Statistical Bureau: Thimphu, Bhutan, 2013.

2.   National Soil Service Center and the Policy and Planning Division. *Land Cover Assessment Report*; National Soil Service Center and the Policy and Planning Division, Ministry of Agriculture and Forest: Thimphu, Bhutan, 2010.

3.   National Biodiversity Center. *Draft National Strategic Action Plan for Cereals*; National Biodiversity Center, Ministry of Agriculture and Forest: Thimphu, Bhutan, 2014.

4.   Bellon, M.R. Conceptualizing interventions to support on-farm genetic resource conservation. *World Dev.* **2004**, *32*, 159–172.

5.   Food and Agriculture Organization (FAO). Smallholders and Family Farmers. Available online: http://www.fao.org/fileadmin/templates/nr/sustainability_pathways/docs/Factsheet_SMALLHOLDE RS.pdf (accessed on 11 April 2014).

6.   Morton, J.F. The impact of climate change on smallholder and subsistence agriculture. *Proc. Natl. Acad. Sci. USA* **2007**, *104*, 19680–19685.

7.   National Environment Commission. *National Adaptation Programme of Action*; National Environment Commission, Royal Government of Bhutan: Thimphu, Bhutan, 2006.

8.   Bhutan Climate Summit. *National Action Plan. Biodiversity Persistence and Climate Change*; Bhutan Climate Change Secretariat, Ministry of Agriculture and Forests: Thimphu, Bhutan, 2011.

9.   Wood, S.A.; Jina, A.S.; Kristjanson, P.; DeFries, R.S. Smallholder farmer cropping decisions related to climate variability across multiple regions. *Glob. Environ. Change* **2014**, *25*,163–172.

10. Ministry of Agriculture and Forests. *Sector Adaptation Plan of Action*; Ministry of Agriculture and Forests: Thimphu, Bhutan, 2013.

11. National Biodiversity Center, Ministry of Agriculture and Forests. *Biodiversity Action Plan for Bhutan*; National Biodiversity Center, Ministry of Agriculture and Forests: Thimphu, Bhutan, 2009.

12. Platform for Agro-biodiversity Research. *Coping with Climate Change: The Use of Agro-biodiversity by Indigenous and Rural Communities*; Platform for Agro-biodiversity Research, Bioversity International: Rome, Italy, 2010.

13. Jarvis, D.I.; Brown, A.H.D.; Cuong, P.H.; Collado-Panduro, L.; Latournerie-Moreno, L.; Gyawali, S.; Tanto, T.; Sawadogo, M.; Mar, I.; Sadiki, M. A global perspective of the richness and evenness of traditional crop-variety diversity maintained by farming communities. *Proc. Natl. Acad. Sci. USA* **2008**, *105*, 5326–5331.

14. Kiwuka, C.; Bukenya-Ziraba, R.; Namaganda, M.; Mulumba, J.W. Assessment of common bean cultivar diversity in selected communities of central Uganda. *Afr. Crop Sci. J.* **2012**, *20*, 239–249.

15. Fowler, C.; Money, P. *The Threatened Gene: Food, Politics and Loss of Genetic Diversity*; Lutworth Press: Cambridge, UK, 1990.

16. Thrupp, L.A. Linking agricultural biodiversity and food security: The valuable role of agro-biodiversity for sustainable agriculture. *Int. Aff.* **2000**, *76*, 265–281.

17. Abay, F.; de Boef, W.; Bjørnstad, A. Network analysis of barley seed flows in Tigray, Ethiopia: Supporting the design of strategies that contribute to on-farm management of plant genetic resources. *Plant Genet. Resour.* **2011**, *9*, 495–505.

18. Policy and Planning Division. *Renewable Natural Resources Statistics*; Ministry of Agriculture and Forest: Thimphu, Bhutan, 2012.

19. Maxted, N.; Guarino, L.; Myer, L.; Chiwona, E.A. Towards a methodology for on-farm conservation of plant genetic. *Genet. Resour. Crop Evol.* **2002**, *49*, 31–46.

20. Mulumba, J.W.; Nankya, R.; Adokorach, J.; Kiwula, C.; Fadda, C.; de Santis, P.; Jarvis, D.I. A risk minimizing argument for traditional crop varietal diversity use to reduce pest and disease damage in agricultural ecosystems of Uganda. *Agric. Ecosyst. Environ.* **2012**, *157*, 70–86.

21. Katwal, T.B.; Wangchuk, D.; Dorji, L.; Wangdi, N.; Choney, R. Evaluation of gray leaf spot tolerant genotypes from CIMMYT in the highland maize production Eco-systems of Bhutan. *J. Life Sci.* **2013**, *7*, 443–452.

22. Bhutan Climate Summit. *Impacts of Climate Change on Food Security*; Ministry of Agriculture and Forests: Thimphu, Bhutan, 2011.

23. Oxfam Novib, ANDES, CTDT, SEARICE, CGN-WUR. *Building on farmers' perceptions and traditional knowledge: Biodiversity management for climate change adaptation strategies*; Oxfam Novib: Hague, Netherland, 2013.

24. Human Wildlife Conflict Management Task Force. *Human-Wildlife Conflict Management Implementation Plan*; Ministry of Agriculture and Forest: Thimphu, Bhutan, 2013.

25. Baldinelli, G.M. Agro-biodiversity conservation as a coping strategy: Adapting to climate change in the Northern Highlands of Bolivia. *Consilience* **2014**, *11*, 153–166.

26. Chettri, G.B.; Gurung, T.R.; Roder, W. *A Review on Seed Sector Development*; Department of Agriculture, Ministry of Agriculture and Forest: Thimphu, Bhutan, 2006.

27. Sthapit, B.R.; Upadhyay, M.P.; Shrestha, P.K.; Jarvis, D.I. On-farm conservation of agricultural biodiversity in Nepal. Volume II. Managing diversity and promoting its benefits. In Proceedings of the Second National Workshop, Nagarkot, Nepal, 25–27 August 2004.

28. Rana, R.B.; Garforth, C.J.; Sthapit, B.R.; Jarvis, D.I. Farmers' rice seed selection and supply system in Nepal: Understanding a critical process for conserving crop diversity. *Int. J. Agric. Sci. Vol.* **2011**, *1*, 852–872.

29. Tin, H.Q.; Cuc, N.H.; Be, T.T.; Ignacio, N.; Berg, T. Impacts of seed clubs in ensuring local seed systems in the Mekong Delta, Vietnam. *J. Sustain. Agric.* **2011**, *35*, 840–854.

30. Bellon, M.R.; Hodson, D.; Hellin, J. Assessing the vulnerability of traditional maize seed systems in Mexico to climate change. *Proc. Natl. Acad. Sci. USA* **2011**, *108*, 13432–13437.

31. Hansen, J.W.; Mason, S.J.; Sun, L.; Tall, A. Review of seasonal climate forecasting for agriculture in Sub-saharan Africa. *Exp. Agric.* **2011**, *47*, 205–240.

10

Pine Woodchip Biochar Impact on Soil Nutrient Concentrations and Corn Yield in a Silt Loam in the Mid-Southern U.S.

Katy E. Brantley, Mary C. Savin *, Kristofor R. Brye and David E. Longer

Department of Crop, Soil, and Environmental Sciences, University of Arkansas, 115 Plant Sciences, Fayetteville, AR 72701, USA; E-Mails: kebrantley89@gmail.com (K.E.B.); kbrye@uark.edu (K.R.B.); dlonger@uark.edu (D.E.L.)

* Author to whom correspondence should be addressed; E-Mail: msavin@uark.edu

Academic Editor: Bin Gao

Abstract: Biochar has altered plant yields and soil nutrient availability in tropical soils, but less research exists involving biochar additions to temperate cropping systems. Of the existing research, results vary based on soil texture, crop grown, and biochar properties. The objective of this study was to determine the effects of pine (*Pinus* spp.) woodchip biochar at 0, 5, and 10 Mg·ha^{-1} rates combined with urea nitrogen (N) on soil chemical properties and corn (*Zea mays* L.) yield under field conditions in the first growing season after biochar addition in a silt-loam alluvial soil. Biochar combined with fertilizer numerically increased corn yields, while biochar alone numerically decreased corn yields, compared to a non-amended control. Corn nitrogen uptake efficiency (NUE) was greater with 10 Mg·ha^{-1} biochar compared to no biochar. There were limited biochar effects on soil nutrients, but biochar decreased nitrate, total dissolved N, and Mehlich-3 extractable sulfur and manganese concentrations in the top 10 cm. Pine woodchip biochar combined with N fertilizer has the potential to improve corn production when grown in silt-loam soil in the mid-southern U.S. by improving NUE and increasing yield. Further research will be important to determine impacts as biochar ages in the soil.

Keywords: biochar; soil microorganisms; temperate agroecosystem; corn production

1. Introduction

Fertile and carbon-rich soils have been discovered throughout the Amazon River basin, an area where soils are typically nutrient leached and weathered [1,2]. Terra Preta soils, or Amazonian dark earth soils, occur in locations classified as Oxisols and Ultisols with similar mineralogical properties as surrounding soils. However, the Terra Preta soils are differentiated by their darker color due to large amounts of organic matter (reportedly nearly 90 $g \cdot kg^{-1}$ in the surface horizon or 250 $Mg \cdot ha^{-1} \cdot m^{-1}$ compared to around 30 $g \cdot kg^{-1}$ or 100 $Mg \cdot ha^{-1} \cdot m^{-1}$ in surrounding Oxisols), charcoal, and A horizons ranging from 30 to 60 cm opposed to the typical 10- to 15-cm depths in adjacent soils [1,3–5].

In light of the observed fertility of Amazonian dark earth soils, presumably due in part to the presence of charcoal, research is ongoing to determine the effects of charcoal addition in different soils and charcoal's agronomic impact as a soil amendment. Biochar, or charcoal added to soils for the purpose of improving agronomic soil properties, can be produced by pyrolysis. Pyrolysis is the thermal conversion of biomass under no or minimal oxygen conditions with temperatures generally between 300 and 700 °C [6–9]. The wide range of biochar products produced through various combinations of biomass types and production conditions can lead to different results when applied to various soils and at different application rates.

With a focus beyond tropical soils, research has been conducted to investigate the effects of soil application of various biochar products in temperate regions. For example, birch (*Betula* spp.) wood biochar applied at 20 $Mg \cdot ha^{-1}$ to a sandy loam (Typic Hapludalf) in Denmark did not increase oat (*Avena sativa* L.) biomass or yield, but barley (*Hordeum vulgare* L.) grown the following year experienced significant biomass increases with biochar addition [10]. In a sandy clay loam (Eutric Cambisol) in Wales, woodchip biochar applied at 25 and 50 $Mg \cdot ha^{-1}$ had no effect on corn (*Zea mays* L.) growth or nutrient concentration, although hay grass (*Dactylis glomerata* L.) grown the year after corn experienced increased foliar N with 50 $Mg \cdot ha^{-1}$ biochar compared to the control and increased biomass in the third year of the study [11]. Poultry litter biochar addition to an Alfisol in Australia increased radish (*Raphanus sativus* L.) yield with increasing biochar addition from rates of 10, 25, to 50 $Mg \cdot ha^{-1}$ in a pot trial [12]. In a sandy loam in Belgium, there was a general reduction in soil nitrate availability and nitrogen-use efficiency as well as reduced biomass in radish and spring barley with the addition of 10 $g \cdot kg^{-1}$ willow (*Salix* spp.) or pine (*Pinus* spp.) biochar, with greater reduction in soil nitrate with the willow biochar compared to the pine biochar in the pot study [13]. Considering these and other observations, the results regarding biochar addition have been mixed based on the biochar products used, soil textures, and the specific crops grown in these temperate region studies.

Corn is an important commodity crop in the United States, with nearly 35.5 million ha harvested for grain, and grain production at a record high of almost 0.49 billion m^3 across the country in 2013 [14]. Corn requires substantial N inputs, with 0.45 kg N expected to be taken up for every 25 kg corn grain produced (bu corn) [15]. If biochar can enhance soil fertility of corn production systems in temperate agroecosystems, there is potential to improve soil quality characteristics, increase yields, and reduce commercial fertilizer-N inputs, thereby improving the sustainability of production systems. Corn yield but not nutrient uptake increased in field research in Iowa after the addition of mixed hardwood biochar (96 $Mg \cdot ha^{-1}$) to a Typic Hapludoll [16]. Peanut (*Arachis hypogaea* L.) hull biochar increased soil N concentration but did not affect corn tissue N when applied at 11.2 and 22.4 $Mg \cdot ha^{-1}$ to loamy sand in

Georgia, while pine woodchip biochar did not increase soil N or tissue N [17]. Corn yield was reduced with the addition of 11.2 Mg·ha^{-1} of the peanut hull biochar, but the application of 22.4 Mg·ha^{-1} produced yields similar to those of the unamended control. The pine woodchip biochar decreased corn yield with increasing application rate in the first year after biochar addition, but the second corn crop experienced increased grain yields with biochar application compared to the control [17]. Corn biomass increased with giant reed (*Arundo donax* L.) biochar irrespective of biochar concentration (0.1, 0.2, and 0.5 g·kg^{-1}) when added to a silt loam in a pot experiment in China [18]. Plant residue biochars tended to increase corn biomass when applied at rates of 2.6 and 6.5 Mg·ha^{-1} to a Glossoboric Hapludalf in a greenhouse study in New York, with minimal differences at greater applications (*i.e.*, 26 and 91 Mg·ha^{-1}) [19]. Poultry manure biochar application produced a similar pattern, while food waste, paper mill waste, and dairy manure applied at 26 and 91 Mg·ha^{-1} tended to decrease corn biomass [19].

Although long-term effects of biochar on soil properties and processes may differ from short-term, there are a range of mechanisms by which biochar may impact soils and agronomic production in the first growing season. Biochar has decreased organic N release but increased nitrification [20]. Maize biochar to loamy sand initially sorbed released ammonium, but also showed short-term increases in N mineralization and nitrification [21]. Through effects on soil properties and alteration of N cycling processes, biochar may interact with fertilizer to increase N availability and N uptake efficiency in plants. Giant reed biochar reduced N leaching and increased nitrogen utilization efficiency, or the amount of corn biomass produced per unit N, in pot experiments [18].

In Arkansas, nearly 60% of the state, or 7.5 million ha (18.6 million ac), is commercial timberland [22]. Thus, pine woodchip biochar could be a potential use of regional wood waste from the forestry industry. Additionally, to the authors' knowledge, there are no published studies conducted in Arkansas regarding biochar application to soils. The objective of this study was to determine the effects of pine woodchip biochar in combination with varying amounts of inorganic N fertilizer on soil biological and chemical properties and corn yield under field conditions in the first growing season after biochar addition. Specifically, it was hypothesized that adequate corn yields and nutrient uptake would occur with biochar addition that could reduce inorganic fertilizer inputs because of increased N uptake efficiency. Despite any sorption or immobilization that may be caused by biochar, more N would be taken up by plants in the presence of biochar and fertilizer than in the absence of biochar.

2. Results

2.1. Initial Biochar and Soil Properties

The pine woodchip biochar had an alkaline pH, EC over 5 dS·m^{-1}, and a C:N ratio of 366:1 (Table 1). The surface texture of the soil was confirmed to be silt loam with percentages of sand, silt, and clay of 26, 65, and 9%, respectively (Table 2). The soil possessed a near-neutral pH of 6.4 and EC of 0.16 dS·m^{-1}. Ideal soil pH for corn growth ranges from 5.8–7 [23]. Since the initial soil pH fell within this range, no additional liming was necessary. All initial soil property variables except for Mehlich-3 extractable soil P were statistically similar among all plots (Table 2). Mehlich-3 soil P ranged from 29.3 to 35.2 μg·g^{-1}. Soil bulk density differed among fertilizer-biochar treatment combinations when analyzed during the middle of the experiment (data not shown). However, a clear pattern was lacking between fertilizer rate and biochar rate in terms of how treatments affected soil bulk density, which ranged from 1.18 to 1.33 $g \cdot cm^{-3}$.

Table 1. Initial mean (± standard error (SE)) pH, electrical conductivity (EC), total carbon (C), total nitrogen (N), C:N ratio, and total recoverable mineral concentrations determined using a nitric acid digest for pine (*Pinus* spp.) woodchip biochar ($n = 2$).

Biochar Property	Mean (± SE)
pH [a]	8.7 (0.03)
EC ($dS \cdot m^{-1}$) [a]	5.3 (0.2)
Total C ($mg \cdot g^{-1}$)	244.5 (21)
Total N ($mg \cdot g^{-1}$)	0.7 (0.2)
C:N ratio	366:1 (64)
Potassium ($mg \cdot g^{-1}$)	2.1 (0.1)
Calcium ($mg \cdot g^{-1}$)	10.1 (0.5)
Magnesium ($mg \cdot g^{-1}$)	2.7 (0.2)
Phosphorus ($\mu g \cdot g^{-1}$)	770.5 (17)
Sulfur ($\mu g \cdot g^{-1}$)	128.5 (1.5)
Sodium ($\mu g \cdot g^{-1}$)	321.5 (13)
Iron ($\mu g \cdot g^{-1}$)	868.0 (57)
Manganese ($\mu g \cdot g^{-1}$)	420.5 (30)
Copper ($\mu g \cdot g^{-1}$)	6.5 (0.04)
Boron ($\mu g \cdot g^{-1}$)	10.4 (0.7)
Zinc ($\mu g \cdot g^{-1}$) [b]	0.01 (0)

[a] pH and EC were determined using a 1:2 (wt:vol) soil:water mixture; [b] Zinc in the woodchip biochar was below the detection limit of the method. Therefore, the detection limit of 0.01 was used for statistical analysis.

Table 2. Mean (± standard error (SE)) soil properties for the Razort silt loam prior to treatment ($n = 36$).

Soil Property	Mean (± SE)
Particle-size distribution ($g \cdot g^{-1}$)	
Sand	0.3 (0.3)
Silt	0.6 (0.3)
Clay	0.1 (0.2)
pH [a]	6.4 (0.03)
Electrical conductivity ($dS \cdot m^{-1}$) [a]	0.2 (0.1)
Organic matter ($mg \cdot g^{-1}$)	27.1 (0.4)
Dissolved organic carbon (C) ($\mu g \cdot g^{-1}$)	33.8 (1.2)
Microbial biomass C ($\mu g \cdot g^{-1}$)	46.1 (2.3)
Microbial biomass nitrogen (N) ($\mu g \cdot g^{-1}$)	8.8 (0.3)
Microbial biomass C:N ratio	5.3:1 (0.2)
Dissolved total N ($\mu g \cdot g^{-1}$)	9.2 (0.2)
Nitrate-N ($\mu g \cdot g^{-1}$)	7.6 (0.2)
Ammonium-N ($\mu g \ g^{-1}$)	0.02 (0.02)

Table 2. *Cont.*

Soil Property	Mean (± SE)
Inorganic N ($\mu g \cdot g^{-1}$)	7.6 (0.2)
Dissolved organic N ($\mu g \cdot g^{-1}$)	1.6 (0.1)
Acid phosphatase activities ($\mu g \cdot g^{-1} \cdot h^{-1}$)	271.9 (14)
Alkaline phosphatase activities ($\mu g \cdot g^{-1} \cdot h^{-1}$)	102.8 (7.2)
Water soluble phosphorus ($\mu g \cdot g^{-1}$)	5.3 (0.3)
Mehlich-3 extractable phosphorus ($\mu g \cdot g^{-1}$)	31.7 (1.1)
Mehlich-3 extractable potassium ($\mu g \cdot g^{-1}$)	104.4 (4.7)
Mehlich-3 extractable calcium ($\mu g \cdot g^{-1}$)	923.4 (13)
Mehlich-3 extractable magnesium ($\mu g \cdot g^{-1}$)	46.4 (0.8)
Mehlich-3 extractable sulfur ($\mu g \cdot g^{-1}$)	5.5 (0.2)
Mehlich-3 extractable iron ($\mu g \cdot g^{-1}$)	51.1 (1.7)
Mehlich-3 extractable manganese ($\mu g \cdot g^{-1}$)	161.7 (3.7)
Mehlich-3 extractable copper ($\mu g \cdot g^{-1}$)	1.9 (0.1)

[a] pH and EC were determined using a 1:2 (wt:vol) soil:water mixture.

2.2. Post-Harvest Soil Characteristics

In terms of treatment effects on various post-harvest soil characteristics, biochar had minimal influence. Neither fertilizer nor biochar as an interaction or main effect had an impact on microbial biomass C, N, or C:N (Table 3). However, alkaline phosphatase enzyme activities differed among fertilizer-biochar treatment combinations. Alkaline phosphatase enzyme activities were greater in the no fertilizer with 5 $Mg \cdot ha^{-1}$ biochar treatment combination and the 112 $kg \cdot ha^{-1}$ fertilizer with 0 $Mg \cdot ha^{-1}$ biochar treatment combination than those in the 10 $Mg \cdot ha^{-1}$ biochar with 224 $kg \cdot ha^{-1}$ fertilizer combination and the no fertilizer with 0 or 10 $Mg \cdot ha^{-1}$ biochar treatment combinations (Figure 1). Water-soluble P also differed among fertilizer-biochar treatment combinations (Table 3), with similar concentrations across fertilizer rates with 10 $Mg \cdot ha^{-1}$ biochar, but with greater WSP in the 224 $kg \cdot ha^{-1}$ fertilizer and no biochar treatment combination compared to other treatments without biochar (Figure 2). Among the 5 $Mg \cdot ha^{-1}$ biochar treatments, WSP concentrations were lowest with the 112 $kg \cdot ha^{-1}$ fertilizer rate.

Table 3. Analysis of variance summary of the effects of fertilizer, biochar, and their interaction (Fert × BC) in post-harvest soil and on corn plant variables.

Soil Variable	*p*-Value		
	Fertilizer	Biochar	Fert × BC
Soil moisture	0.754	0.065	0.335
pH	0.248	0.105	0.417
Electrical conductivity	0.639	0.308	0.865
Soil organic matter	0.587	0.104	0.671
Dissolved organic carbon (C)	0.041 *	0.356	0.241
Dissolved total nitrogen (N)	0.093	0.022 *	0.727
Microbial biomass C	0.257	0.778	0.435
Microbial biomass N	0.255	0.768	0.477
Microbial biomass C:N	0.385	0.577	0.419

Table 3. *Cont.*

Soil Variable	*p*-Value		
	Fertilizer	**Biochar**	**Fert × BC**
Nitrate-N	0.031 *	0.044 *	0.570
Ammonium-N	1.000	1.000	1.000
Inorganic N	0.031 *	0.044 *	0.570
Dissolved organic N	0.002 *	0.644	0.203
Acid phosphatase activities	0.513	0.603	0.882
Alkaline phosphatase activities	0.143	0.235	0.046 *
Water soluble phosphorus	0.015 *	0.767	0.045 *
Mehlich-3 extractable phosphorus	0.103	0.403	0.661
Mehlich-3 extractable potassium	0.708	0.697	0.785
Mehlich-3 extractable calcium	0.094	0.146	0.064
Mehlich-3 extractable magnesium	0.641	0.580	0.903
Mehlich-3 extractable sulfur	0.582	0.008 *	0.519
Mehlich-3 extractable iron	0.035 *	0.453	0.520
Mehlich-3 extractable manganese	0.150	0.045 *	0.105
Mehlich-3 extractable copper	0.952	0.226	0.942
Yield	<0.001 *	0.439	0.011 *
Grain total N	0.014 *	0.491	0.477
NUE	0.170	0.003 *	0.300
Ear-leaf weight	0.005 *	0.548	0.333
Ear-leaf N	<0.001 *	0.555	0.710

* $p < 0.05$.

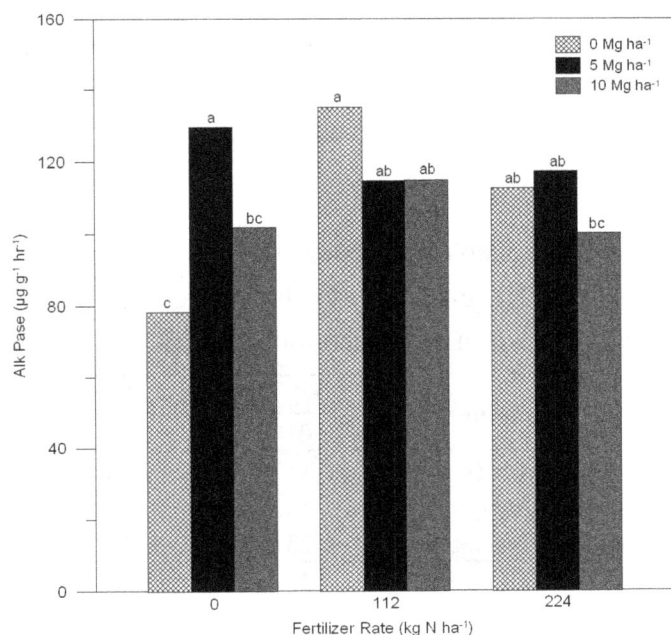

Figure 1. Alkaline phosphatase (Alk Pase) enzyme activity as influenced by pine woodchip biochar and fertilizer rates. The fertilizer rates are 0, 112, and 224 kg·nitrogen (N)·ha^{-1} rates. The biochar treatments are displayed in the shaded boxes at rates of 0, 5, and 10 Mg·ha^{-1}. Bars with different letters are statistically different from each other ($p < 0.05$).

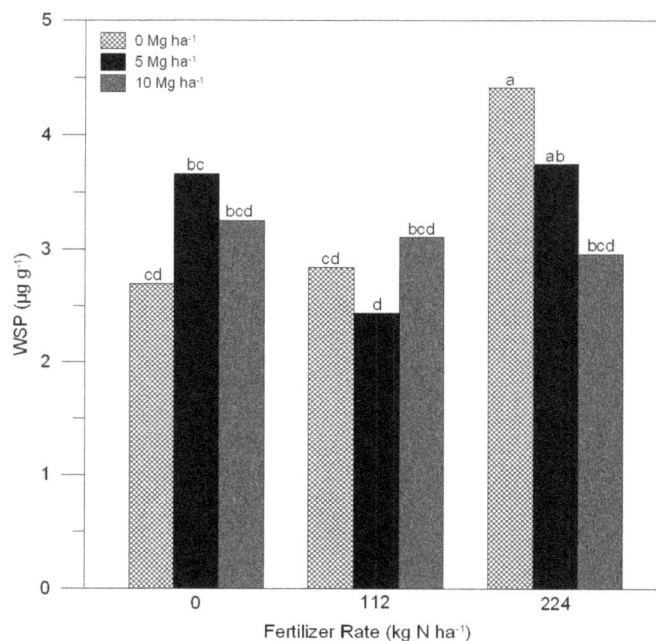

Figure 2. Water soluble phosphorus (WSP) concentrations as influenced by pine woodchip biochar and fertilizer rates. The fertilizer rates are 0, 112, and 224 kg·nitrogen (N)·ha^{-1} rates. The biochar treatments are displayed in the shaded boxes at rates of 0, 5, and 10 Mg·ha^{-1}. Bars with different letters are statistically different from each other ($p < 0.05$).

Nitrate-N and dissolved organic N concentrations differed among fertilizer rates (Table 3). Nitrate increased with the 224 kg·ha^{-1} fertilizer treatment compared to the 0 kg·ha^{-1} fertilizer treatment, while DON was lower with the 224 kg·ha^{-1} fertilizer treatment compared to the 0 and 112 kg·ha^{-1} fertilizer treatments (Table 4). Nitrate and DTN concentrations differed among biochar rates (Table 3). The addition of biochar, irrespective of rate, decreased dissolved total N concentrations compared to no biochar addition, while nitrate concentrations decreased with the 10 Mg·ha^{-1} biochar treatment compared to no biochar addition (Table 5).

Table 4. Mean corn grain total nitrogen (N), ear-leaf weight, and ear-leaf N concentration, and soil dissolved organic N (DON), nitrate (NO$_3^-$-N), inorganic N, dissolved organic C (DOC), and Mehlich-3 extractable soil iron concentration as affected by fertilizer rate.

Plant/Soil	Variable	Fertilizer Rate (kg·N·ha^{-1})		
		0	112	224
Plant	Grain total N (mg·g^{-1})	11.1 b	11.4 b	12.4 a
	Ear-leaf weight (g)	14.0 b	15.7 a	16.7 a
	Ear-leaf N (mg·g^{-1})	23.1 b	26.7 a	28.0 a
Soil	DON (µg·g^{-1})	2.0 a	2.1 a	1.6 b
	NO$_3^-$-N (µg·g^{-1})	2.4 b	2.8 ab	3.9 a
	Inorganic N (µg·g^{-1})	2.4 b	2.8 ab	3.9 a
	DOC (µg·g^{-1})	8.6 b	12.4 a	11.0 ab
	Iron (µg·g^{-1})	25.3 b	28.5 ab	30.6 a

Means followed by different letters in the same row are statistically different ($p < 0.05$).

Table 5. Mean nitrogen uptake efficiency (NUE) in corn and soil dissolved total N (DTN), nitrate (NO_3^--N), inorganic N, and Mehlich-3 extractable soil sulfur and manganese concentrations as affected by biochar rate.

Plant/Soil	Variable	Biochar Rate (Mg·ha^{-1})		
		0	5	10
Plant	NUE (%)	12.2 b	21.8 b	44.4 a
Soil	DTN (µg·g^{-1})	5.7 a	4.8 b	4.4 b
	NO_3^--N (µg·g^{-1})	3.8 a	2.8 ab	2.5 b
	Inorganic N (µg·g^{-1})	3.8 a	2.8 ab	2.5 b
	Sulfur (µg·g^{-1})	12.3 a	11.4 a	9.4 b
	Manganese (µg·g^{-1})	80.6 a	78.2 ab	72.3 b

Means followed by different letters in the same row are statistically different ($p < 0.05$).

Dissolved organic C concentrations increased (Table 3) with the 112 kg·ha^{-1} compared to the 0 kg·ha^{-1} fertilizer treatment (Table 4). Mehlich-3 extractable soil S and Mn differed among biochar rates (Table 3). Soil S decreased with the 10 Mg·ha^{-1} biochar treatment compared to the 0 and 5 Mg·ha^{-1} treatments (Table 5). Manganese decreased with the 10 Mg·ha^{-1} biochar treatment compared to the no biochar treatment. Conversely, Fe increased (Table 3) with the 224 kg·ha^{-1} fertilizer treatment compared to the no fertilizer treatment (Table 4).

2.3. Corn Characteristics

Corn yield differed among fertilizer-biochar treatment combinations (Table 3), with the 224 kg·N·ha^{-1} fertilizer and 10 Mg·ha^{-1} biochar treatment combination resulting in a greater yield than those produced by treatments with no fertilizer or no biochar (Figure 3). Yields were similar between the 10 Mg·ha^{-1} biochar treatment combined with either the 112 or 224 kg·ha^{-1} fertilizer N. Yields produced with the 5 Mg·ha^{-1} biochar treatment combined with the 112 or 224 kg·ha^{-1} fertilizer treatment were similar to yields produced with all treatment combinations involving fertilizer, except the 112 kg·ha^{-1} fertilizer and 5 Mg·ha^{-1} biochar treatment combination produced lower yields than the 224 kg·ha^{-1} fertilizer and 10 Mg·ha^{-1} biochar combination. When no fertilizer was applied, the 10 Mg·ha^{-1} biochar treatment reduced yield compared to the no fertilizer and no biochar treatment combination.

In contrast to the differences in corn yield among fertilizer-biochar treatment combinations, grain N concentrations differed among fertilizer rates (Table 3). The 224 kg·ha^{-1} fertilizer treatment produced greater grain N than the 112 kg·ha^{-1} fertilizer treatment and the no fertilizer treatment (Table 4). The addition of fertilizer at either rate increased ear-leaf weight and ear-leaf N concentrations compared to the no fertilizer treatment. While biochar did not significantly affect (Table 3) grain N concentrations, ear-leaf weight, or ear-leaf N, NUE was greatest (Table 3) with the 10 Mg·ha^{-1} biochar treatment compared to the 5 or 0 Mg·ha^{-1} biochar treatments (Table 5).

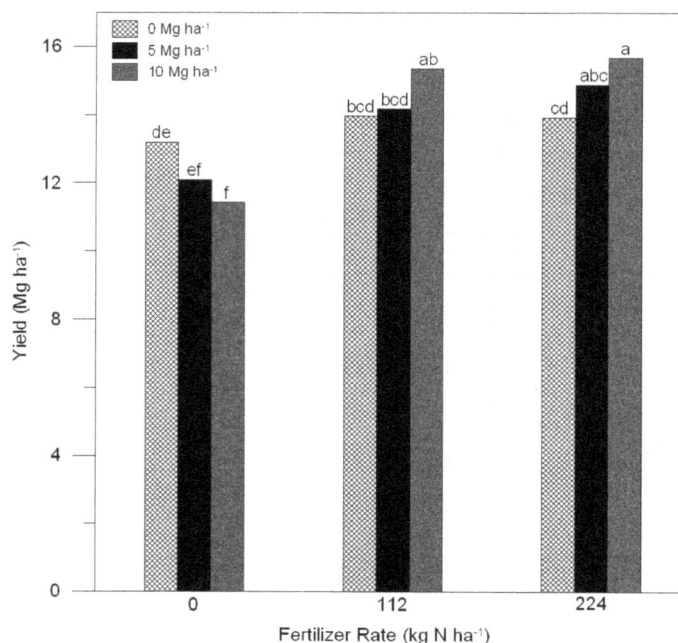

Figure 3. Corn yield as influenced by pine woodchip biochar and fertilizer rates. The fertilizer rates are 0, 112, and 224 kg·nitrogen (N)·ha^{-1} rates. The biochar treatments are displayed in the shaded boxes at rates of 0, 5, and 10 Mg·ha^{-1}. Bars with different letters are statistically different from each other ($p < 0.05$).

3. Discussion

Average corn yields in Arkansas for 2012 and 2013 were close to 11.3 Mg·ha^{-1} and consistent with the lowest yields in this study [14]. The above-state-average yields with the no fertilizer and no biochar treatment were potentially due to residual N in the soil. Furthermore, variety testing in 2013 of irrigated corn in silt loam soil using the specific corn variety used in this study, DEKALB DKC64-69, yielded 15.8 Mg·ha^{-1} when grown at three other University of Arkansas Experiment stations (Stuttgart, Rohwer, and Bell Farm) [24]. The greatest yields from this study were consistent with those reported from the variety testing in the same growing year. In this field experiment, the addition of biochar combined with fertilizer improved crop production to reach the yield potential achieved in the variety testing trials.

Corn yields actually decreased with the addition of 10 Mg·ha^{-1} biochar in the no fertilizer treatment compared to the no fertilizer and no biochar treatment combination. There was the possibility of biochar increasing microbial immobilization or sorption of soil N in the treatments lacking fertilizer. Three Wisconsin cropped soils, a Typic Hapludalf (clay loam), Typic Udipamment (loamy sand), and Typic Argiudoll (silt loam), all experienced increased microbial biomass and activity with mixed manure and pine biochar application in a soil microcosm experiment; microbial biomass and activity increased with increasing biochar concentration (0, 10, 25, 50, and 100 g·kg^{-1}) and extractable-N decreased [25]. Streubel *et al.* [26] reported decreased N mineralization in a silt loam with wood biochar in Washington. Immobilization of soil or biochar N from microbial decomposition of the pine woodchip biochar may have occurred in this study based on the wide C:N ratio of the biochar, but there was no significant biochar effect on microbial biomass C, N, or C:N ratio, thus failing to provide support that changes in the microbial community size explain lower yield in the absence of fertilizer, or explain differences in soil N concentrations or NUE of corn with biochar treatments.

With the addition of fertilizer, corn yield increased numerically with biochar additions, so any reduction of N availability by biochar was presumably not negatively impacting yield when inorganic N fertilizer was applied. Biochar has increased ammonium-N sorption and decreased inorganic N leaching by increasing water holding capacity of soil, N immobilization, and ammonium sorption [18]. Various studies have suggested mechanisms by which biochar may retain N in soil. Zheng *et al.* [18] reported reduced nitrate leaching after nitrate or ammonium fertilizer addition as well as increased microbial activity in giant reed biochar-amended soil. Increased N immobilization in the presence of biochar could have temporarily retained the N in the organic form and reduced the potential for inorganic N leaching [18]. In a study by Güereña *et al.* [27], there was less N leaching with biochar addition combined with the recommended N fertilizer rate than with the fertilizer addition alone. The increases in soil N retention were thought to be due to increases in microbial biomass and thus N immobilization, lower gaseous and erosion losses, and increased retention of organic N on biochar surfaces. Biochar may increase N cycling and availability by providing habitat for microbes and altering food web dynamics [20]. A hardwood biochar decreased N leached from a manure-amended Typic Hapludoll [28], while increasing total soil N [29].

Cycling of N between organic and inorganic pools may be decoupled in the presence of biochar and fertilizer, promoting retention of organic C and N in soil, while increasing plant availability and uptake utilization of fertilizer N [20]. Gajić and Koch [30] found hydrochar reduced early beet growth in a pot and field trial, but that N fertilizer could alleviate yield reductions. They suggested that early plant growth was inhibited by increased microbial immobilization of N especially with use of higher C:N char. Nitrogen may be mineralized later in the growing season, but fertilizer was necessary to compensate for low nutrient availability during early plant growth stages [30].

Biochar decreased soil N concentrations which may have resulted in part from N uptake, or could have resulted from loss mechanisms such as volatilization or erosion which were not directly investigated in this study. With the application of urea fertilizer, there was the potential for N loss, such as by ammonia volatilization [31,32]. However, since biochar did not affect soil water content or pH, and ammonia volatilization was not quantified, there was no evidence to suggest that ammonia volatilization differed among biochar treatments and thus does not explain differences between soil N concentrations based on biochar treatments. Further investigation is warranted to understand the mechanisms responsible for increased yield in the combined presence of fertilizer N and high C:N biochar.

Despite increases in yield in the combined presence of fertilizer and biochar, wood biochar did not affect plant N concentrations, with only fertilizer influencing grain and ear-leaf N concentrations. In another wood biochar field study, increases in soybean (*Glycine max* L.) yields and doubling of forage biomass with 3.9 Mg·ha^{-1} hardwood biochar were observed in a high-P clay loam in Quebec [33]. However, hardwood biochar applied at rates of 0, 25, and 50 Mg·ha^{-1} did not affect growth performance (corn height or biomass) and did not affect grain N concentrations in a sandy clay loam (Eutric Cambisol) in a field trial in Wales [11].

Although changes in N concentrations in the grain and ear leaf were not influenced by biochar treatments, there was an increase in NUE with the 10 Mg·ha^{-1} biochar application compared to the 0 or 5 Mg·ha^{-1} biochar treatments. Increased radish NUE with greenwaste biochar occurred in an acidic,

Australian Alfisol, attributed in part to increases in pH, extractable soil P and K, and field capacity water content in a pot trial [34]. However, increases in pH and extractable soil P and K do not seem to be reasons for improvement in this study. Soil K was not significantly different across treatments in this experiment. Biochar did not alter soil pH despite the biochar's alkaline pH. In this experiment, the soil was initially similar, except for soil P concentrations, despite attempts at homogenizing the soil and uniform field preparation. However, by the end of the experiment, there were no differences based on treatment effects with Mehlich-3 soil P. The changes in WSP based on fertilizer and biochar addition did not seem to have a clear pattern, perhaps impacted by the differences in initial Mehlich-3 soil P and increases in alkaline phosphatase activities by the end of the experiment. The alkaline phosphatase activities were greatest with 5 $Mg \cdot ha^{-1}$ biochar application without fertilizer, possibly related to the greatest residual P in the initial soil later amended with the 5 $Mg \cdot ha^{-1}$ biochar treatment. With fertilizer addition, the activity seemed to level out across biochar treatments without significant differences among fertilizer treatments. Field capacity water content was not analyzed.

In a silt loam in an Italian field study, durum wheat (*Triticum durum* L.) grain production increased with biochar addition (30 and 60 $Mg \cdot ha^{-1}$) compared to no biochar addition, while there was no biochar effect on grain N concentration [35]. Since grain N concentrations were unaffected by biochar addition, there was no N-dilution effect corresponding with the increase in yield, suggesting greater fertilizer NUE with biochar treatments than without [35]. The lack of a N-dilution effect also resulted in a field study conducted in a Typic Hapludoll in Iowa with hardwood biochar applied at five rates between 0 and 95.8 $Mg \cdot ha^{-1}$, in which corn yield increased with biochar addition, but biochar did not affect plant tissue N concentrations [16]. Zheng *et al.* [18] reported lower concentrations of N absorbed by corn seedlings in a three-month pot experiment but also greater biomass production with the amount of N that was absorbed compared to the no biochar treatment. Therefore, it was suggested that, based on the lower N accumulation efficiency but greater NUE with compared to without biochar, biochar could have enhanced N availability to the corn plants and therefore decreased N fertilizer demand [18].

A variety of mechanisms may retain N in soil and decrease N fertilizer demand. Laird *et al.* [28] suggested that biochar adsorbed ammonium onto CEC sites and reduced nitrification and in turn the potential for nitrate leaching. Prendergast-Miller *et al.* [36,37] observed nitrate retained by biochar, presumably in solution in biochar pores rather than on CEC sites, and proposed that nitrate could potentially be released from biochar pores by diffusion gradients [36]. Prendergast-Miller *et al.* [36,37] suggested that biochar localized nitrate in the rhizosphere soil and that roots grew preferentially towards biochar particles. Deciduous wood biochar applied at 20 and 60 $Mg \cdot ha^{-1}$ produced longer wheat roots that increased the root-soil contact compared to non-biochar amended soil and could have increased root N uptake in biochar treatments [36]. A more extensive root system could increase plant ability to access nutrients in the soil [18], thus providing an explanation for the decrease in dissolved total N, nitrate, and Mehlich-3 extractable soil S and Mn concentrations with the 10 $Mg \cdot ha^{-1}$ biochar treatment compared to the no biochar treatment in this experiment. Additional belowground sampling would be necessary in future studies to better understand the effects of pine woodchip biochar on rhizosphere soil and corn roots.

4. Experimental Section

4.1. Biochar Characteristics

Pine woodchip biochar (Waste to Energy Solutions Inc., Destin, FL, USA), which was produced through pyrolysis at 500 °C, was selected for this field study. Biochar was dried for 48 h at 70 °C then ground to pass a 40-mesh screen before pH and EC were determined potentiometrically on a 1:2 (wt:vol) sample:water mixture. Total nitrogen and total carbon concentrations were determined by combustion with Elementar Variomax (Elementar Americas, Inc., Mt. Laurel, NJ, USA). The C:N ratio was calculated from the total N and C concentrations. Total recoverable minerals (*i.e.*, P, K, Ca, Mg, sulfur (S), sodium (Na), Fe, Mn, Zn, Cu, and boron (B)) were determined from acid digestion [38] using an ARCOS inductively coupled plasma (ICP) spectrophotometer (SPECTRO Analytical Instruments Inc., Mahwah, NJ, USA).

4.2. Site Description and Experimental Design

The field experiment was conducted at the University of Arkansas Agricultural Research and Extension Center in Fayetteville, Arkansas in summer 2013. The Global Positioning System data for the four corners of the field were collected using World Geodetic System 1984 in latitude, longitude format (36.09780719° N, 94.16717458° W; 36.09780275° N, 94.16708997° W; 36.09846935° N, 94.16705646° W; 36.09846342° N, 94.16713982° W). Annual precipitation from 30-year normal data [39] was 126.5 cm, average annual maximum air temperature was 20.2 °C, and average annual minimum air temperature was 8.7 °C. The soil within the field was classified as a Razort silt loam, occasionally flooded (fine-loamy, mixed, active, mesic Mollic Hapludalf) [40]. There was a slight positive slope from north to south. The 0.26-ha field was planted the previous two years in cotton (*Gossypium hirsutum* L.). The old cotton stalks had been mowed in fall 2012.

The experimental design was a full factorial randomized complete block. There were 36 plots, each 6 m long and 3.6 m wide and consisting of 4 rows. The total field was 73 m by 11 m. A random number generator was used to place the treatments in each block. Pine woodchip biochar was added at rates of 0, 5, and 10 Mg·ha^{-1}. For the 5 Mg·ha^{-1} biochar application rate, 11.2 kg of biochar were used per plot, and 22.5 kg were used per plot for the 10 Mg·ha^{-1} biochar application rate. Biochar was manually applied on 28 May and was incorporated with mechanical tillage into approximately the top 5 cm before rows were bedded and knocked down for planting.

Corn, DEKALB hybrid DKC64-69 with the Genuity VT Triple PRO value-added trait, was planted 74,100 seeds·ha^{-1}, which equated to 7 to 10 seeds·m^{-1}, on 29 May with a four-row planter. Full emergence occurred after one week, and no thinning of seedlings was required. The rows were watered by furrow irrigation as needed with the use of the Arkansas online irrigation scheduler [41].

Nitrogen fertilizer was applied at 0, half, and full rates in a split application. The full rate was chosen to achieve a theoretical corn yield of 12.5 Mg·ha^{-1} [23]. Therefore, 224 kg·N·ha^{-1} were added as the full rate and 112 kg·N·ha^{-1} were added as the half rate of fertilizer, both in the form of urea (46-0-0). Based on soil analyses conducted prior to experiment initiation (data not shown), no other fertilizer amendments were applied. The first urea application was manually applied 20 June, and the split application was applied 9 July. Herbicide application consisted of 1.3 L·ha^{-1} of broadcasted Cornerstone herbicide on 20 June and a directed spray of 1.7 L·ha^{-1} Atrazine plus 1.7 L·ha^{-1} Cornerstone on 10 July.

4.3. Soil Analyses

Soil was sampled at the 0- to 10-cm depth prior to treatment application to document initial properties and assess initial plot variability. Soil sampling was also conducted at the end of the growing season to assess potential treatment effects. After moist soil was sieved through a 2-mm mesh screen, soil was analyzed for dissolved organic C (DOC), dissolved total N (DTN), ammonium (NH$_4^+$-N), and nitrate (NO$_3^-$-N) using a single extraction approach [42]. A Skalar segmented-flow autoanalyzer (Skalar Inc., Norcross, GA, USA) colorimetrically determined NH$_4^+$ following the salicylate hypochlorite procedure and NO$_3^-$ following a modification of Griess-Ilosvay cadmium-copper reduction of NO$_3^-$ to NO$_2^-$ procedure [43]. Using the chloroform-fumigation method, microbial biomass C and N were quantified by calculating the difference between fumigated and unfumigated samples for both C and N [44]. Fumigated and unfumigated soils were extracted and analyzed for DOC and DTN on a Shimadzu TOC-V PC-controlled total organic carbon with attached total nitrogen analyzer (Shimadzu, Columbia, MD, USA). Inorganic N (N$_i$) was calculated by summing concentrations of NO$_3^-$ and NH$_4^+$. Dissolved organic nitrogen (DON) concentration was calculated by subtracting N$_i$ from DTN [42].

Water-soluble phosphorus (WSP) was obtained from extractions from 2-g moist soil samples using a 1:10 (wt:vol) soil:water ratio [45] and analyzed by a Skalar Sans-plus segmented-flow autoanalyzer (Skalar Inc., Norcross, GA, USA) using the ascorbic acid method [46]. Acid and alkaline phosphatase activities were measured using the colorimetric estimation of *p*-nitrophenol produced by phosphatase enzyme activity after soil incubation in buffered sodium *p*-nitrophenyl phosphate solution [47].

After drying soil at 70 °C for at least 48 h, soil pH and EC were measured using a 1:2 (wt:vol) soil:water mixture. Organic matter was determined by loss-on-ignition using a muffle furnace. Mehlich-3 extractable soil nutrients (*i.e.*, P, K, Ca, Mg, S, Fe, Mn, Cu) [48] were determined by ICP spectrometry.

One soil core per plot, 4.8 cm in diameter, was collected at the 0- to 10-cm depth on 29 July. After drying at 70 °C for at least 48 h, soil cores were weighed for bulk density determinations. Oven-dried soil was sieved through a 2-mm mesh screen and particle-size analysis was conducted using an adaptation of the 12-h hydrometer method [49].

4.4. Corn Analyses

Ear leaves were harvested at tasseling from the outer two rows of each plot for leaf tissue-N analysis. Ear leaves were dried at 65 °C, ground to pass a 40-mesh screen, and weighed before ear-leaf total N was determined by combustion [50] using a Model Rapid N III (Elementar Americas, Inc., Mt. Laurel, NJ, USA). The harvested yield area was the center 1.5 m of the center two rows in each plot. Grain was harvested on 28 September once physiological maturity had been reached, and yield was calculated based on grain dry weight. Grain samples were ground prior to analysis for total N by combustion. Nitrogen uptake efficiency (NUE) was calculated using the difference method, where NUE in grain was equal to the difference between N removed in grain and the N removed in the unamended control grain divided by the fertilizer-N applied [51]. The N removed in grain was calculated by multiplying the N concentration in the grain by the mass of grain (yield), assuming 720 kg grain per m^3.

4.5. Data Analyses

A two-way analysis of variance (ANOVA) was performed using SAS (version 9.2, SAS Institute, Inc., Cary, NC, USA) to determine any initial differences in the plots for soil pH, EC, OM, DOC, DTN, microbial C, N, and C:N ratio, NO_3^--N, NH_4^+-N, N_i, DON, WSP, acid and alkaline phosphatase activities, and Mehlich-3 extractable soil nutrient concentrations (*i.e.*, P, K, Ca, Mg, S, Fe, Mn, and Cu). A two-way ANOVA was also performed using the data from the soil harvested at the end of the growing season to determine the effects of biochar, fertilizer, and their interaction on soil characteristics as were analyzed for the initial soil. An additional ANOVA was performed to determine the effect of biochar, fertilizer, and their interaction on soil bulk density and particle-size fractions sand, silt, and clay. A two-way ANOVA was performed to determine the effects of biochar, fertilizer, and their interaction on ear-leaf weight and N, corn yield, grain total N, and NUE. Least significant differences were used to separate treatment means at $\alpha = 0.05$.

5. Summary and Conclusions

Pine woodchip biochar applied at rates of 5 and 10 Mg·ha^{-1} in combination with N fertilizer to a fertile silt loam in northwest Arkansas numerically increased corn yields compared to fertilizer application without biochar. However, biochar addition numerically decreased yields when applied without fertilizer, leading to statistically lower yields with the 10 Mg·ha^{-1} biochar treatment combined with no fertilizer compared to the no fertilizer treatment alone. Nitrogen uptake efficiency was greatest with 10 Mg·ha^{-1} biochar application, while soil nitrate concentrations at the end of the growing season were lowest at the same rate of biochar application. Microbial biomass C and N were statistically similar among all treatments. Biochar potentially altered N availability in soil, although the exact mechanisms regarding biochar-soil-plant nutrient relations need further study to elucidate. Pine woodchip biochar can improve corn NUE even in a fertile, temperate, alluvial soil and can increase corn yields in combination with N fertilizer. Given that results will vary based on soil texture, crop management system, biochar properties, as well as through time as a result of dynamic processes, caution must be exercised generalizing results to other systems. Subsequent research is also important to determine long-term changes in soil resulting from pine woodchip biochar addition.

Acknowledgments

The authors would like to thank the University of Arkansas Division of Agriculture for funding and Jeff Armstrong and Bradley Jackson with Monsanto for the corn seed. Support from the Altheimer laboratory staff, Trent Roberts, Chester Greub, Vaughn Skinner, Ron Cox, Kamela Mitchell, Jade Ford, and the USDA Poultry Waste Lab is gratefully acknowledged.

Author Contributions

Katy Brantley was a graduate student at the time of this research, and she was responsible for managing the study, collecting and analyzing data, and writing the manuscript. She was aided in the implementation of the experimental design, writing, and editing by Mary Savin. David Longer assisted with the experimental design and field work, while Kris Brye contributed to the writing and editing of the manuscript.

Conflicts of Interest

The authors declare no conflict of interest.

References

1. Kern, D.C.; D'aquino, G.; Rodrigues, T.E.; Frazao, F.J.L.; Sombroek, W.; Myers, T.P.; Neves, E.G. Distribution of Amazonian dark earths in the Brazilian Amazon. In *Amazonian Dark Earths: Origin, Properties, Management*; Lehmann, J., Kern, D.C., Glaser, B., Woods, W.I., Eds.; Kluwer Academic Publishers: Dordrecht, The Netherlands, 2003; pp. 51–75.

2. Lehmann, J. Terra Preta de Indio. In *Encyclopedia of Soil Science*; Lal, R., Ed.; Taylor and Francis: Boca Raton, FL, USA, 2006; Volume 1, pp. 1–4.

3. Glaser, B.; Haumaier, L.; Guggenberger, G.; Zech, W. The "Terra Preta" phenomenon: A model for sustainable agriculture in the humid tropics. *Naturwissenschaften* **2001**, *88*, 37–41.

4. Smith, N. Anthrosols and human carrying capacity in Amazonia. *Ann. Assoc. Am. Geogr.* **1980**, *70*, 553–566.

5. Zech, W.; Haumaier, L.; Hempfling, R. Ecological aspects of soil organic matter in tropical land use. In *Humic Substances in Soil and Crop Sciences: Selected Readings*; MacCarthy, P., Clapp, C.E., Malcolm, L., Bloom, P.R., Eds.; Soil Science Society of America: Madison, WI, USA, 1990; pp. 187–202.

6. Demirbaş, A. Biomass resource facilities and biomass conversion processing for fuels and chemicals. *Energy Convers. Manag.* **2001**, *42*, 1357–1378.

7. Gaskin, J.W.; Steiner, C.; Harris, K.; Das, K.C.; Bibens, B. Effect of low-temperature pyrolysis conditions on biochar for agricultural use. *Trans. ASABE* **2008**, *51*, 2061–2069.

8. Kloss, S.; Zehetner, F.; Dellantonio, A.; Hamid, R.; Ottner, F.; Liedtke, V.; Schwanninger, M.; Gerzabek, M.H.; Soja, G. Characterization of slow pyrolysis biochars: Effects of feedstocks and pyrolysis temperature on biochar properties. *J. Environ. Qual.* **2012**, *41*, 990–1000.

9. Skodras, G.; Grammelis, P.; Basinas, P.; Kakaras, E.; Sakellaropoulos, G. Pyrolysis and combustion characteristics of biomass and waste-derived feedstock. *Ind. Eng. Chem. Res.* **2006**, *45*, 3791–3799.

10. Sun, Z.; Bruun, E.W.; Arthur, E.; de Jonge, L.W.; Moldrup, P.; Hauggaard-Nielsen, H.; Elsgaard, L. Effect of biochar on aerobic processes, enzyme activity, and crop yields in two sandy loam soils. *Biol. Fertil. Soils* **2014**, *50*, 1087–1097.

11. Jones, D.L.; Rousk, J.; Edwards-Jones, G.; DeLuca, T.H.; Murphy, D.V. Biochar-mediated changes in soil quality and plant growth in a three year field trial. *Soil Biol. Biochem.* **2012**, *45*, 113–124.

12. Chan, K.Y.; van Zwieten, L.; Meszaros, I.; Downie, A.; Joseph, S. Using poultry litter biochars as soil amendments. *Aust. J. Soil Res.* **2008**, *46*, 437–444.

13. Nelissen, V.; Ruysschaert, G.; Müller-Stöver, D.; Bodé, S.; Cook, J.; Ronsse, F.; Shackley, S.; Boeckx, P.; Hauggaard-Nielsen, H. Short-term effect of feedstock and pyrolysis temperature on biochar characteristics, soil and crop response in temperate soils. *Agronomy* **2014**, *4*, 52–73.

14. National Agricultural Statistics Service. Crop Production: 2013 Summary. Available online: http://usda.mannlib.cornell.edu/usda/nass/CropProdSu//2010s/2014/CropProdSu-01-10-2014.pdf (accessed on 2 December 2014).

15. International Plant Nutrition Institute (IPNI). *Nutri-Facts. Agronomic Fact Sheets on Crop Nutrients: Nitrogen. Ref. #1 #14024*; International Plant Nutrition Institute: Peachtree Corners, GA, USA.

16. Rogovska, N.; Laird, D.A.; Rathke, S.J.; Karlen, D.L. Biochar impact on Midwestern Mollisols and maize nutrient availability. *Geoderma* **2014**, *230*, 340–347.

17. Gaskin, J.W.; Speir, R.A.; Harris, K.; Das, K.C.; Lee, R.D.; Morris, L.A.; Fisher, D.S. Effect of peanut hull and pine chip biochar on soil nutrients, corn nutrient status, and yield. *Agron. J.* **2010**, *102*, 623–633.

18. Zheng, H.; Wang, Z.; Deng, X.; Herbert, S.; Xing, B. Impacts of adding biochar on nitrogen retention and bioavailability in agricultural soil. *Geoderma* **2013**, *206*, 32–39.

19. Rajkovich, S.; Enders, A.; Hanley, K.; Hyland, C.; Zimmerman, A.R.; Lehmann, J. Corn growth and nitrogen nutrition after additions of biochars with varying properties to a temperate soil. *Biol. Fertil. Soils* **2012**, *48*, 271–284.

20. Prommer, J.; Wanek, W.; Hofhansl, F.; Trojan, D.; Offre, P.; Urich, T.; Schleper, C.; Sassmann, S.; Kitzler, B.; Soja, G.; *et al.* Biochar decelerates soil organic nitrogen cycling but stimulates soil nitrification in a temperate arable field trial. *PLoS One* **2014**, *9*, doi:10.1371/journal.pone.0086388.

21. Nelissen, V.; Rütting, T.; Huygens, D.; Staelens, J.; Ruysschaert, G.; Boeckx, P. Maize biochars accelerate short-term soil nitrogen dynamics in a loamy sand soil. *Soil Biol. Biochem.* **2012**, *55*, 20–27.

22. Pelkki, M.H. *An Economic Assessment of Arkansas' Forest Industries: Challenges and Opportunities for the 21st Century*; Arkansas Forest Resources Center Series 007; Arkansas Agricultural Experiment Station: Fayetteville, AR, USA, 2005.

23. Espinoza, L.; Ross, J. Fertilization and liming. In *Corn Production Handbook*; University of Arkansas Cooperative Extension Service: Little Rock, AR, USA, 2003; pp. 23–27.

24. Bond, R.D.; Dombek, D.G.; Still, J.A.; Pryor, R.M. *Arkansas Corn and Grain Sorghum Performance Tests 2013*; Arkansas Agricultural Experiment Station: Fayetteville, AR, USA, 2013.

25. Kolb, S.E.; Fermanich, K.J.; Dornbush, M.E. Effect of charcoal quantity on microbial biomass and activity in temperate soils. *Soil Sci. Soc. Am. J.* **2009**, *73*, 1173–1181.

26. Streubel, J.D.; Collins, H.P.; Garcia-Perez, M.; Tarara, J.; Granatstein, D.; Kruger, C.E. Influence of contrasting biochar types on five soils at increasing rates of application. *Soil Sci. Soc. Am. J.* **2011**, *75*, 1402–1413.

27. Güereña, D.; Lehmann, J.; Hanley, K.; Enders, A.; Hyland, C.; Riha, S. Nitrogen dynamics following field application of biochar in a temperate North American maize-based production system. *Plant Soil* **2013**, *365*, 239–254.

28. Laird, D.; Fleming, P.; Wang, B.; Horton, R.; Karlen, D. Biochar impact on nutrient leaching from a Midwestern agricultural soil. *Geoderma* **2010**, *158*, 436–442.

29. Laird, D.A.; Fleming, P.; Davis, D.D.; Horton, R.; Wang, B.; Karlen, D.L. Impact of biochar amendments on the quality of a typical Midwestern agricultural soil. *Geoderma* **2010**, *158*, 443–449.

30. Gajić, A.; Koch, H.-J. Sugar beet (*Beta vulgaris* L.) growth reduction caused by hydrochar is related to nitrogen supply. *J. Environ. Qual.* **2012**, *41*, 1067–1075.

31. Chen, C.R.; Phillips, I.R.; Condron, L.M.; Goloran, J.; Xu, Z.H.; Chan, K.Y. Impacts of greenwaste biochar on ammonia volatilization from bauxite processing residue sand. *Plant Soil* **2013**, *367*, 301–312.

32. Zhao X.; Yan, X.; Wang, S.; Xing, G.; Zhou, Y. Effects of the addition of rice-straw-based biochar on leaching and retention of fertilizer N in highly fertilized cropland soils. *Soil Sci. Plant Nutr.* **2013**, *59*, 771–782.

33. Husk, B.; Major, J. Commercial Scale Agricultural Biochar Field Trial in Québec, Canada over Two Years: Effects of Biochar on Soil Fertility, Biology and Crop Productivity and Quality. Available online: http://www.researchgate.net/publication/237079745_Commercial_scale_agricultural_ biochar (accessed on 2 December 2014).

34. Chan, K.Y.; van Zwieten, L.; Meszaros, I.; Downie, A.; Joseph, S. Agronomic values of greenwaste biochar as a soil amendment. *Aust. J. Soil Res.* **2007**, *45*, 629–634.

35. Vaccari, F.P.; Baronti, S.; Lugato, E.; Genesio, L.; Castaldi, S.; Fornasier, F.; Miglietta, F. Biochar as a strategy to sequester carbon and increase yield in durum wheat. *Euro. J. Agron.* **2011**, *34*, 231–238.

36. Prendergast-Miller, M.T.; Duvall, M.; Sohi, S.P. Localisation of nitrate in the rhizosphere of biochar-amended soils. *Soil Biol. Biochem.* **2011**, *43*, 2243–2246.

37. Prendergast-Miller, M.T.; Duvall, M.; Sohi, S.P. Biochar-root interactions are mediated by biochar nutrient content and impacts on soil nutrient availability. *Eur. J. Soil Sci.* **2013**, *65*, 173–185.

38. Method 3050B: Acid Digestion of Sediments, Sludges, and Soils. Available online: http://www. epa.gov/wastes/hazard/testmethods/sw846/pdfs/3050b.pdf (accessed on 2 December 2014).

39. National Climatic Data Center. NOAA's 1981–2010 Climate Normals. Available online: http://www. ncdc.noaa.gov/oa/climate/normals/usnormals.html (accessed on 2 December 2014).

40. Web Soil Survey—Home. Available online: http://websoilsurvey.sc.egov.usda.gov/App/ HomePage.htm (accessed on 2 December 2014).

41. Division of Agriculture. Irrigation for Agriculture in Arkansas. Available online: http://www.uaex. edu/environment-nature/water/irrigation.aspx (accessed on 2 December 2014).

42. Jones, D.; Willett, V. Experimental evaluation of methods to quantify dissolved organic nitrogen (DON) and dissolved organic carbon (DOC) in soil. *Soil Biol. Biochem.* **2006**, *38*, 991–999.

43. Mulvaney, R.L. Nitrogen: Inorganic forms. In *Methods of Soil Analysis, Part 3. Chemical Methods*; Sparks, D.L., Helmke, P.A., Page, A.L., Loeppert, R.H., Eds.; SSSA Book Series; Soil Science Society of America: Madison, WI, USA, **1996**; pp. 1123–1184.

44. Vance, E.D.; Brookes, P.C.; Jenkinson, D.S. Microbial biomass measurements in forest soils: The use of the chloroform fumigation-incubation method in strongly acid soils. *Soil Biol. Biochem.* **1987**, *19*, 697–702.

45. Self-Davis, M.L.; Moore, P.A.; Joern, B.C. Determination of water- and/or dilute salt-extractable phosphorus. In *Methods of Phosphorus Analysis for Soils, Sediments, Residuals, and Waters*; Pierzynski, G.M., Ed.; Southern Cooperative Series Bulletin; North Carolina State University: Raleigh, NC, USA, **2000**; pp. 24–26.

46. Kuo, S. Phosphorus. In *Methods of Soil Analysis, Part 3. Chemical Methods*; Sparks, D.L., Page, A.L., Helmke, P.A., Loeppert, R.H., Eds.; SSSA Book Series; Soil Science Society of America: Madison, WI, USA, 1996; pp. 869–920.

47. Tabatabai, M.A. Soil enzymes. In *Methods of Soil Analysis, Part 2. Microbiological and Biochemical Properties*; Bottomley, P.S., Angle, J.S., Weaver, R.W., Eds.; SSSA Book Series; Soil Science Society of America: Madison, WI, USA, 1994; pp. 775–833.

48. Tucker, M.R. Determination of phosphorus by Mehlich 3 extractant. In *Reference Soil and Media Diagnostic Procedure for the Southern Region of the United States*; Donohue, S.J., Ed.; Southern Cooperative Series Bulletin; Va. Agric. Exp. Station: Blacksburg, VA, USA, 1992; pp. 9–12.

49. Gee, G.W.; Bauder, J.W. Particle-size analysis. In *Methods of Soil Analysis, Part 1. Physical and Mineralogical Methods*; Klute, A., Ed.; Agronomy Monograph; American Society of Agronomy: Madison, WI, USA, 1986; pp. 383–411.

50. Plank, C.O. *Plant Analysis Reference Procedures for the Southern Region of the United States*; Southern Cooperative Series Bulletin; The University of Georgia: Athens, GA, USA, 1992.

51. Pomares-Garcia, F.; Pratt, P.F. Recovery of ^{15}N-labeled fertilizer from manured and sludge-amended soil. *Soil Sci. Soc. Am. J.* **1978**, *42*, 717–720.

11

Emerging Perspectives on the Natural Microbiome of Fresh Produce Vegetables

Colin R. Jackson [1,*], Bram W. G. Stone [1] and Heather L. Tyler [2]

[1] Department of Biology, the University of Mississippi, University, MS 38677, USA; E-Mail: bstone2@go.olemiss.edu

[2] Crop Production Systems Research Unit, USDA Agricultural Research Service, Stoneville, MS 38776, USA; E-Mail: Heather.Tyler@ars.usda.gov

* Author to whom correspondence should be addressed; E-Mail: cjackson@olemiss.edu

Academic Editor: Pascal Delaquis

Abstract: Plants harbor a diverse microbiome existing as bacterial populations on the leaf surface (the phyllosphere) and within plant tissues (endophytes). The composition of this microbiome has been largely unexplored in fresh produce vegetables, where studies have tended to focus on pathogen detection and survival. However, the application of next-generation 16S rRNA gene sequencing approaches is beginning to reveal the diversity of this produce-associated bacterial community. In this article we review what is known about the composition of the microbiome of fresh produce vegetables, placing it in the context of general phyllosphere research. We also demonstrate how next-generation sequencing can be used to assess the bacterial assemblages present on fresh produce, using fresh herbs as an example. That data shows how the use of such culture-independent approaches can detect groups of taxa (anaerobes, psychrophiles) that may be missed by traditional culture-based techniques. Other issues discussed include questions as to whether to determine the microbiome during plant growth or at point of purchase or consumption, and the potential role of the natural bacterial community in mitigating pathogen survival.

Keywords: microbiome; bacteria; salad produce; herbs; phyllosphere; endophytes; 16S rRNA

1. Introduction

The last few decades have seen rapid growth in the fields of microbial ecology and environmental microbiology, such that natural microbial communities are now recognized as being much more diverse than previously thought. While much of the early research in this field focused on using molecular techniques to characterize the diversity of microbial assemblages in soils and waters; from the late 1990s onwards there has been an increased awareness of the natural microbiota associated with plants. Just as there has been a research focus on determining the make-up of the complex microbiome of animals (and especially mammals) and its role in health and disease, the importance of the plant microbiome is beginning to be examined, whether in the context of its role in plant growth and disease resistance, or in the context of human consumption of plant material (e.g., [1]).

The natural bacterial populations associated with plants are present on both aboveground (stem, leaves, flowers, fruits, *etc.*) and belowground (roots, tubers, *etc.*) structures. That roots have an associated microbial community (the rhizosphere) has long been recognized, and aboveground portions of plants (primarily leaves) also have such a community (the phyllosphere). While agricultural scientists are likely aware of the potential for microorganisms to be found naturally on the plant surface, the idea that plants also contain many microbial populations living inside of them (endophytes) is less well known. However, virtually every study that has tested for the presence of endophytes inside of plant samples has found them, so it is likely that every plant species is colonized by at least one endophytic bacterial species. While most endophytes are likely to be non-pathogenic to humans, a number of pathogenic bacteria can become internalized as at least temporary endophytes within leaves. These and other endophytes present an interesting problem from the viewpoint of consumer safety as no amount of washing or vegetable preparation will remove them, and while this is not a potential issue with the consumption of cooked vegetables, it may be with the consumption of raw vegetables in salads. Here we review the state of our current knowledge of the microbial assemblages associated with produce vegetables and present perspectives on how these assemblages relate to consumer health. We also demonstrate how next-generation sequencing technologies can provide valuable insights into the composition of these assemblages, paying particular attention to an understudied vegetable type, fresh herbs.

2. General Characteristics of Phyllosphere Bacterial Communities

Leaves of plants present an area that is available for potential microbial colonization and growth, whether they are on the leaf surface (the phylloplane) or within the leaf itself as endophytes. As such, microbial populations in or on leaves can be abundant, and the phyllosphere community can be taxonomically diverse [2]. The leaf surface itself can be a harsh environment for microbial growth, with microorganisms being potentially exposed to extreme fluctuations in moisture availability, ultraviolet variation, temperature, and nutrient availability [2,3]. However, phyllosphere adapted bacteria may be capable of modifying their local environment, increasing the leakage of nutrients from the host plant and producing extracellular polysaccharides to resist desiccation and help attachment [4]. Studies of the phyllosphere using traditional culture-based approaches have been a part of agricultural and environmental microbiology since at least the 1950s, although for obvious reasons many studies have tended to focus on the detection of known plant pathogens (e.g., species of *Xanthomonas*, *Erwinia* and *Pseudomonas*) or human pathogens from a food safety standpoint. However, the development and application of molecular (largely 16S ribosomal RNA) approaches in microbial ecology, has led to a more thorough understanding of the diversity and complexity of leaf-associated bacterial communities.

Building upon Carl Woese's work demonstrating that ribosomal RNA (rRNA) gene sequences could be used to determine bacterial phylogeny [5], microbial ecology underwent a dramatic shift in the 1980s and 1990s in that it became feasible to directly obtain bacterial DNA from environmental samples, polymerase chain reaction (PCR)-amplify bacterial 16S rRNA genes, and then to use those genes to examine bacterial diversity either through DNA sequencing or through techniques such as denaturing gradient gel electrophoresis (DGGE) or terminal restriction fragment length polymorphism (T-RFLP) analysis. Such analyses revealed that microbial communities were much more complex than previously believed and that perhaps as few as 1% of bacteria had been identified using traditional culture techniques [6,7]. Initially, these molecular microbial ecology studies focused on marine systems (likely because of easier analysis), with a greater emphasis on soils from the mid-1990s onwards [8]. However, despite an increased interest in the microbial diversity of soils from an agricultural standpoint, even including the plant rhizosphere [9–13]; initially, only a few studies applied this concept to examine the natural microbial diversity on the aboveground structures of plants.

Yang *et al.* [14] used DGGE to examine the natural bacterial communities in the phyllosphere of seven agriculturally important plant species: cotton (*Gossypium hirsutum*), corn (*Zea mays*), sugar beet (*Beta vulgaris*), green bean (*Phaseolus vulgaris*), Valencia and navel oranges (both *Citrus sinensis*), and the grapefruit-pomelo hybrid OroBlanco (*C. grandis* × *C. paradisi*). DGGE banding patterns indicated that leaves sampled from different individuals of the same plant species tended to show similar bacterial community profiles, but the profiles for the different species were generally unique. While the phyllosphere communities on the two orange varieties examined (Valencia, navel) were similar, they were very different from those on the other citrus variety (OroBlanco) [14]. Sequence analysis of 17 of the dominant DGGE bands obtained from the Valencia orange samples showed that only four (identified as *Enterobacter agglomerans*, *Bacillus pumilus*, *Acinetobacter* sp., and a member of the Cytophagales) represented established phyllosphere taxa. The other sequences included representatives of the Alpha-, Gamma-, and Deltaproteobacteria that had not previously been reported to inhabit the phyllosphere. This was in contrast to the results of the same analysis conducted on bacteria growing in mixed culture in BIOLOG plates, which were predominantly members of *Pseudomonas*, *Erwinia*, *Acinetobacter*, and *Enterobacter* [14]; all genera that are recognized phyllosphere inhabitants based on traditional culture techniques. Thus, using culture-independent 16S rRNA based methods revealed a phyllosphere community that was very different, and presumably more diverse, than the one that would have been determined through culture-dependent methodologies.

The last 10–15 years has seen a rapid increase in the application of rRNA-based methods to describe leaf-associated assemblages. DGGE and other approaches have been used to characterize the bacterial community in the phyllosphere of food crops, such as corn (*Zea mays*) [15], pepper (*Capsicum annuum*) [16], cucumber (*Cucumis sativus*) [17], and spinach (*Spinacia oleracea*), celery (*Apium graveolens*), broccoli (*Brassica oleracea var. italica*), and cauliflower (*Brassica oleracea var. botrytis*) [18]. Similarly, the diversity and structure of the bacterial community in the phyllosphere of a number of tree species has been described [19–22]. More than basic descriptions of community structure,

16S rRNA techniques have also been used to examine how the phyllosphere community is influenced by environmental factors, such as ultraviolet light [15] and rainfall [23], as well as pesticide [24,25] or biological control agent [16,26] application. At an ecological level, both spatial [20] and temporal [21] patterns in the composition of the phyllosphere have been examined. Thus, while our understanding of the phyllosphere initially lagged behind our knowledge of soil or rhizosphere microbiology, these, and other, culture-independent studies are leading us to insights into both the composition of leaf-associated microbial communities, and how these communities interact with the host plant [27,28].

3. Bacterial Communities Associated with Leaves of Salad Produce Vegetables

Fresh produce vegetables such as lettuce, spinach, *etc.* present an example of leaves that may be consumed by humans without the use of preparation methods that would remove or kill the associated bacterial community. As such, studies of phyllosphere or endophytic bacterial assemblages on these plants are interesting from both a basic scientific and applied aspect. From an applied viewpoint, the presence of potential human pathogens is of interest given that these are not likely to be removed prior to consumption. Most studies of bacteria in produce have used culture-dependent methods to determine the culturable pathogenic populations present, or have examined the recovery of bacteria following intentional exposure to pathogens such as *Escherichia coli* O157:H7, *Salmonella*, or *Listeria* [29–33]. Direct visualization of bacterial cells in the phyllosphere through microscopy has also been used to examine the microbial colonization of leafy vegetables [29,32–34]. Together, these studies indicate that pathogenic bacteria are capable of colonizing leaf surfaces and internal tissues (*i.e.*, they can be part of both phyllosphere and endophytic assemblages), and thus are not necessarily removed by washing. Both *Salmonella* and *E. coli* have been found to be capable of becoming endophytic and colonize the interior tissues of plants [30,33,35–38], although the ability to colonize and persist within the leaf varies between specific strains of bacteria. The ability of these human pathogens to survive and persist on the leaf surface is also less than that of the naturally occurring bacterial populations that reside there. For example, *Salmonella enterica* serovar Thompson has been shown to be capable of colonizing the phyllosphere of the herb cilantro (*Coriandrum sativum*), but was not as resilient to environmental conditions on the leaf surface as native bacterial epiphytes [32]. Similarly, plant-associated *E. coli* strains that were specifically isolated from cabbage roots were more proficient at colonizing the surface of alfalfa sprouts than an O157:H7 clinical isolate [39]. These examples highlight the importance of considering the naturally occurring microbiome on leafy produce, as it may be more persistent than, and even interact with, potential pathogens.

Most of the earlier studies on the presence and persistence of potential human pathogens on salad produce relied on the use of selective microbiological culture media and biochemical tests for the identification of bacterial isolates [29,32]. However, as with general microbial diversity studies, this has been supplemented by culture-independent molecular approaches using 16S rRNA gene amplification and subsequent community profiling by techniques, such as DGGE, T-RFLP, DNA sequencing, and/or microarrays [34,40]. More recently, metagenomic or next-generation sequencing approaches have been used [1,41–43].

An interesting question regarding the microorganisms present on consumable produce is when to sample, or even broader, at what time point do we say that the leaf-associated bacteria that are present are the representative bacterial community? The naturally occurring microbiome that is associated with the plant (*i.e.*, the true phyllosphere and endophyte community) should probably be defined as the microbial populations that are present on or in vegetables growing in the field. However, from the standpoint of the consumer, the microbial populations present at the point of sale, or even the time of consumption, may be more relevant, and both phyllosphere and endophytic microorganisms may differ at these different time points. DGGE of 16S rRNA genes showed reduced bacterial diversity in the phyllosphere of lettuce (*Lacutuca sativa var. capitata*) samples collected from grocery stores compared to samples taken directly from a farm field site, suggesting a simplification of the phyllosphere community during processing of salad vegetables [44]. Furthermore, commercially pre-bagged, refrigerated versions of the same lettuce samples showed evidence of the presence of additional bacterial populations (notably *Pseudomonas libaniensis*) [44] indicating the potential for either contamination or the selective growth of specific bacterial populations during processing and storage. A similar observation has been made for spinach (*Spinacia oleracea*), for which numbers of culturable Pseudomonadaceae and Enterobacteriaceae were found to increase after 12 days of cold storage [29]. The latter study is probably a more reliable assessment of changes that occur post-harvest and following storage, as changes in DGGE profiles or in the composition of 16S rRNA gene clone libraries represent a proportional change in community composition not a change in the absolute number of organisms. Thus, for example, while the presence of *Pseudomonas libaniensis* in clone libraries generated from lettuce samples following refrigerated storage [44] could indicate the growth of that species, it could also indicate a reduction in overall bacterial numbers following refrigeration, especially of more dominant taxa, so that *P. libaniensis* could only then be detected. Regardless, it seems reasonable to assume that the microbiota on vegetables are likely to change following harvest, and while the microbiome present in fresh produce vegetables at the time of purchase and consumption may not entirely be indicative of that present in the growing plant, it does represent the microbial populations to which consumers are potentially exposed, and may be more relevant to study.

Two recent studies took that viewpoint and used 16S rRNA techniques to survey the bacterial populations present in the phyllosphere [41] or in both the phyllosphere and internalized within the leaves [1] of vegetables obtained directly from grocery stores/supermarkets. Both report 16S rRNA gene sequences identified as representing members of the Gammaproteobacteria to be the most abundant sequence types detected, with the Enterobacteriaceae being particularly prevalent. Sequences identified as Enterobacteriaceae were proportionally much more abundant on fresh salad vegetables such as lettuce and spinach, than they were on apples, peaches or grapes [41]. Clustering 16S rRNA gene sequences together using a 97% similarity criterion to define operational taxonomic units (OTUs; a surrogate for species), suggested the presence of around 40–70 different OTUs or species on the lettuce or spinach leaf surface [1,41]. Thus, while not as diverse as soils or natural waters, the leaf-associated community of consumable fresh produce is quite diverse, and consumers are likely being exposed to 50 or more species of bacteria during salad consumption. While many of these populations are likely to be plant symbionts or pathogens, some of the bacterial genera and species detected in those studies include strains that can be human pathogens [1].

Taken together, these studies suggest that there are some families and genera of bacteria that are commonly found in the phyllosphere of leafy salad vegetables. Members of the Enterobacteriaceae have been consistently detected, and are often among the most abundant members of the phyllosphere community [34,40–43]. Members of the genus *Pseudomonas* have been found to be among the dominant bacterial populations in spinach [42] and lettuce [40,43], as well as in bagged salad mixes containing lettuce, red cabbage, and carrots [34]. However, species of *Pseudomonas* are not always dominant members of the phyllosphere, as shown in the survey of Leff and Fierer [41] which found that the entire family Pseudomonadaceae accounted for less than 5% of the bacterial 16S rRNA gene sequences obtained from spinach or lettuce, compared to Enterobacteriaceae which accounted for 38% (lettuce) or 58% (spinach) of the sequences obtained. This same variability is seen in the less abundant bacterial populations detected in phyllosphere studies, where certain bacterial genera can account for >1% of the community in some studies but are not detected at all in others. Thus, despite the generally widespread occurrence of the Enterobacteriaceae and, to a lesser extent, the genus *Pseudomonas*, there does not appear to be a consistent core set of taxa that are found in the phyllosphere of all produce vegetables.

While there may be no consistent core taxa, environmental conditions during produce growth and storage likely impact the leaf-associated bacterial community. Although they are almost always abundant, Enterobacteriaceae have been reported as being more prevalent in the phyllosphere of conventionally grown spinach and lettuce compared to those under organic cultivation [41]. At least that was the case for produce sampled at the point of purchase: the opposite trend was reported when lettuce was sampled after being freshly harvested, when organically grown plants showed higher numbers of Enterobacteriaceae than conventionally grown ones [45]. However, these two studies may not be entirely comparable as they differed in the style of analysis used (culture-dependent *vs.* molecular) as well as in the time of sampling. A comparison of the plant-associated bacterial communities on six varieties of organically and conventionally grown salad produce sampled at point of purchase revealed no consistent differences in community composition between the two growth approaches [1]. Regardless of the influence of cultivation style, other environmental variables during growth such as soil moisture, organic content, or nutrient (fertilizer) availability could also affect microbial community composition.

As previously mentioned, storage undoubtedly has an influence on the leaf-associated microbial community. When fresh-cut spinach was stored at 10 °C, numbers of both Enterobacteriaceae and Pseudomonadaceae increased at least 1000-fold over 12 days [29]. Storage under refrigeration also lowered the diversity and richness of the phyllosphere community on spinach [42], and the temperature of storage influenced how much the community changes over the storage period. After 15 days of storage, spinach samples that were held at 15 °C harbored bacterial communities that were more similar to the original microbiome than samples that were stored at 10 °C [42]. Changes in the bacterial community associated with bagged lettuce mixes have also been reported following storage at 10 °C, with an increase in the relative abundance of Enterobacteriaceae and a decrease in the relative abundance of *Pseudomonas* [34]. Interestingly, when the bagged lettuce mixes were stored at refrigerator temperature (4 °C), the decrease in *Pseudomonas* was less pronounced and this genus was still the dominant taxonomic group, as it was prior to storage [34]. Thus, refrigerated storage might help retain the natural microbiome, while extended storage at cool, but not cold, temperatures might be more likely to promote shifts in the phyllosphere community, and potentially favor pathogenic strains.

4. Bacterial Communities Associated with Consumable Herbs: A Demonstration of Next-Generation Sequencing Approaches

Studies of microbial diversity, even those using culture-independent molecular techniques, are limited by the ability of the technique used to adequately sample the microbiome in question. With thousands to millions of bacteria cells present on a leaf surface, even 16S rRNA gene based methods such as DGGE or cloning and Sanger sequencing of amplified 16S rRNA gene fragments only reveal a miniscule subset of the bacteria that are present. The last few years, however, have seen the emergence of next-generation sequencing technologies that allow a far more in-depth profiling of bacterial assemblages to occur. Such approaches can readily yield thousands or tens of thousands of partial 16S rRNA gene sequences from a single sample of plant material (often from just 0.1–0.5 g of material), providing much more coverage of the community present, and increasing the ability to detect less common taxa. Such approaches were used by the previously mentioned studies of fresh produce at the point of consumer purchase [1,41] suggesting that those studies may well be better assessments of the bacterial component present on fresh vegetables than those before. That said, even those studies only assessed around 2000 [1] or 200 (this lower amount because more vegetables were sampled) [41] sequences, largely because they were based around the Roche 454 next-generation sequencing platform. Illumina next-generation sequencing has replaced 454 as the procedure of choice, and emerging protocols on that platform now facilitate more in-depth sampling of microbiome structure [46,47]. This approach is beginning to be used to characterize various plant-associated bacterial communities, with studies being reported for the phyllosphere of neo-tropical rainforest trees [48] and wetland plants [49]. However, other than a recent study examining how pesticide application can change the phyllosphere of tomatoes [50], there have been few, if any, studies that have used Illumina next-generation sequencing to characterize the bacterial communities on fresh fruit or vegetables. Here, we demonstrate the use of this platform to describe the natural bacterial assemblages present on a generally ignored subset of fresh vegetables, fresh herbs, and suggest some guidelines for using next-generation sequencing approaches in this context.

4.1. Sample Collection, Processing, and 16S rRNA Gene Illumina Sequencing

Samples of six fresh herbs were purchased from an Oxford, Mississippi, USA, grocery store in September 2014. Herbs were basil (*Ocimum basilicum*), chives (*Allium schoenoprasum*), dill (*Anethum graveolens*), mint (*Mentha spicata*), rosemary (*Rosmarinus officinalis*), and thyme (*Thymus vulgaris*). All were packaged in polymer film-sealed plastic containers, labeled as products of the USA. For each herb type, three representative leaves/fronds were sampled aseptically and 0.1 g weighed and used for DNA extraction. DNA was extracted from the leaf sample using a PowerPlant DNA Isolation kit (Mo Bio Laboratories, Carlsbad, CA, USA). Thus, 16S rRNA gene sequences obtained reflect those of bacterial populations in the phyllosphere and potential endophytes. An alternative approach would have been to remove bacterial cells from the leaf surface prior to extracting DNA from them, and this has been done for some studies [41–44]. While that approach has the advantage of minimizing the co-amplification of chloroplast and mitochondrial sequences (which typically can amplify with bacterial specific primers), it prevents the sampling of endophyte populations, which may be as relevant as phyllosphere populations in regards to food safety. Furthermore, for this specific study, removing bacteria from the leaf surface would have been difficult given the small size of the particular leaf samples used. Given the large number of sequences per sample that can be generated using next-generation sequencing, we recommend extracting DNA from the intact plant material rather than leaf washes, as the ability to detect the entirety of the bacterial community associated with the plant (*i.e.*, both phyllosphere and endophyte populations) likely outweighs the potential loss of some data to chloroplast or mitochondrial sequence reads.

A dual-index barcoding approach was used for Illumina next-generation sequencing whereby each sample was amplified with bacterial specific 16S rRNA gene primers, each tagged with a specific 8-nucleotide barcode [47]. These primers amplify a 250 nucleotide long region of the V4 variable region of the 16S rRNA gene. The specific amplification approach used involves a single round of PCR, thereby limiting the risk of amplification artifacts [47]. Following amplification, amplicons from all samples were pooled, and the assembled library spiked with 5% PhiX to increase nucleotide base diversity prior to sequencing (a necessary step for bacterial community analysis by this method). The final library was sequenced on an Illumina MiSeq instrument, via two index sequencing reads, at the University of Mississippi Medical Center (UMMC) Molecular and Genomics Core Facility. Illumina MiSeq-based sequencing was chosen as it reflects a suitable balance of a large number of sequence reads per run (and therefore per unit cost), the ability to obtain high quality data, and the ability to obtain reads in excess of 200 nucleotides long, thereby facilitating more accurate sequence classification. While the Illumina HiSeq or NextSeq systems can provide higher sample throughput, the MiSeq system is currently more commonly used for 16S rRNA gene studies, and at a cost at or below U.S. $100,000 is more suitable for individual investigators or testing laboratories [46,47].

4.2. Data Processing and Bioinformatics Pipeline

Raw sequence data (fastq files) were downloaded and accessed via the 16S rRNA bioinformatics software package mothur [51,52]. Sequences were processed along a bioinformatics pipeline following the general guidelines of Kozich *et al.* [47], which are optimized for Illumina MiSeq-generated 16S rRNA gene data. Briefly, contigs were assembled using forward and reverse reads and screened to only include those with a maximum length of 275 bp and no base ambiguities (*i.e.*, sequence contigs had to have identical base calls for both the forward and reverse read). Sequences were aligned against the SILVA 16S rRNA database [53] and misaligned sequences deleted. Sequences were clustered together by roughly 1% sequence similarity to account for potential amplification and sequencing errors, and chimeras removed using UCHIME [54]. Valid sequences were classified using the Greengenes [55] 16S rRNA classification scheme, and erroneous (archaeal, eukaryotic) sequences removed. Remaining bacterial sequences were grouped into operational taxonomic units (OTUs) based on >97% sequence similarity. As stated, these procedures follow published recommendations for this type of data [47], which are also routinely updated online [56]. These procedures were developed after a series of studies that assessed issues such as the impacts of clustering level, chimera removal procedure, alignment and classification reference, and OTU similarity criterion on bacterial community analysis [56] and we have found them suitable for sequences amplified from a range of samples including plant material, soils, and waters.

4.3. Taxonomic Composition of Bacterial Communities on Fresh Herbs

The microbiome of each of the herbs examined was generally dominated by members of the Firmicutes, Bacteroidetes, and Proteobacteria (Table 1). Of the latter, Alpha- and Gamma- sub-phyla were the most prevalent. However, the proportional abundance of each of these phyla in the sequence dataset varied between herb species with members of the Firmicutes being more prevalent in/on basil, rosemary, and thyme, and the Proteobacteria (especially Gammaproteobacteria) being more prevalent in/on dill and, to a lesser extent, chives (Table 1). A substantial proportion of the sequences recovered from mint were also classified as Proteobacteria, although these were Alphaproteobacteria and their proportional abundance was highly variable between samples (Table 1). Other major bacterial groups that were detected in relatively common proportions (accounting for >1% of the sequence dataset) were the Bacteroidetes (accounting for an average of 6.1%–17.9% of sequences, depending upon the herb species) and Actinobacteria (an average of 0.3%–6.5% of sequences).

Table 1. Major phyla and subphyla of bacteria associated with prepackaged fresh herbs sampled from a grocery store in Mississippi, USA, as determined from high throughput 16S rRNA gene sequencing. Numbers represent mean (+/− SE) percentage of total sequence reads obtained from 3 samples per herb that were identified as that phylum.

Phylum	Basil	Chives	Dill	Mint	Rosemary	Thyme
Acidobacteria	0.02 (0.01)	0.07 (0.07)	0.10 (0.04)	0.04 (0.03)	0.01 (0.00)	0.01 (0.01)
Actinobacteria	0.26 (0.04)	1.00 (0.13)	6.52 (1.46)	0.74 (0.10)	0.36 (0.05)	0.67 (0.07)
Bacteroidetes	6.1 (1.26)	13.7 (7.14)	6.1 (2.73)	17.9 (3.34)	12.2 (1.28)	13.1 (1.21)
Firmicutes	84.2 (1.87)	51.6 (24.19)	13.4 (5.33)	48.8 (23.94)	82.2 (1.02)	81.5 (0.74)
Lentisphaerae	0.02 (0.01)	0.08 (0.04)	-	0.13 (0.07)	0.07 (0.01)	0.10 (0.02)
Proteobacteria	9.2 (3.08)	28.9 (26.74)	73.6 (4.32)	31.99 (27.29)	5.05 (1.05)	4.40 (0.54)
Alpha-	1.41 (0.48)	0.22 (0.17)	2.08 (0.52)	25.56 (22.82)	2.36 (0.58)	0.15 (0.05)
Beta-	0.28 (0.04)	0.78 (0.29)	1.12 (0.36)	2.65 (1.60)	0.46 (0.09)	0.51 (0.04)
Gamma-	7.29 (2.63)	27.70 (26.94)	70.33 (4.10)	3.40 (3.01)	1.89 (0.51)	3.40 (0.58)
Delta-	0.17 (0.02)	0.23 (0.12)	0.02 (0.00)	0.37 (0.14)	0.33 (0.04)	0.34 (0.07)

The total number of valid bacterial sequences obtained from each individual sample averaged 18,140, but ranged from 5129 to 47,548. Compared to prior molecular surveys of bacterial assemblages on fresh vegetables, this represents a substantial increase in sequencing depth, but raises the question as to how much sequence data is needed to accurately assess the phyllosphere and endophyte communities associated with such samples. From analyses of Roche 454 next-generation sequencing datasets, 1000 valid sequences have been found to capture 90% of the patterns in community structure between samples (beta-diversity), while 5000 sequences were necessary to accurately predict total diversity within a sample (alpha-diversity) [57]. However, those numbers were derived from analyses of sediment samples, and bacterial communities in sediments are generally more diverse than those associated with plants. Thus, while we recommend a minimum of 5000 sequences to be analyzed from a sample in order to accurately assess the bacterial community present, in practice 2000–3000 sequences is likely to be enough to obtain a reasonable determination of community composition for plant samples. Rarefaction curves of the number of distinct OTUs observed as a function of sampling effort suggested that similar sequencing depth (2000–5000 sequences per sample) was sufficient to describe the bacterial assemblages associated with fresh salad produce [1]. However, the more reads per sample the greater the ability to detect rare taxa, and if the goal is to detect specific bacterial populations (e.g., pathogens) that may account for a minor proportion of the total bacterial community, then greater sequencing depth may be needed.

The dominant specific bacterial populations (as determined from proportional abundance of each OTU) associated with the fresh herb samples included various members of the Firmicutes, Bacteroidetes, and Proteobacteria (Table 2). An unclassified member of the Bacillaceae represented by far the most numerous sequence type followed by sequences identified as *Streptococcus* sp. and *Prevotella copri*. Other major OTUs included *Pseudomonas* sp., *Bacillus cereus*, and *Bacteroides* sp. (Table 2). While some sequences could clearly be identified to the species level, others could only be identified to genus, or even less resolved (e.g., the dominant unclassified Bacillaceae, as well as unclassified members of the Enterobacteriaceae and Planococcaceae). This highlights a potential limitation of next-generation sequencing in that sequence read lengths are currently limited to a few hundred bases (in this case 250 bp). However, even with that limitation some patterns are apparent. For example, a number of well-represented OTUs were identified as taxa that are obligate anaerobic bacteria (e.g., *Prevotella copri*, *Bacteroides* sp., *Phascolarctobacterium* sp., *Catenibacterium* sp.). These taxa represent bacteria that are typically found in the human large intestine, and were initially described after isolation from fecal samples [58–60]. Thus, their presence in prepackaged fresh herbs suggests that those herbs show evidence of past fecal contamination. Regular microbiological culture methods would not necessarily have detected these organisms because of their anaerobic culture requirements, demonstrating that while culture-independent methods, such as next-generation sequencing may have limitations, they can be a useful tool in detecting the presence of intestinal microbial populations.

Table 2. Identities of the most prevalent bacterial populations (OTUs) associated with prepackaged fresh herbs (basil, chives, dill, mint, rosemary, thyme) sampled from a grocery store in Mississippi, USA, as determined from high throughput 16S rRNA gene sequencing. Number of reads represents the total number of sequence reads identified as that taxon, out of a total of 392,272 reads.

OTU	Phylum (Class)	Number of Reads
Unclassified Bacillaceae	Firmicutes (Bacilli)	204,483
Streptococcus sp.	Firmicutes (Bacilli)	22,984
Prevotella copri	Bacteroidetes (Bacteroidia)	22,597
Pseudomonas sp.	Proteobacteria (Gamma)	15,955
Bacillus cereus	Firmicutes (Bacilli)	12,823
Bacteroides sp.	Bacteroidetes (Bacteroidia)	10,045
Phascolarctobacterium	Firmicutes (Clostridia)	6124
Bacteroides sp.	Bacteroidetes (Bacteroidia)	5231
Psychrobacter sp.	Proteobacteria (Gamma)	4293
Unclassified Enterobacteriaceae	Proteobacteria (Gamma)	4240
Faecalibacterium prausnitzii	Firmicutes (Clostridia)	4232
Unclassified Planococcaceae	Firmicutes (Bacilli)	3485

Table 2. *Cont.*

Methylobacterium adhaesivum	Proteobacteria (Alpha)	2821
Catenibacterium sp.	Firmicutes (Erysipelotrichi)	2738
Pseudomonas veronii	Proteobacteria (Gamma)	2702
Sphingomonas sp.	Proteobacteria (Alpha)	2572
Agrobacterium sp.	Proteobacteria (Alpha)	2546
Sphingomonas sp.	Proteobacteria (Alpha)	2445
Sphingomonas sp.	Proteobacteria (Alpha)	2327
Unclassified Rikenellaceae	Bacteroidetes (Bacteroidia)	1950
Arthrobacter psychrolactophilus	Actinobacteria (Actinobacteria)	1943
Portiera aleyrodidarum	Proteobacteria (Gamma)	1868
Exiguobacterium sp.	Firmicutes (Bacilli)	1733
Sutterella sp.	Proteobacteria (Beta)	1633
Alistipes putredinis	Bacteroidetes (Bacteroidia)	1600

Three taxa (*Psychrobacter* sp., *Arthrobacter psychrolactophilus*, *Exiguobacterium* sp.) detected as important OTUs were bacterial species that are psychrophilic or psychrotolerant (*i.e.*, cold tolerant). The presence of sequences affiliated with these groups likely arises from the refrigerated storage of the produce sampled, as they are unlikely to be abundant in the natural herb phyllosphere. This confirms the accepted idea that the bacterial community associated with produce likely does change during processing and storage, with an increased prevalence of psychrotolerant groups following refrigeration. Indeed, given that the herbs sampled were packaged in plastic containers sealed with polymer file, the high prevalence of OTUs that would be classified as anaerobic likely also reflects microaerophilic storage conditions, and while those OTUs may have originated from fecal contamination, their relative abundance appears to have increased during storage. This provides some support for the argument to sample at point of sale, rather than at point of growth, if the goal is to determine the bacterial community present on produce from a consumer standpoint [1,41]. Further supporting the idea that changes in the herb phyllosphere and endophytic communities must have occurred during processing and storage is the finding that relatively few of the dominant OTUs (essentially just species of *Methylobacterium*, *Agrobacterium*, and *Sphingomonas*) would be regarded as primarily plant-associated bacteria.

4.4. Overall Summary and Conclusions of the Example Study

Overall, the results of the herb study demonstrate that next-generation sequencing technologies such as Illumina 16S rRNA gene sequencing can serve a useful function in the analysis of bacterial communities associated with fresh produce such as herbs. These approaches are rapidly becoming more affordable and easier to perform, and in terms of labor intensity tend to be more efficient than culture-based approaches. In terms of time expenditure, processing and DNA extraction from all herb samples was accomplished with 1–2 days of purchase, and could realistically be completed within a few hours by a dedicated technician. Illumina 16S rRNA gene library preparation takes another few days, with the actual sequencing run taking 24–48 h. Thus, with a dedicated laboratory, it would be possible to generate complete community data in <2 weeks. In this particular study, the bacterial assemblages associated with herbs were found to be dominated by members of the Firmicutes and Gammaproteobacteria, although there was some variation between the particular herbs sampled.

Plant-associated taxa were relatively scarce, supporting previous findings [34,42] that phyllosphere and endophyte communities can change during storage, whether from refrigeration or being packaged in modified atmospheres. These changes appeared to result in a higher presence than expected of 16S rRNA sequences identified as being from psychrophilic organisms, and a substantial number of sequences classified as coming from anaerobic taxa. The latter finding is particularly alarming as many of the genera identified (e.g., *Bacteroides*, *Catenibacterium*, *Phascolarctobacterium*, *Prevotella*) are associated with the human large intestine so are typically only detected in fecal samples. Thus, the herbs likely were subject to fecal contamination at some point, and these bacterial populations increased during the storage period. Given the obligate anaerobic nature of these taxa, they would not have been detected by routine microbiological culture, demonstrating that the use of culture-independent approaches can provide useful insights into bacteria present on fresh produce.

5. Conclusions

Assessing the natural microbiome of consumable plants is important, as natural microbial assemblages may reduce the likelihood of pathogen colonization or survival. For example, reduced levels of *Salmonella enterica* colonization have been observed in lettuce that has a more diverse endophyte community [35]. The mechanism for this pathogen reduction with increased endophyte diversity is not conclusively known, but may well be an increased likelihood of antagonists to *S. enterica* being present with increased overall diversity. Additional support for this concept comes from a study examining the viability of *E. coli* O157:H7 on romaine lettuce, where the phyllosphere bacterial diversity in plants that had culturable *E. coli* O157:H7 cells differed from that on plants where the *E. coli* was no longer viable [61]. That native plant-associated microorganisms can act as competitors to potential human pathogens such as *Salmonella* species and *E. coli* O157:H7 has been shown in lettuce and alfalfa sprouts [62,63], suggesting that even in the absence of specific antagonistic interactions, natural phyllosphere and endophytic communities may limit the presence and abundance of pathogenic bacteria by simply outcompeting them in the living plant. Determining the structure of these communities might therefore provide insights into produce-borne outbreaks of disease, and even lead to the development of tools to assess the likelihood of these outbreaks occurring [1]. Intentional addition of competitive native microbiota has even been proposed as a potential method to reduce enteropathogen contamination of fresh produce [64].

Whether it is most important to determine the microbiome of fresh produce vegetables in the field or at point of sale/consumption is debatable. From the viewpoint of the potential for the natural microbial community to mitigate the growth and persistence of pathogenic bacteria, then the composition of the phyllosphere community in the field may be most important. However, a number studies have shown that the composition and diversity of this community changes during processing, handling, and storage [29,34,42,44], so it could be questioned as to whether the field community is as important if exposure to pathogens occurs post-harvesting. From the viewpoint of the consumer, the bacterial populations that are in or on the produce at point of sale, or more specifically the point of consumption, are the most relevant, as these are the bacteria to which they are being exposed. Interestingly, there have been few, if any, studies that have determined the changes that may occur in produce-associated microbial communities in the time between purchase and consumption, even though that time could well span a number of days. More in-depth tracking of the changes that can occur over the entire period from growth through harvesting, processing, storage, purchase, and final consumption is certainly merited.

The methodologies used to assess the composition of the bacterial community on produce need to be current. The use of culture-dependent methods is certainly justified when the goal is to determine the presence or viable abundance of specific bacterial populations such as pathogens or indicator species, whose culture requirements are known, and for which specific selective and differential growth media exist. However, culture-independent molecular methods allow the entire bacterial community to be examined, facilitating a more thorough examination of the microbiome present. Emerging technologies such as next-generation sequencing can be used to detect populations that may be missed by standard culture approaches (such as the anaerobic taxa detected on the herb samples described above), and as they become more affordable are likely to compete with traditional culture approaches in routine assessment and monitoring of produce. That said, when we have used culture-dependent approaches and culture-independent next-generation sequencing to analyze the same produce samples, we found that while sequencing revealed the presence of more bacterial taxa, the dominant taxa in our sequence libraries were the same ones detected by the culturing approach [1]. Thus the two approaches are best viewed as complementary and, ideally, both would be used in analyses of fresh produce vegetables.

Just as microbial ecologists have come to understand the complexity and diversity of naturally occurring bacterial assemblages in soils and waters, we are now beginning to grasp the diversity of the phyllosphere and endophytic communities in agricultural crops. Determining this diversity is especially important for crops such as fresh salad produce as they are not extensively processed or cooked prior to consumption. Understanding the complexity of these communities could lead to the development of new pathogen control mechanisms [64] and a greater understanding of pathogen survival and persistence. Regardless of potential future developments, it's becoming clear that leafy salad vegetables harbor a diverse set of phyllosphere and endophytic populations, which consumers must continually be being exposed to. The impact of this native microbiome on consumer health, whether positive or negative, has been largely unexplored.

Raw bacterial sequence data (fastq files) from the herb samples analyzed as part of this manuscript have been deposited in the NCBI Sequence Reads Archive (SRA) under Project accession number SRP052782. Individual herb samples have SRA Sample accession numbers SRS826210, SRS826225, SRS826710, SRS826711, SRS826712, SRS826713, SRS826714, SRS826715, SRS826716, SRS826717, SRS826718, SRS826719, SRS826720, SRS826721, SRS826722, SRS826723, SRS826724, SRS826726, SRS826727, SRS826728, SRS826729, SRS826730, SRS826731, and SRS826732.

Acknowledgments

Colin R. Jackson's contribution was in part supported by award R01AT007042 from the U.S. National Institutes of Health and also by the Department of Biology at the University of Mississippi. The work performed through the UMMC Molecular and Genomics Facility is supported, in part, by funds from the National Institute of General Medical Sciences of the National Institutes of Health, including Mississippi INBRE (P20GM103476), Center for Psychiatric Neuroscience (CPN)-COBRE (P30GM103328) and Obesity, Cardiorenal and Metabolic Diseases-COBRE (P20GM104357). The content of the manuscript is solely the responsibility of the authors and does not necessarily represent the official views of the National Institutes of Health. Mention of trade names or commercial products is solely for the purpose of providing specific information and does not imply recommendation or endorsement by the US Department of Agriculture (USDA).

Author Contributions

Colin R. Jackson reviewed the literature and wrote the initial draft of the manuscript with assistance from Heather L. Tyler; Bram W. G. Stone performed the next-generation sequencing experiments in Section 4; Colin R. Jackson performed the bioinformatics analyses on the sequencing data and generated Tables 1 and 2. All three authors reviewed and edited the final submission.

Conflicts of Interest

The authors declare no conflict of interest.

References

1. Jackson, C.R.; Randolph, K.C.; Osborn, S.L.; Tyler, H.L. Culture dependent and independent analysis of bacterial communities associated with commercial salad leaf vegetables. *BMC Microbiol.* **2013**, *13*, 274, doi:10.1186/1471-2180-13-274.

2. Lindow, S.E.; Brandl, M.T. Microbiology of the phyllosphere. *Appl. Environ. Microbiol.* **2003**, *69*, 1875–1883.

3. Hirano, S.S.; Upper, C.D. Bacteria in the leaf ecosystem with emphasis on *Pseudomonas syringiae*— A pathogen, ice nucleus, and epiphyte. *Microbiol. Mol. Biol. Rev.* **2000**, *64*, 624–653.

4. Beattie, G.A.; Lindow, S.E. Bacterial colonization of leaves: A spectrum of strategies. *Phytopathology* **1999**, *89*, 353–359.

5. Woese, C.R. Bacterial evolution. *Microbiol. Rev.* **1987**, *51*, 221–271.

6. Pace, N.R. A molecular view of microbial diversity and the biosphere. *Science* **1997**, *276*, 734–740.

7. Hugenholtz, P. Exploring prokaryotic diversity in the genomic era. *Genome Biol.* **2002**, *3*, doi:10.1186/gb-2002-3-2-reviews0003.

8. Rappé, M.S.; Giovannoni, S.J. The uncultured microbial majority. *Annu. Rev. Microbiol.* **2003**, *57*, 369–394.

9. Duineveld, B.M.; Rosado, A.S.; van Elsas, J.D.; van Veen, J.A. Analysis of the dynamics of bacterial communities in the rhizosphere of the chrysanthemum via denaturing gradient gel electrophoresis and substrate utilization patterns. *Appl. Environ. Microbiol.* **1998**, *64*, 4950–4957.

10. Smalla, K.; Wachtendorf, U.; Heuer, H.; Liu, W.T.; Forney, L. Analysis of BIOLOG GN substrate utilization patterns by microbial communities. *Appl. Environ. Microbiol.* **1998**, *64*, 1220–1225.

11. Kim, J.S.; Sakai, M.; Hosoda, A.; Matsuguchi, T. Application of DGGE analysis to the study of bacterial community structure in plant roots and in nonrhizosphere soil. *Soil Sci. Plant Nutr.* **1999**, *45*, 493–497.

12. Smit, E.; Leeflang, P.; Glandorf, B.; van Elsas, J.D.; Wernars, K. Analysis of fungal diversity in the wheat rhizosphere by sequencing of cloned PCR-amplified genes encoding 18S rRNA and temperature gradient gel electrophoresis. *Appl. Environ. Microbiol.* **1999**, *65*, 2614–2621.

13. Yang, C.H.; Crowley, D.E. Rhizosphere microbial community structure in relation to root location and plant iron nutritional status. *Appl. Environ. Microbiol.* **2000**, *66*, 345–351.

14. Yang, C.H.; Crowley, D.E.; Borneman, J.; Keen, N.T. Microbial phyllosphere populations are more complex than previously realized. *Proc. Nat. Acad. Sci. USA* **2001**, *98*, 3889–3894.

15. Kadivar, H.; Stapleton, A.E. Ultraviolet radiation alters maize phyllosphere bacterial diversity. *Microb. Ecol.* **2003**, *45*, 353–361.

16. Zhang, B.; Bai, Z.; Hoefel, D.; Tang, L.; Yang, Z.; Zhuang, G.; Yang, Z.; Zhang, H. Assessing the impact of the biological control agent *Bacillus thuringiensis* on the indigenous microbial community within the pepper plant phyllosphere. *FEMS Microbiol. Lett.* **2008**, *284*, 102–108.

17. Suda, W.; Nagasaki, A.; Shishido, M. Powdery mildew-infection changes bacterial community composition in the phyllosphere. *Microbes Environ.* **2009**, *24*, 217–223.

18. Zhang, B.; Bai, Z.; Hoefel, D.; Wang, X.; Zhang, L.; Li, Z. Microbial diversity within the phyllosphere of different vegetable species. In *Current Research, Technology and Education Topics in Applied Microbiology and Microbial Biotechnology*; Méndez-Vilas, A., Ed.; Formatex: Badajoz, Spain, 2010; Volume 2, pp. 1067–1077.

19. Lambais, M.R.; Crowley, D.E.; Curry, J.C.; Büll, J.C.; Rodrigues, R.R. Bacterial diversity in tree canopies of the Atlantic forest. *Science* **2006**, *312*, doi:10.1126/science.1124696.

20. Redford, A.J.; Bowers, R.M.; Knight, R.; Linhart, Y.; Fierer, N. The ecology of the phyllosphere: Geographic and phylogenetic variability in the distribution of bacteria on tree leaves. *Environ. Microbiol.* **2010**, *12*, 2885–2893.

21. Jackson, C.R.; Denney, W.C. Annual and seasonal variation in the phyllosphere bacterial community associated with leaves of the Southern Magnolia (*Magnolia grandiflora*). *Microb. Ecol.* **2011**, *61*, 113–122.

22. Kim, M.; Singh, D.; Lai-Hoe, A.; Go, R.; Rahim, R.A.; Ainuddin, A.N.; Chun, J.; Adams, J.M. Distinctive phyllosphere bacterial communities in tropical trees. *Microb. Ecol.* **2012**, *63*, 674–681.

23. Jackson, E.F.; Echlin, H.E.; Jackson, C.R. Changes in the phyllosphere community of the resurrection fern, *Polypodium polypodioides*, associated with rainfall and wetting. *FEMS Microbiol. Ecol.* **2006**, *58*, 236–246.

24. Zhang, B.; Bai, Z.; Hoefel, D.; Tang, L.; Wang, X.; Li, B.; Li, Z.; Zhuang, G. The impacts of cypermethrin pesticide application on the non-target microbial community of the pepper plant phyllosphere. *Sci. Total Environ.* **2009**, *407*, 1915–1922.

25. Gu, L.; Bai, Z.; Jin, B.; Hu, Q.; Wang, H.; Zhuang, G.; Zhang, H. Assessing the impact of fungicide enostroburin application on bacterial community in wheat phyllosphere. *J. Environ. Sci.* **2010**, *22*, 134–141.

26. Alberghini, S.; Battisti, A.; Squartini, A. Monitoring a genetically modified *Pseudomonas* sp. released on pine leaves reveals concerted successional patterns of the bacterial phyllospheric community. *Antonie van Leeuwenhoek* **2008**, *94*, 415–422.

27. Vorholt, J.A. Microbial life in the phyllosphere. *Nat. Rev. Microbiol.* **2012**, *10*, 828–840.

28. Müller, T.; Ruppel, S. Progress in cultivation-independent phyllosphere microbiology. *FEMS Microbiol. Ecol.* **2014**, *87*, 2–17.

29. Babic, I.; Roy, S.; Watada, A.E.; Wergin, W.P. Changes in microbial populations on fresh cut spinach. *Int. J. Food Microbiol.* **1996**, *31*, 107–119

30. Franz, E.; Visser, A.A.; van Diepeningen, A.D.; Klerks, M.M.; Termorshuizen, A.J.; van Bruggen, A.H.C. Quantification of contamination of lettuce by GFP-expressing *Escherichia coli* O157:H7 and *Salmonella enterica* serovar Typhimurium. *Food Microbiol.* **2007**, *24*, 106–112.

31. Jablasone, J.; Warriner, K.; Griffiths, M. Interactions of *Escherichia coli* O157:H7, *Salmonella typhimurium* and *Listeria monocytogenes* plants cultivated in a gnotobiotic system. *Int. J. Food Microbiol.* **2005**, *99*, 7–18.

32. Brandl, M.T.; Mandrell, R.E. Fitness of *Salmonella enterica* serovar Thompson in the cilantro phyllosphere. *Appl. Environ. Microbiol.* **2002**, *68*, 3614–3621.

33. Solomon, E.B.; Yaron, S.; Matthews, K.R. Transmission of *Escherichia coli* O157:H7 from contaminated manure and irrigation water to lettuce plant tissue and its subsequent internalization. *Appl. Environ. Microbiol.* **2002**, *68*, 397–400.

34. Rudi, K.; Flateland, S.L.; Hanssen, J.F.; Bengtsson, G.; Nissen, H. Development and evaluation of a 16S ribosomal DNA array-based approach for describing complex microbial communities in ready-to-eat vegetable salads packed in a modified atmosphere. *Appl. Environ. Microbiol.* **2002**, *68*, 1146–1156.

35. Klerks, M.M.; Franz, E.; van Gent-Pelzer, M.; Zijlstra, C.; van Bruggen, A.H.C. Differential interaction of *Salmonella enterica* serovars with lettuce cultivars and plant microbe factors influencing the colonization efficiency. *ISME J.* **2007**, *1*, 620–631.

36. Guo, X.; Chen, J.; Brackett, R.E.; Beuchat, L.R. Survival of Salmonellae on and in tomato plants from the time of inoculation at flowering and early stages of fruit development through fruit ripening. *Appl. Environ. Microbiol.* **2001**, *67*, 4760–4764.

37. Wachtel, M.R.; Whitehand, L.C.; Mandrell, R.E. Association of *Escherichia coli* O157: H7 with preharvest leaf lettuce upon exposure to contaminated irrigation water. *J. Food Prot.* **2002**, *65*, 18–25.

38. Deering, A.J.; Mauer, L.J.; Pruitt, R.E. Internalization of *E. coli* O157:H7 and *Salmonella* spp. in plants: A review. *Food Res. Int.* **2012**, *45*, 567–575.

39. Barak, J.D.; Whitehand, L.C.; Charkowski, A.O. Differences in attachment of *Salmonella enterica* serovars and *Escherichia coli* O157:H7 to alfalfa sprouts. *Appl. Environ. Microbiol.* **2002**, *68*, 4758–4763.

40. Hunter, P.J.; Hand, P.; Pink, D.; Whipps, J.M.; Bending, G.D. Both leaf properties and microbe-microbe interactions influence within-species variation in bacterial population diversity and structure in the lettuce (*Lactuca* species) phyllosphere. *Appl. Environ. Microbiol.* **2010**, *76*, 8117–8125.

41. Leff, J.W.; Fierer, N. Bacterial communities associated with the surfaces of fresh fruits and vegetables. *PLoS ONE* **2013**, *8*, e59310, doi:10.1371/journal.pone.0059310.

42. Lopez-Velasco, G.; Welbaum, G.E.; Boyer, R.R.; Mane, S.P.; Ponder, M.A. Changes in spinach phylloepiphytic bacteria communities following minimal processing and refrigerated storage described using pyrosequencing of 16S rRNA amplicons. *J. Appl. Microbiol.* **2011**, *110*, 1203–1214.

43. Rastogi, G.; Sbodio, A.; Tech, J.J.; Suslow, T.V.; Coaker, G.L.; Leveau, J.H.J. Leaf microbiota in an agroecosystem: Spatiotemporal variation in bacterial community composition on field-grown lettuce. *ISME J.* **2012**, *6*, 1812–1822.

44. Handschur, M.; Pinar, G.; Gallist, B.; Lubitz, W.; Haslberger, A.G. Culture free DGGE and cloning based monitoring of changes in bacterial communities of salad due to processing. *Food Chem. Toxicol.* **2005**, *43*, 1595–1605.

45. Oliveira, M.; Usall, J.; Viñas, I.; Anguera, M.; Gatius, F.; Abadias, M. Microbiological quality of fresh lettuce from organic and conventional production. *Food Microbiol.* **2010**, *27*, 679–684.

46. Caporaso, J.G.; Lauber, C.L.; Walters, W.A.; Berg-Lyons, D.; Huntley, J.; Fierer, N.; Owens, S.M.; Betley, J.; Fraser, L.; Bauer, M.; *et al.* Ultra-high-throughput microbial community analysis on the Illumina HiSeq and MiSeq platforms. *ISME J.* **2012**, *6*, 1621–1624.

47. Kozich, J.J.; Westcott, S.L.; Baxter, N.T.; Highlander, S.K.; Schloss, P.D. Development of a dual-index sequencing strategy and curation pipeline for analyzing amplicon sequence data on the MiSeq Illumina sequencing platform. *Appl. Environ. Microbiol.* **2013**, *79*, 5112–5120.

48. Kembel, S.W.; O'Connor, T.K.; Arnold, H.K.; Hubbell, S.P.; Wright, S.J.; Green, J.L. Relationships between phyllosphere bacterial communities and plant functional traits in a neotropical forest. *Proc. Nat. Acad. Sci. USA* **2014**, *111*, 13715–13720.

49. Xie, W.Y.; Su, J.Q.; Zhu, Y.G. Phyllosphere bacterial community of floating macrophytes in paddy soil environments as revealed by Illumina high-throughput sequencing. *Appl. Environ. Microbiol.* **2015**, doi:10.1128/AEM.03191-14.

50. Ottesen, A.R.; Gorham, S.; Pettengill, J.B.; Rideout, S.; Evans, P.; Brown, E. The impact of systemic and copper pesticide applications on the phyllosphere microflora of tomatoes. *J. Sci. Food Agric.* **2015**, doi:10.1002/jsfa.7010.

51. Schloss, P.D.; Westcott, S.L.; Ryabin, T.; Hall, J.R.; Hartmann, M.; Hollister, E.B.; Lesniewski, R.A.; Oakley, B.B.; Parks, D.H.; Robinson, C.J.; *et al.* Introducing mothur: Open-source, platform-independent, community-supported software for describing and comparing microbial communities. *Appl. Environ. Microbiol.* **2009**, *75*, 7537–7541.

52. Schloss, P.D.; Gevers, D.; Westcott, S.L. Reducing the effects of PCR-amplification and sequencing artifacts on 16S rRNA-based studies. *PLoS ONE* **2011**, doi:10.1371/journal.pone.0027310.

53. Quast, C.; Pruesse, E.; Yilmaz, P.; Gerken, J.; Schweer, T.; Yarza, P.; Peplies, J.; Glöckner, F.O. The SILVA ribosomal RNA gene database project: Improved data processing and web-based tools. *Nucleic Acids Res.* **2013**, *41*, 590–596.

54. Edgar, R.C.; Haas, B.J.; Clemente, J.C.; Quince, C.; Knight, R. UCHIME improves sensitivity and speed of chimera detection. *Bioinformatics* **2011**, *27*, 2194–2200.

55. DeSantis, T.Z.; Hugenholtz, P.; Larsen, N.; Rojas, M.; Brodie, E.L.; Keller, K.; Huber, T.; Dalevi, D.; Hu, P.; Andersen, G.L. Greengenes, a chimera-checked 16S rRNA gene database and workbench compatible with ARB. *Appl. Environ. Microbiol.* **2006**, *72*, 5069–5072.

56. MiSeq SOP. Available online: http://www.mothur.org/wiki/MiSeq_SOP (accessed on 3 October 2015).

57. Lundin, D.; Severin, I.; Logue, J.B.; Östman, Ö.; Andersson, A.F.; Lindström, E.S. Which sequencing depth is sufficient to describe patterns in bacterial α- and β-diversity? *Environ. Microbiol. Rep.* **2012**, *4*, 367–372.

58. Hayashi, H.; Shibata, K.; Sakamoto, M.; Tomita, S.; Benno, Y. *Prevotella copri* sp. nov. and *Prevotella stercorea* sp. nov., isolated from human faeces. *Int. J. Syst. Evol. Microbiol.* **2007**, *57*, 941–946.

59. Kageyama, A.; Benno, Y. *Catenibacterium mitsuokai* gen. nov., sp. nov., a gram-positive anaerobic bacterium isolated from human faeces. *Int. J. Syst. Evol. Microbiol.* **2000**, *50*, 1595–1599.

60. Watanabe, Y.; Nagai, F.; Morotomi, M. Characterization of *Phascolartcobacterium succinatutens* sp. nov., and asaccharolytic, succinate-utilizing bacterium isolated from human feces. *Appl. Environ. Microbiol.* **2012**, *78*, 511–518.

61. Williams, T.R.; Moyne, A.L.; Harris, L.J.; Marco, M.L. Season, irrigation, leaf age, and *Escherichia coli* inoculation influence the bacterial diversity in the lettuce phyllosphere. *PLoS ONE*, **2013**, *8*, e68642, doi:10.1371/journal.pone.0068642.

62. Matos, A.; Garland, J.L. Effects of community *versus* single strain inoculants on the biocontrol of *Salmonella* and microbial community dynamics in alfalfa sprouts. *J. Food Prot.* **2005**, *68*, 40–48.

63. Cooley, M.B.; Chao, D.; Mandrell, R.E. *Escherichia coli* O157:H7 survival and growth on lettuce is altered by the presence of epiphytic bacteria. *J. Food Prot.* **2006**, *69*, 2329–2335.

64. Heaton, J.C.; Jones, K. Microbial contamination of fruit and vegetables and the behavior of enteropathogens in the phyllosphere: A review. *J. Appl. Microbiol.* **2007**, *104*, 613–626.

12

Evaluation of Single or Double Hurdle Sanitizer Applications in Simulated Field or Packing Shed Operations for Cantaloupes Contaminated with *Listeria monocytogenes*

Cathy C. Webb *, Marilyn C. Erickson, Lindsey E. Davey and Michael P. Doyle

Center for Food Safety, Department of Food Science and Technology, 1109 Experiment Street, University of Georgia, Griffin, GA 30223, USA; E-Mails: mericks@uga.edu (M.C.E.); ledavey@uga.edu (L.E.D.); mdoyle@uga.edu (M.P.D.)

* Author to whom correspondence should be addressed; E-Mail: ccwebb@uga.edu

Academic Editor: Pascal Delaquis

Abstract: *Listeria monocytogenes* contamination of cantaloupes has become a serious concern as contaminated cantaloupes led to a deadly outbreak in the United States in 2011. To reduce cross-contamination between cantaloupes and to reduce resident populations on contaminated melons, application of sanitizers in packing shed wash water is recommended. The sanitizing agent of 5% levulinic acid and 2% sodium dodecyl sulfate (SDS) applied as a single hurdle in either a simulated dump or dip treatment significantly reduced *L. monocytogenes* to lower levels at the stem scar compared to a simulated dump treatment employing 200 ppm chlorine; however pathogen reductions on the rind tissue were not significantly different. Double hurdle approaches employing two sequential packing plant treatments with different sanitizers revealed decreased reduction of *L. monocytogenes* at the stem scar. In contrast, application of sanitizers both in the field and at the packing plant led to greater *L. monocytogenes* population reductions than if sanitizers were only applied at the packing plant.

Keywords: levulinic acid; sodium dodecyl sulfate; cantaloupes; stem scar; *Listeria monocytogenes*

1. Introduction

Listeria monocytogenes has been identified as a pathogen of concern for the cantaloupe industry. In 2011, Colorado (USA) grown cantaloupes were determined to be contaminated with *L. monocytogenes* present on harvesting equipment and in the packing facility [1]. An ensuing outbreak resulted in 147 illnesses, 1 miscarriage, and 33 deaths [2]. Its widespread prevalence is another cause for concern. A study of five farms in New York State revealed the prevalence of *L. monocytogenes* in produce fields was 15% compared to 4.6% and 2.7% for *Salmonella* and Shiga toxin-producing *Escherichia coli*, respectively [3]. *L. monocytogenes* has been detected in sewage, water, soil, vegetation, silage, in domestic and wild animals, and in food processing plants [4].

Given the potential for *L. monocytogenes* contamination in cantaloupe packing facilities, the application of sanitizers in such environments is critical. Currently, many processing facilities utilize sanitizers (primarily chlorine or chlorine dioxide) in dump tanks that may contain more than several thousand liters of water. The water in these tanks serves both as a vehicle for unloading and moving the melons onto conveyor belts as well as to wash debris from the product's surface. However, to prevent cross-contamination of contaminated to uncontaminated melons during this stage of processing, it is recommended that the water be chlorinated at a rate of 150 ppm of free chlorine [5]. Reduction of pathogens on cantaloupe surfaces may occur during exposure of the melons to chlorine for nearly 10 min in these tanks, but it is minimal (1 to 2 log). Hence, alternative sanitizers have been evaluated. Rodgers *et al.* [6] reported greater than 5-log colony forming units (CFU)/g reductions of *L. monocytogenes* on inoculated cantaloupes treated for 5 min with peracetic acid (80 ppm), chlorinated trisodium phosphate (100 and 200 ppm), chlorine dioxide (3 and 5 ppm), and ozone (3 ppm). In another study, a chlorine dioxide gas (5 ppm) treatment for 10 min reduced *L. monocytogenes* levels by 4.3 log CFU/5 cm^2 [7]. Unfortunately, sanitizers such as chlorine dioxide and ozone must be generated on site and used in concentrations less than 3 ppm to ensure worker safety in processing facilities [8]. Hence, plant-derived antimicrobials, such as carvacrol, thymol, β-resorcylic acid, and caprylic acid, that would not jeopardize worker safety have been studied for their potential as a sanitizing agent in the presence or absence of hydrogen peroxide and have been shown to reduce *L. monocytogenes* levels from 2.5 to 6 log CFU/cm^2 on cantaloupe rinds [9]. Unfortunately, application in a commercial dump tank could require large quantities of the antimicrobial. Insertion of a smaller volume dip tank (<1000 liters) into the packing line where the melons would be exposed to the antimicrobial for a short period of time prior to their being sorted and packed could be a long-term, cost-effective alternative and one that would not reduce the packing line speed. However, additional up-front costs would be required for purchasing and incorporating these dip tanks into packing lines and an alternative mode for unloading melons in the absence of a dump tank would need to be devised and implemented.

Levulinic acid and sodium dodecyl sulfate (SDS) are generally recognized as safe food additives by the United States Food and Drug Administration (FDA) for specific purposes and have the potential to be produced in large quantities at low cost [10,11]. Previous studies have revealed that these combined chemicals can reduce pathogenic bacteria on lettuce, alfalfa seed, and tomatoes [12–15]. Based on the potential low cost and demonstrated efficacy of these chemicals, they have recently been investigated as a sanitizing agent in cantaloupe wash water. In that investigation, 2% levulinic acid in combination with 0.2% SDS in cantaloupe wash water reduced *Salmonella* Poona populations by 3.4 and 4.5 log CFU/g on netted rind tissue after a 6 min simulated dump tank or dump tank with brushing treatment, respectively, compared to reductions of 1.5 and 2.6 log CFU/g when chlorine (120 ppm) was used as the sanitizing agent under those same treatment conditions [16].

Although single hurdle interventions would be desirable from the standpoint of reduced cost and processing time, adoption of double hurdle interventions could be potentially advantageous if their implementation were to generate additive or synergistic reductions to pathogen populations on contaminated cantaloupes. Therefore, the purpose of this study was to determine the fate of *L. monocytogenes* on stem scar and netted rind tissues when cantaloupes were exposed to sanitizers in single or double hurdle applications. Single hurdle approaches included a simulated dump tank or 1-min dip treatment, whereas the double hurdle approach included either: (1) a simulated dump tank treatment followed by a simulated dip treatment; or (2) a field-applied injection and spray treatment followed by the simulated dump tank treatment. During these treatments, several sanitizers (*i.e.*, chlorine, chlorine dioxide, and levulinic acid and SDS) were compared for their efficacy.

2. Results and Discussion

2.1. Effectiveness of Sanitizers for Reduction of L. monocytogenes from Cantaloupe Stem Scar and Rind Tissue

Sanitizers (chlorine and chlorine dioxide) commonly used in dump tanks by cantaloupe growers and processors were compared with 5% levulinic acid/2% SDS for their efficacy in reducing *L. monocytogenes* populations on cantaloupe surface tissue. *L. monocytogenes* populations in stem scar and rind tissue of cantaloupes exposed for 8 min to chlorine (200 ppm) or chlorine dioxide (3 ppm) in a simulated dump tank were not statistically different from non-treated (control) melons (Tables 1 and 2). In contrast, *L. monocytogenes* populations on both stem scar and netted rind tissues decreased significantly by 2.4 log CFU/sample when cantaloupes were treated in a simulated dump tank containing 5% levulinic acid/2% SDS compared to the non-treated control melons (Tables 1 and 2). In a previous study, greater reductions had been observed with levulinic acid and SDS compared to chlorine for netted rind tissue contaminated with *Salmonella* Poona, but in that case, the effectiveness of both chemicals in reducing *Salmonella* populations was much greater than seen here with *L. monocytogenes* [16].

2.2. Fate of Surviving L. monocytogenes during Storage at 4 °C Following Sanitizer Treatment

Storage at 4 °C slows or inhibits the growth of most bacteria; however, the psychrotrophic characteristics of *L. monocytogenes* allow it to grow in cold temperatures [17,18]. Hence, cantaloupes treated with chlorine, chlorine dioxide, and levulinic acid/SDS were analyzed for *L. monocytogenes* after short-term (3-day period representing transport time to market) and long-term (15-day period representing melons held refrigerated for duration of shelf life) storage. Growth of *L. monocytogenes* did not occur in any of the treated or non-treated cantaloupes during storage. In fact, populations on treated and non-treated Day 3 stem scar samples were statistically less than found on Day 0 samples and treated Day 3 rind samples were statistically less than Day 0 samples (Tables 1 and 2). With continued storage to 15 days, further decreases in populations were statistically significant for treated rind samples but not for stem scar samples. As population decreases during storage were greater when the cantaloupes had been treated with either chlorine or levulinic acid/SDS, it may be presumed that for many of the cells not inactivated by these chemicals, the tolerance of the pathogen to stresses encountered during storage was reduced and contributed to further inactivation of *L. monocytogenes*.

Table 1. *Listeria monocytogenes* populations on cantaloupe stem scar surface tissue after sanitizer treatment and storage at 4 °C for 0, 3 or 15 days.

Sanitizer [b]	log CFU/sample [a] (log reduction compared to Day 0 control)		
	Day 0	Day 3	Day 15
None (control) [c]	6.5 ± 0.6 A	4.6 ± 0.7 CD (1.9)	5.6 ± 0.3 A–C (0.9)
Chlorine, 200 ppm	5.6 ± 1.0 A–C (0.9)	3.5 ± 1.0 E (3.0)	4.2 ± 0.4 DE (2.3)
Chlorine dioxide, 3 ppm	6.3 ± 0.4 AB (0.2)	4.8 ± 0.2 CD (1.7)	5.4 ± 0.4 BC (1.1)
5% Levulinic acid/2% SDS	4.1 ± 1.5 DE (2.4)	2.3 ± 0.8 F (4.2)	2.3 ± 1.8 F (4.2)

[a] Mean ± SD per 2.5 cm² sample. Each value was derived from two replicate trials and in each trial $n = 3$. Mean ± SD not followed by the same letter are statistically different ($p < 0.05$); [b] Cantaloupes exposed to sanitizer for 8 min in simulated dump tank; [c] Control, non-treated.

Table 2. *Listeria monocytogenes* populations on cantaloupe rind surface tissue after sanitizer treatment and storage at 4 °C for 0, 3 or 15 days.

Sanitizer [b]	log CFU/sample [a] (log reduction compared to Day 0 control)		
	Day 0	Day 3	Day 15
None (control) [c]	4.4 ± 1.9AB	3.4 ± 1.7 BC (1.0)	2.1 ± 1.7 C–E (2.3)
Chlorine, 200 ppm	3.6 ± 1.2 B (0.8)	2.2 ± 1.3 CE (2.2)	0.1 ± 0.2 G (4.3)
Chlorine dioxide, 3 ppm	5.2 ± 0.4 A (−0.8)	3.3 ± 0.8 B–D (1.1)	1.6 ± 1.0 EF (2.8)
5% Levulinic acid/2% SDS	2.0 ± 1.4 DE (2.4)	0.1 ± 0.3 G (4.3)	0.4 ± 0.6 FG (4.0)

[a] Mean ± SD per 2.5 cm² sample. Each value was derived from two replicate trials and in each trial $n = 3$. Mean ± SD not followed by the same letter are statistically different ($p < 0.05$); [b] Cantaloupes exposed to sanitizer for 8 min in simulated dump tank; [c] Control, non-treated.

2.3. Detection of L. monocytogenes in Cantaloupe Flesh after Simulated Dump Tank Treatment

The immersion of cantaloupes into dump tanks after harvest has been described by Richards and Beuchat [19] as a potential route for infiltration of pathogens into internal tissues. Consequently, the grower practice of sanitizing cantaloupes in tanks using ground water (20 to 22 °C) plus sanitizer was simulated to determine if surface-inoculated *L. monocytogenes* could infiltrate stem scar and netted rind tissue. *L. monocytogenes* was isolated in both positive controls (inoculated but non-immersed) as well as cantaloupes immersed in sanitizer solutions with more *Listeria* found in internal stem scar tissue (17%–28%) than in internal tissue beneath the rind (6%–11%, Table 3). Only 3 of 144 surface samples (2.0%) tested *L. monocytogenes*-positive by enrichment culture, but in all those cases, the corresponding flesh sample tested *L. monocytogenes*-negative (Table 3). Hence, steaming was considered an effective method for inactivating pathogens on the surface and preventing cross-contamination during sampling of internal flesh samples. In terms of the mode by which the pathogen reaches the internal tissues, the similar frequency of internalized contamination observed between non-immersed and immersed melons for each type of tissue argues against immersion in dump tanks as being a likely scenario in this study (Table 3). Hence, contamination of internal tissue likely occurred during inoculation through passive diffusion of the liquid inoculum. Moreover, the increased prevalence of infiltration of *Listeria* into stem scar tissue compared to rind tissue could likely be attributed to the greater porosity of the stem scar tissue compared to the rind tissue. Following harvest of cantaloupes, physiological changes by the cantaloupe tissue may occur as the fruit encounters different temperatures. Thus during the commercial practice of holding melons in shaded staging areas, the pore size may shrink, in which case its subsequent transfer to the dump tank could lead to a decreased possibility of infiltration of pathogens through the stem scar.

Table 3. Presence of *Listeria monocytogenes* in cantaloupe flesh after simulated dump tank treatment and storage for 0, 3, and 15 days at 4 °C [a,b].

Sanitizer [c]	# *L. monocytogenes* Positive by Enrichment Culture/# Samples Analyzed			
	Stem scar tissue w/o surface layer [d]	Stem scar surface layer [d]	Rind tissue w/o surface layer [d]	Rind surface layer [d]
None (control)	5/18	0/18	2/18	0/18
Chlorine, 200 ppm	3/18	2/18 [e]	2/18	0/18
Chlorine dioxide, 3 ppm	5/18	1/18 [e]	1/18	0/18
5% Levulinic acid/2% SDS	5/18	0/18	1/18	0/18

[a] Two replicate trials were performed with $n = 18$. Day 0, 3, and 15 day storage data combined; [b] No positive values were obtained for non-inoculated, non-treated controls, data not shown; [c] Cantaloupes treated for 8 min in sanitizer solution, then held for 5 min at room temperature before either storing at 4 °C or analyzing the sample; [d] Cantaloupes were subjected to a 6-min steam treatment to inactivate pathogens on the surface. Stem scar and rind pieces were then immediately removed using a sterile knife and divided into flesh and surface samples of 3.13 cm^3 each; [e] *L. monocytogenes*-positive surface samples in this column corresponded to *L. monocytogenes*-negative flesh samples, indicating no cross contamination during sample preparation.

2.4. Comparison of Single Hurdle to Double Hurdle Approaches Incorporating Sanitizer Treatments

Effectiveness of both single and double hurdle sanitizer treatments was greater on rind tissue than stem scar tissue in that *L. monocytogenes* populations in treated stem scar samples were still large enough to be detected by plate count enumeration but the pathogen could only be detected by enrichment culture in treated rind samples (Table 4). Similar levels of prevalence of the pathogen was found in treated rind samples implying that double and single hurdle treatments were equally effective. In the case of stem scar samples, the chlorine tank treatment (200 ppm) resulted in only a 1-log CFU/sample reduction of *L. monocytogenes* compared to non-treated cantaloupes. However, in comparison to non-treated cantaloupes, treatments incorporating 5% levulinic acid/2.5% SDS reduced *L. monocytogenes* in stem scar tissue by 3.4 and 1.4 log CFU/sample in the single and double hurdle approaches, respectively (Table 4). The decreased effectiveness of the double hurdle approach suggests that chlorine altered the pathogen or tissue in some manner as to diminish the effectiveness of the levulinic acid and SDS dip treatment. One possibility could include activation of the pathogen's defenses by chlorine. Alternatively, chlorine may react with the stem scar tissue to make it less porous and accessible for levulinic acid and SDS to penetrate and reach the pathogen. In any event, incorporation of a levulinic acid/SDS dip treatment as a second hurdle would likely have only a minimal impact on reducing *Listeria* on cantaloupes in commercial operations employing chlorine dump tanks. As was the case with stem scar tissue, there was no advantage to employing a double hurdle rather than a single hurdle approach to eliminating the pathogen from rind tissue.

Table 4. *Listeria monocytogenes* recovered from cantaloupe stem scar and rind tissue [a] following treatment with sanitizers in single hurdle versus double hurdle approach [b].

Hurdle	Sanitizer [d]	Log *L. monocytogenes* CFU/ sample [c] or # positive by enrichment culture/# samples (log reduction compared to control)	
		Stem Scar [e]	Netted Rind [f]
None	None (control)	4.4 ± 1.0 C	4.4 ± 1.4
Single	200 ppm chlorine in dump tank	3.3 ± 1.5 BC (1.1)	38/41
	5% levulinic acid/2.5% SDS in dip tank	1.0 ± 1.4 A (3.4)	34/41
Double	200 ppm chlorine in dump tank followed by 5% levulinic acid/2.5% SDS in dip tank	3.0 ± 0.8 B (1.4)	35/41

[a] Stem scar and netted rind samples are 3.13 cm^3, two replicate trials performed; [b] No positive values were obtained for non-inoculated, non-treated controls, data not shown; [c] Mean ± SD, stem scar means not followed by the same letter are significantly different ($p < 0.05$); [d] Dump tank treatments were for 10 min, whereas dip treatments were for 1 min; [e] The number of samples analyzed was $n = 6$ (control) and $n = 8$ (single/double hurdle treatments); [f] The number of samples analyzed was $n = 24$ (control) and $n = 41$ (single/double hurdle treatments).

2.5. Stem Scar Injection and Spray Treatment of Cantaloupes Prior to Inoculation and Simulated Dump Tank Treatment

An injection treatment of cantaloupe stem scar tissue with 200 μL 7.5% levulinic acid/0.5% SDS combined with a spray sanitizer treatment of 30 mL 7.5% levulinic acid/0.5% SDS administered to the entire cantaloupe surface immediately after harvest was evaluated as a means to prevent subsequent cross-contamination of freshly harvested cantaloupes during transport to the packing shed. Therefore, after application of the field sanitizer treatments, cantaloupes were spot- or soil-inoculated to simulate contact with contaminated liquids or surfaces on transport trailers. Moreover, in this set of experiments, cultures were prepared from solid media compared to liquid media as studies conducted by Uesegi *et al.* and Theofel and Harris [20,21] and preliminary studies conducted in our laboratory have demonstrated that cultures prepared from solid media were more resistant to desiccation than liquid cultures. Once inoculated, the cantaloupes were held for a short period before subjecting them to the dump tank washing hurdle, typically applied by growers in the southeastern United States, using either 200 ppm chlorine or 5% levulinic acid/2% SDS.

L. monocytogenes populations on rind samples of soil-inoculated cantaloupes were not significantly different for melons receiving either the chlorine or levulinic acid/SDS treatment as their second hurdle (Table 5). In contrast, both rind samples from spot-inoculated cantaloupes and stem scar samples from either spot- or soil-inoculated cantaloupes were significantly less when the second hurdle employed a levulinic acid/SDS treatment as opposed to a chlorine treatment ($p < 0.05$). To determine whether any antimicrobial activity had occurred by the first hurdle employing the levulinic acid/SDS injection and spraying, comparisons of the net reductions in this experiment to the previous experiments employing a levulinic acid/SDS (Table 1) or a chlorine (Tables 1 and 4) dump tank were made. Stem scar and rind samples averaged net reductions of more than 1.5 and 2.0 fold greater, respectively, for the double hurdle approach compared to the single hurdle. These comparisons would therefore suggest that both stem scar injection and a field spray treatment with levulinic acid/SDS did exert some antimicrobial activity. These results are in agreement where *Salmonella* inactivation was observed following vacuum diffusion of sanitizers into tomato stem scar tissue [22].

Table 5. *Listeria monocytogenes* populations on stem scar and rind tissues of non-treated cantaloupes and cantaloupes treated by injection of stem scar and spraying of cantaloupes [a] prior to inoculation [b] and then exposing the cantaloupes to a sanitizer as a second hurdle [c].

2nd Hurdle Sanitizer [e]	Log CFU/sample [d] (log reduction compared to control)			
	Spot Inoculation		Soil Inoculation	
	Stem Scar	Rind	Stem Scar	Rind
None (control) [f]	6.7 ± 0.2 A	7.2 ± 0.3 A	5.8 ± 0.4 A	5.5 ± 0.5 A
200 ppm Chlorine	5.3 ± 0.2 B (1.4)	2.6 ± 1.4 B (4.6)	3.9 ± 1.0 B (1.9)	0.4 ± 0.9 B (5.1)
5% levulinic acid/2% SDS	3.9 ± 1.2 C (2.8)	1.4 ± 1.5 C (5.8)	2.2 ± 1.6 C (3.6)	0.3 ± 0.7 B (5.2)

[a] Stem scars were injected with 200 μL of 7.5% levulinic acid/1.0% SDS after which the entire cantaloupe was sprayed with *ca.* 30 mL of 7.5% levulinic acid/0.5% SDS; [b] Solid media preparation used for inoculation; [c] No positive values were obtained for non-inoculated, non-treated controls; [d] Mean ± S.D per 3.13 cm^3 sample. Within each column, means not followed by the same letter are significantly different ($p < 0.05$). Four replicate trials occurred with $n = 5$ for each trial; [e] Cantaloupes were treated in simulated dump tanks for 10 min; [f] Positive controls were inoculated but did not receive a first or second hurdle treatment.

3. Experimental Section

3.1. Cantaloupes

Freshly harvested Eastern variety cantaloupes (*Cucumis melo* L. var. *reticulatus* cv. Athena) were obtained from a commercial grower in Tifton, GA. The melons were chosen from the transport trailers prior to washing and packing, to be of similar maturity, size, degree of netting, and free of any visible blemishes.

3.2. Bacterial Strains

Five strains of *Listeria monocytogenes* (2011L-2624, 2011L-2625, 2011L-2626, 2011L-2663, and 2011L-2676) isolated from patients who had consumed cantaloupe involved in a 2011 outbreak, were obtained from the Centers for Disease Control and Prevention. Each strain was transformed with plasmid pNF8 [23] containing genes to produce green fluorescence and erythromycin resistance [24]. The resulting *L. monocytogenes* pNF8 strains LD22 (2011L-2625), LD23 (2011L-2626), LD24 (2011L-2663), and LD25 (2011L-2676) produced bright green colonies under a Dark Reader trans-illuminator at ~500 nm (Clare Chemical, Dolores, CO, USA). *L. monocytogenes* wild type and pNF8 strains were stored at −80 °C in brain heart infusion broth (BHIB) (Neogen, Lansing, MI, USA) with 25% glycerol. Strains from frozen stock were struck on to brain heart infusion agar (BHIA) (Neogen) or BHIA supplemented with 8 μg/mL erythromycin (MP Biomedicals, LLC, Santa Ana, CA, USA), BHIAE, and incubated at

37 °C for 18–21 h. *L. monocytogenes* wild-type or brightly glowing green colonies were re-streaked on BHIA or BHIAE, respectively, and incubated at 37 °C for 18–21 h.

3.3. Preparation of Liquid- and Soil-based Inoculum

Wild-type *L. monocytogenes* strains were used for the simulated dump tank storage studies due to the potential instability of the pNF8 plasmid during storage where potential proliferation could occur [23], whereas *L. monocytogenes* pNF8-containing strains were used in all non-storage studies for ease of detection and limited plasmid segregation pressure after short-term exposure to cantaloupe surfaces.

Preliminary experiments revealed that *L. monocytogenes* inoculum prepared from solid media plates was more resistant to sanitizer treatment after a 2-h incubation prior to sampling, therefore solid media-prepared cultures were used for spot and soil inoculation in the field spray treatment study (Table 5). Liquid-media prepared cultures were used for all other studies because preliminary experiments revealed no difference in the two culture preparations inoculated on cantaloupes and incubated for 16 to 18 h prior to treatment and sampling (data not shown).

Inoculum preparation was initiated by taking either one or two colonies of each wild-type strain to individually inoculate 50 mL of BHIB (liquid media) or one or two *L. monocytogenes* pNF8 colonies to inoculate either 50 mL of BHIB supplemented with 8 µg/mL erythromycin, BHIBE (liquid media) or 3 BHIAE plates (solid media). The liquid media cultures were incubated at 37 °C with agitation (150 rpm) for 21–24 h, and the solid media plates were incubated for 24 h at 37 °C. After incubation, 4 mL of 0.1% peptone water was added to each solid media plate, the colonies were gently removed with a glass spreader, collected in a 50-mL centrifuge tube, and the process was repeated to dislodge remaining cells from the plate. Both liquid media cultures and suspended solid media cultures were sedimented by centrifugation ($4193 \times g$ for 15 min at 4 °C) and washed two times in sterile 0.1% peptone water. *L. monocytogenes* wild-type strains were resuspended in 45 mL of sterile water, combined in equal proportions, and an additional 50 mL of sterile water was added to make a 5-strain mixture of *ca.* 9 log CFU/mL. *L. monocytogenes* pNF8 strains were resuspended in 3 mL of 0.1% peptone water and combined in equal proportions to make a 5-strain mixture of *ca.* 10 log CFU/mL.

Soil was collected from the cantaloupe farm and sifted to remove rocks and other large debris. A portion (200 g) was placed in a 3.07 L Glad container (The Clorox Company, Oakland, CA, USA) and 4 mL of 10 log CFU/mL of the *L. monocytogenes* pNF8 (solid media culture) mixture was applied in a fine mist spray. The inoculated soil was mixed for 1 min with a spoon, covered, and held 18 h in the dark at room temperature for pathogen acclimation. *L. monocytogenes* concentration in the inoculated soil ranged from 7.11 to 8.46 log CFU/g.

3.4. Preparation of Sanitizing Solutions

Chlorine (200 ppm) was prepared by adding 40 mL of sodium hypochlorite solution containing 5% available chlorine (Ricca Chemical Company, Arlington, TX, USA) with 10 L of sterile deionized water. The chlorine solution was adjusted to pH 7.0 with sulfuric acid (Sigma-Aldrich, St. Louis, MO, USA). Free chlorine was determined with the Hach digital titrator using the DPD (*N,N*-diethyl-p-phenylenediamine)-ferrous ethylenediammonium sulfate titration cartridge (Hach Co., Loveland, CO, USA).

A chlorine dioxide stock solution was prepared by dissolving 4 g of Aqua-Tab (Beckart Environmental, Inc., Kenosha, WI, USA) in 1L of sterile deionized water. The chlorine dioxide stock was further diluted in sterile deionized water to prepare three 10-L batches of a 3 ppm solution, pH 4.36, per trial. Chlorine dioxide concentrations were measured using a Chlorine Dioxide Pocket Colorimeter™ II (Hach, Co. Loveland, CO, USA).

Levulinic acid (98%, Acros Organics, Fair Lawn, NJ, USA) and sodium dodecyl sulfate (SDS, 20%, Acros Organics) were combined with sterile deionized water to make 10 liters of 5% levulinic acid/2.0% SDS, pH 2.84, for dump tank treatments; 4 liters of 5.0% levulinic acid/2.5% SDS, pH 2.85, for dip treatments; 2 liters of 7.5% levulinic acid/0.5% SDS, pH 2.70, for spray treatments; and 200 mL of 7.5% levulinic acid/0.5% SDS, pH 2.71, for injection treatments.

All sanitizer solutions were prepared daily for each type of treatment and for each replicate trial. The temperature of all sanitizer solutions was 20 to 22 °C at the time of treatment.

3.5. Simulated Dump Tank Sanitizer Treatment of Cantaloupes and Subsequent Storage

Cantaloupes were spot inoculated, using a wild-type *L. monocytogenes* liquid medium cultured mixture by applying 100 μL of either an 8-log CFU/mL stock inoculum within a 2-cm diameter circle drawn on the netted rind, or a 7-log CFU/mL stock inoculum on the stem scar. After inoculation, the melons were held for 16 to 18 h at 22 °C. Inoculated melons for treatment were floated in 10 liters of 200 ppm chlorine, 3 ppm chlorine dioxide or 5% levulinic acid/2.0% SDS in a 53-L (51 × 35 × 30 cm) storage box (Roughneck, Rubbermaid Home Products, Fairlawn, OH, USA). The melons were held for a total of 8 min with the inoculation zones on stem scar and rind surface submerged the entire time, placed in 354-mL foam bowls (Walmart, Bentonville, AR, USA), and held for 5 min at 22 °C before analysis or storage. Positive control samples were not submerged in a sanitizer solution but were inoculated under conditions similar to treated cantaloupes. Two subsets of treated and non-treated melons were stored for 3 and 15 days at 4 °C and then analyzed. Two replicate trials were performed for each treatment. No positive values were obtained for non-inoculated, non-treated controls.

3.6. Sanitizer Treatment of Cantaloupes for the Single Hurdle or Double Hurdle Approach

Cantaloupes held at 4 °C for 48 and 72 h after harvest (replicate 1 and 2, respectively) were spot inoculated, using a *L. monocytogenes* pNF8 liquid medium cultured mixture, by applying 10 μL of either a 9-log CFU/mL stock inoculum within a 2-cm diameter circle drawn on the netted rind or an 8-log CFU/mL stock inoculum on the stem scar tissue. Inoculated cantaloupes were held for 16 h at 22 °C before treatment. Dip-treated cantaloupes were submerged for 1 min in 4 L of 5.0% levulinic acid/2.5% SDS held in 11.35-L pails and then placed in a foam bowl. Dump tank-treated melons were floated in 10 L of 200 ppm chlorine, as described in section 3.5, but with a 10-min treatment time. Cantaloupes designated for sequential dump tank and dip treatments first received the 10-min tank treatment in 200 ppm chlorine immediately followed by a 1 min dip treatment in 5.0% levulinic acid/2.5% SDS. All treated cantaloupes were held for 1 h prior to sample analysis. Non-treated melons (positive controls) were also inoculated and analyzed after holding for 16 h. Two replicate trials were performed for each treatment.

3.7. Stem Scar Injection and Spray Treatment of Cantaloupes Prior to Inoculation and Subsequent Simulated Dump Tank Treatment

Cantaloupes held at 22 °C for 20–24 h after harvest were injected at the stem scar with 0.2 mL of 7.5% levulinic acid/0.5% SDS at 60 psi using a P50 Microdose needle free injector fitted with a 3-stream nozzle (Pulse NeedleFree Systems, Lenexa, KS, USA) and attached to a CO_2 compressed gas canister. The stem scar-treated melons were immediately sprayed for 10 s with *ca.* 30 mL of 7.5% levulinic acid/0.5% SDS using a hand-held commercial garden sprayer (Flo Master Yard and Garden Sprayer, model 1401, Root-Lowell Manufacturing Co., Lowell, MI, USA). Stem scar- and spray-treated cantaloupes were placed in foam bowls and held at 22 °C for 2 h before either spot inoculation of cantaloupes with 100 μL of a 9 log (netted rind) and an 8 log (stem scar) CFU/mL *L. monocytogenes* pNF8 solid medium cultured mixture, or inoculation of cantaloupes with *L. monocytogenes* pNF8 solid medium cultured inoculated soil (2.5 g) pressed onto the stem scar and netted rind tissue (2-cm diameter circle) using waxed weighing paper (Thermo Fisher Scientific Inc., Waltham, MA, USA). Inoculated cantaloupes were held for 2 h at 22 °C prior to dump tank treatment of either 200 ppm of chlorine or 5% levulinic acid/2.0% SDS using the protocols described in Sections 3.5 and 3.6. The treated cantaloupes were held for 2 h at 22 °C prior to sample analysis. Non-treated melons (positive controls) were also inoculated and analyzed. Four replicate trials were performed for each treatment.

3.8. Sample Collection

Treatment and inoculation procedures were staggered to maintain equal time intervals before analysis of each melon. Control and treated melons (storage study only, Table 3) in which flesh beneath inoculated stem scar and netted rind tissue was to be analyzed, were first steam treated for 6 min in a 12.06 L stainless steel food steamer (Secura Inc., Brookfield, WI, USA) to remove residual wild-type *Listeria monocytogenes* from cantaloupe surfaces. Steamed cantaloupes were then placed in a sterile 3.79- or 7.58-L- Ziploc® bag (SC Johnson, Racine, WI, USA) containing 225 mL of 0.1% peptone water for non-treated control melons or sanitizer neutralizing buffers (0.1% buffered peptone water (Neogen, Lansing, MI, USA) for levulinic acid/SDS-treated melons or 0.1% peptone water supplemented with 0.01 g/liter of sodium thiosulfate (Sigma-Aldrich, St. Louis, MO, USA) for chlorine or chlorine dioxide-treated melons). The melons were massaged in the sealed bags by hand for 1 min, then were removed for sampling and the remaining buffer was collected for enrichment culture. Sampling of stored and non-stored cantaloupes involved placing cantaloupes on a sterile cutting board and using a stainless steel coring knife to cut out cores from the rind and stem scar inoculation sites. In the case of samples from non-stored cantaloupes, both the rind and stem scar samples were trimmed with a sterile knife into 3.13 cm³ samples (2.5 × 2.5 × 0.5 cm) which included surface tissue and flesh beneath the inoculation site before placing in a Whirl-Pak bag (Nasco, Fort Atkinson, WI, USA). In the case of samples from stored cantaloupes, the cored pieces were first separated into surface (2.5 cm² squares) or flesh (3.13 cm³) tissue, before placing in sterile Whirl-Pak bags. The surface and flesh samples collected during the storage study were weighed, and either sterile 0.1% peptone water or sanitizer neutralizing buffer was added at a 1:9 (wt/vol) ratio. Rind and stem scar surface samples collected in non-storage studies were combined either with 25 mL of 0.1% peptone water or neutralizing buffer.

3.9. Processing and Analysis of Cantaloupe Rind and Flesh Samples

Each sample was macerated in a Stomacher 400C (Seward Laboratory Systems, Port Saint Lucie, FL, USA) for 1 min at 260 rpm. The sample was directly plated, or a portion of the sample was diluted (1:9 wt/vol portions) in 0.1% peptone water and either 100 or 250 μL was plated in duplicate for enumeration of *L. monocytogenes* on modified Oxford agar (MOX) supplemented with 100 μg/mL sodium pyruvate (Fisher Bioreagents, Fair Lawn, NJ, (MOXP)) or MOXP supplemented with 8 μg/mL erythromycin (MOXPE) for wild-type *L. monocytogenes* and *L. monocytogenes* pNF8 strains, respectively. The MOX agar consisted of Oxford Listeria base (Neogen) and modified Oxford antimicrobic supplement (Becton, Dickinson and Company, Franklin Lakes, NJ, USA). The plates were incubated at 37 °C for 48 h. The limit of detection for directly plated samples was 2 log CFU/sample.

Wild-type *L. monocytogenes* samples (storage studies) were enriched according to the FDA Bacterial Analytical Manual protocol [25] and 2× BHIBE was added to *L. monocytogenes* pNF8 samples (non-storage studies) and incubated 24 h at 37 °C. The enrichment cultures (20 μL) were plated on MOXP or MOXPE plates and incubated at 37 °C for 48 h. The limit of detection by enrichment was 1 CFU per sample. Representative presumptive colonies of wild-type *L. monocytogenes* were streaked on tryptic soy agar (Neogen) and analyzed with API *Listeria* test kits (bioMérieux, Inc., Marcy l'Etoile, France) for confirmation.

3.10. Statistical Analysis

Data were analyzed by analysis of variance (ANOVA) using StatGraphics Centurion XVI statistical software package (Statpoint, Inc., Herndon, VA, USA). When statistical differences were observed ($p < 0.05$) with ANOVA, differences among sample means were determined using the least significant difference test at $p < 0.05$.

4. Conclusions

Using a single hurdle approach, treatment of cantaloupes with 5% levulinic acid/2% SDS either in a dump or dip tank provided greater reductions of *L. monocytogenes* than 200 ppm chlorine for stem scar tissue. In contrast, no significant differences were observed with these sanitizers on rind tissue samples. Regardless of sanitizer, it was revealed that stem scar-attached *L. monocytogenes* was more difficult to eliminate than rind-attached *L. monocytogenes*. This response was attributed to greater infiltration of the pathogen into stem scar tissue than rind tissue. A double hurdle approach employing a 200-ppm chlorine dump tank treatment followed by a 5% levulinic acid/2.5% SDS dip treatment did not provide any additional inactivation of *L. monocytogenes* on contaminated rind samples and actually had reduced efficacy compared to the dip treatment alone on contaminated stem scar samples. In contrast, a double hurdle approach employing field injection or spraying of 7.5% levulinic acid/0.5% SDS into stem scar and rinds, respectively, followed by exposure to either chlorine or levulinic acid/SDS in the packing plant dump tank appeared to have provided additional reductions of *L. monocytogenes* at stem scars and rinds compared to treatment of the cantaloupe with the dump tank sanitizer only. Based on these results, application of levulinic acid and SDS in the field could provide an additional hurdle to ensure the safety of the final product.

Acknowledgments

This research was funded by unrestricted gifts from the food industry. We thank Bill Brim, Ed Walker, Peter Germishuizen, Pablo A. Navia Giné, Philip Grimes, Jane Grimes, Alan Parrish, and Lynda Glenn for providing cantaloupes and invaluable information regarding cantaloupe growing and processing practices. We also thank Charles Hall and the Eastern Cantaloupe Growers Association.

Author Contributions

C.C.W., M.C.E. and M.P.D., conceived and designed the experiments. C.C.W., M.C.E. and L.E.D., performed the experiments. C.C.W. and M.C.E., analysed the data. C.C.W. and L.E.D., contributed reagents, material, and analysis tools. C.C.W. and M.C.E., wrote the paper.

Conflicts of Interest

The authors declare no conflict of interest.

References

1. U.S. Food and Drug Administration. Final Update on Multistate Outbreak of *Listeriosis* Linked to Whole Cantaloupes, 2012. Available online: http://www.fda.gov/Food/RecallsOutbreaks Emergencies/Outbreaks/ucm272372.htm#final (accessed on 7 February 2015).

2. McCollum, J.T.; Cronquist, A.B.; Silk, B.J.; Jackson, K.A.; O'Connor, K.A.; Cosgrove, S.; Gossack, J.P.; Parachini, S.S.; Jain, N.S.; Ettestad, P.; *et al.* Multistate outbreak of *Listeriosis* associated with cantaloupe. *N. Engl. J. Med.* **2013**, *369*, 944–953.

3. Strawn, L.K.; Fortes, E.D.; Bihn, E.A.; Nightingale, K.K.; Grohn, Y.T.; Worobo, R.W.; Wiedmann, M.; Bergholz, P.W. Landscape and meteorological factors affecting prevalence of three food-borne pathogens in fruit and vegetable farms. *Appl. Environ. Microbiol.* **2013**, *79*, 588–600.

4. Ivanek, R.; Groehn, Y.T.; Wiedmann, M. *Listeria monocytogenes* in multiple habitats and host populations: Review of available data for mathematical modeling. *Foodborne Pathog. Dis.* **2006**, *3*, 319–336.

5. Hurst, W.C. Good agricultural practices in the harvest, handling and packing of cantaloupes. In *Cantaloupe and Specialty Melons*; University of Georgia Cooperative Extension: Athens, GA, USA, 2014; pp. 20–23.

6. Rodgers, S.L.; Cash, J.N.; Siddiq, M.; Ryser, E.T. A comparison of different chemical sanitizers for inactivating *Escherichia coli* O157:H7 and *Listeria monocytogenes* in solution and on apples, lettuce, strawberries, and cantaloupe. *J. Food Prot.* **2004**, *67*, 721–731.

7. Mahmoud, B.S.M.; Vaidya, N.A.; Corvalan, C.M.; Linton, R.H. Inactivation kinetics of inoculated *Escherichia coli* O157:H7, *Listeria monocytogenes* and *Salmonella* Poona on whole cantaloupe by chlorine dioxide gas. *Food Microbiol.* **2008**, *25*, 857–865.

8. U.S. Food and Drug Administration. Part 173.300—Secondary Direct Food Additives Permitted in Food for Human Consumption. In *CFR-Code of Federal Regulations Title 21*; U.S. Government Printing Office: Washington, DC, USA, 2014. Available online: http://www.accessdata.fda.gov/scripts/cdrh/cfdocs/cfcfr/CFRSearch.cfm?fr=173.300 (accessed on 8 February 2015).

9. Upadhyay, A.; Upadhyaya, I.; Mooyottu, S.; Kollanoor-Johny, A.; Venkitanarayanan, K. Efficacy of plant-derived compounds combined with hydrogen peroxide as antimicrobial wash and coating treatment for reducing *Listeria monocytogenes* on cantaloupes. *Food Microbiol.* **2014**, *44*, 47–53.

10. Bozell, J.J.; Moens, L.; Elliott, D.C.; Wang, Y.; Neuenscwander, G.G.; Fitzpatrick, S.W.; Bilski, R.J.; Jarnefeld, J.L. Production of levulinic acid and use as a platform chemical for derived products. *Resour. Conserv. Recycl.* **2000**, *28*, 227–239.

11. Fang, Q.; Hanna, M.A. Experimental studies for levulinic acid production from whole kernel grain sorghum. *Bioresour. Technol.* **2002**, *81*, 187–192.

12. Zhao, T.; Zhao, P.; Cannon, J.L.; Doyle, M.P. Inactivation of *Salmonella* in biofilms and on chicken cages and preharvest poultry by levulinic acid and sodium dodecyl sulfate. *J. Food Prot.* **2011**, *74*, 2024–2030.

13. Zhao, T.; Zhao, P.; Doyle, M.P. Inactivation of *Salmonella* and *Escherichia coli* O157:H7 on lettuce and poultry skin by combinations of levulinic acid and sodium dodecyl sulfate. *J. Food Prot.* **2009**, *72*, 928–936.

14. Zhao, T.; Zhao, P.; Doyle, M.P. Inactivation of *Escherichia coli* O157:H7 and *Salmonella* Typhimurium DT 104 on alfalfa seeds by levulinic acid and sodium dodecyl sulfate. *J. Food Prot.* **2010**, *73*, 2010–2017.

15. Zhao, T.; Zhao, P.; Doyle, M.P. Inactivation of foodborne pathogens on tomatoes by levulinic acid plus sodium dodecyl sulfate. In Proceedings of the 12th ASEAN Food Conference 2011, BITEC Bangna, Bangkok, Thailand, 16–18 June 2011; pp. 416–424.

16. Webb, C.C.; Davey, L.E.; Erickson, M.C.; Doyle, M.P. Evaluation of levulinic acid and sodium dodecyl sulfate as a sanitizer for use in processing Georgia-grown cantaloupes. *J. Food Prot.* **2013**, *76*, 1767–1772.

17. Junttila, J.R.; Niemela, S.I.; Him, J. Minimun growth temperatures of *Listeria monocytogenes* and non-haemolytic *Listeria*. *J. Appl. Bacteriol.* **1988**, *65*, 321–327.

18. Walker, S.J.; Stringer, M.F. Growth of *Listeria monocytogenes* and *Aeromonas hydrophila* at chill temperatures. *J. Appl. Bacteriol.* **1987**, *63*, 20.

19. Richards, G.M.; Beuchat, L.R. Attachment of *Salmonella* Poona to cantaloupe rind and stem scar tissues as affected by temperature of fruit and inoculum. *J. Food Prot.* **2004**, *67*, 1359–1364.

20. Uesugi, A.R.; Danyluk, M.D.; Harris, L.J. Survival of *Salmonella enteritidis* phage type 30 on inoculated almonds stored at −20, 4, 23, and 35 °C. *J. Food Prot.* **2006**, *69*, 1851–1857.

21. Theofel, C.G.; Harris, L.J. Impact of preinoculation culture conditions on the behavior of *Escherichia coli* O157:H7 inoculated onto Romaine lettuce (*Lactuca sativa*) plants and cut leaf surfaces. *J. Food Prot.* **2009**, *72*, 1553–1559.

22. Gurtler, J.B.; Smelser, A.M.; Niemira, B.A.; Jin, T.Z.; Yan, X.; Geveke, D.J. Inactivation of *Salmonella enterica* on tomato stem scars by antimicrobial solutions and vacuum perfusion. *Int. J. Food Microbiol.* **2012**, *159*, 84–92.

23. Ma, L.M.L.; Zhang, G.D.; Doyle, M.P. Green fluorescent protein labeling of *Listeria*, *Salmonella*, and *Escherichia coli* O157:H7 for safety-related studies. *PLoS ONE* **2011**, *6*, e18083, doi:10.1371/journal.pone.0018083.

24. Fortineau, N.; Trieu-Cuot, P.; Gaillot, O.; Pellegrini, E.; Berche, P.; Gaillard, J.L. Optimization of green fluorescent protein expression vectors for *in vitro* and *in vivo* detection of *Listeria monocytogenes*. *Res. Microbiol.* **2000**, *151*, 353–360.

25. Hitchins, A.D.; Jinneman, K. *Detection and enumeration of Listeria monocytogenes in foods*; Food and Drug Administration: Silver Spring, MD, USA, 2011. Available online: http://www.fda.gov/Food/FoodScienceResearch/LaboratoryMethods/ucm071400.htm (accessed on 9 May 2014).

Permissions

List of Contributors

Ognjen Zurovec, Pål Olav Vedeld and Bishal Kumar Sitaula
Department of International Environment and Development Studies (Noragric), Norwegian University of Life Science (NMBU), Universitetstunet 3 1430 Ås, Norway

Dennis Wichelns
P.O. Box 2629, Bloomington, IN 47402, USA

Shreya Wani, Jagpreet K. Maker, Joseph R. Thompson, Jeremy Barnes and Ian Singleton
School of Biology, Newcastle University, Newcastle upon Tyne, NE1 7RU, UK

Mariangela Diacono
Consiglio per la Ricerca e l'analisi dell'economia Agraria, CRA-SCA, Research Unit for Cropping Systems in Dry Environments, Via Celso Ulpiani 5, 70125, Bari, Italy

Francesco Montemurro
Consiglio per la Ricerca e l'analisi dell'economia Agraria, CRA-SCA, Research Unit for Cropping Systems in Dry Environments (Azienda Sperimentale Metaponto), SS 106 Jonica, km 448.2, 75010, Metaponto (MT), Italy

Saskia M. van Ruth
RIKILT Wageningen UR, P.O. Box 230, 6700 EV Wageningen, The Netherlands
Food Quality and Design Group, Wageningen University, P.O. Box 17, 6700 AA Wageningen, The Netherlands
Ries de Visser
IsoLife B.V., P.O. Box 349, 6700 AH Wageningen, The Netherlands

Nicholas Sadgrove and Graham Jones
Pharmaceuticals and Nutraceuticals Group, Centre for Bioactive Discovery in Health and Ageing, University of New England, S & T McClymont Building UNE, Armidale NSW 2351, Australia

Upendra B. Pradhanang
Shankar Dev Campus, Faculty of Management, Tribhuvan University, Kathmandu 44600, Nepal

Soni M. Pradhanang
Department of Geosciences, University of Rhode Island, Kingston, RI 02881, USA

Arhan Sthapit
Department of Management, Public Youth Campus, Tribhuvan University, Kathmandu 44600, Nepal

Nir Y. Krakauer
Department of Civil Engineering, The City College of New York, City University of New York, New York, NY 10031, USA

Ajay Jha
Department of Horticulture and Landscape Architecture, College of Agricultural Sciences, Horticulture and Landscape Architecture, Colorado State University, Fort Collins, CO 80523, USA

Tarendra Lakhankar
NOAA-Cooperative Remote Sensing Science & Technology (CREST) Center, The City College of New York, City University of New York, New York, NY 10031 USA

Lydia C. Medeiros and Jeffrey T. LeJeune
Food Animal Health Research Program, Ohio Agricultural Research and Development Center, The Ohio State University, 1680 Madison Ave, Wooster, OH 44691, USA

Tirtha Bdr. Katwal
Specialist III-Maize, RNR Research and Development Center, Yusipang, Department of Forest and Park Services, Ministry of Agriculture and Forests, Thimphu, P.O. Box 212, Bhutan

Singay Dorji
National Coordinator, Global Environment Facility-Small Grants Programme, UNDP, Thimphu, P.O. Box 162, Bhutan

Rinchen Dorji and Lhab Tshering
Biodiversity Officers, National Biodiversity Center, Serbithang, Ministry of Agriculture and Forest, Thimphu, P.O. Box 875, Bhutan

Mahesh Ghimiray
Rice Specialist III, RNR Research and Development Center, Bajo, Ministry of Agriculture and Forest, Wangduephodrang, P.O. Box 1263, Bhutan

Ganesh B. Chhetri
Agriculture Specialist II, Department of Agriculture, Ministry of Agriculture and Forest, Thimphu, P.O. Box 392, Bhutan

Tashi Yangzome Dorji
Program Director, National Biodiversity Center, Serbithang, Ministry of Agriculture and Forest, Thimphu, P.O. Box 875, Bhutan

Asta Maya Tamang
Principal Biodiversity Officer, National Biodiversity Center, Serbithang, Ministry of Agriculture and Forest, Thimphu, P.O. Box 875, Bhutan

Katy E. Brantley, Mary C. Savin, Kristofor R. Brye and David E. Longer
Department of Crop, Soil, and Environmental Sciences, University of Arkansas, 115 Plant Sciences, Fayetteville, AR 72701, USA

Colin R. Jackson, Bram W. G. Stone and Heather L. Tyler
Department of Biology, the University of Mississippi, University, MS 38677, USA Crop Production Systems Research Unit, USDA Agricultural Research Service, Stoneville, MS 38776, USA

Cathy C. Webb, Marilyn C. Erickson, Lindsey E. Davey and Michael P. Doyle
Center for Food Safety, Department of Food Science and Technology, 1109 Experiment Street, University of Georgia, Griffin, GA 30223, USA